Statistical Techniques
in
Geographical Analysis

Gareth Shaw

Department of Geography, University of Exeter

and

Dennis Wheeler

Department of Geography, Sunderland Polytechnic

JOHN WILEY & SONS

Chichester . New York . Brisbane . Toronto . Singapore

Copyright © 1985 by John Wiley & Sons Ltd.

All rights reserved.

Library of Congress Cataloguing in Publication Data

Shaw, Gareth.
  Statistical techniques in geographical analysis.
  1. Geography—Statistical methods.
  I. Wheeler, Dennis. II. Title.
  G70.3.S52 1984 910' .212 84–12003
  ISBN 0 471 10317 9
  ISBN 0 471 90538 0 (pbk.)

British Library Cataloguing in Publication Data

Shaw, Gareth
  Statistical techniques in geographical analysis.
  1. Geography—Statistical methods
  I. Title II. Wheeler, Dennis
  519.5'02491 G70.3
  ISBN 0 471 10317 9
  ISBN 0 471 90538 0 Pbk

Phototypeset by D.U.P. Sandymount,
Dublin, Ireland and printed in Great Britain by Page Bros.
(Norwich) Ltd., Norwich.

Contents

Preface

Since the mid 1960s a number of textbooks have been published which have aimed at teaching geography undergraduates the rudiments of quantitative methods. Three general features have characterized the majority of these books. First, they have sensibly avoided delving too deeply into statistical theory and have concentrated on a more pragmatic approach, whereby students are taught how and when to use particular methods, and how to interpret the results. Second, and less justifiably, these books have paid almost exclusive attention to univariate and bivariate methods; that is to the examination of geographical variables either singly or in correlated pairs. Undoubtedly a great deal can be achieved by such studies, but geography is all too often a multivariate discipline in which the features of the geographical phenomenon (however that may be defined) extend over many variables. In how many ways, for example, can we describe a town, a soil sample or a river system? The answer is certainly more than one. The third characteristic common to these books is that they pay no attention to the development of computing facilities in the application of quantitative methods. No recognition is given of the widespread availability of computer hardware and statistical package programs.

Therefore, while we entirely agree with a pragmatic approach which includes only essential theory, we feel that the geographer's training, particularly at an undergraduate level, should pay far more attention to both multivariate methods and the use of computers. Practical application of multivariate methods raises problems not always experienced when only univariate or bivariate statistics are used. In particular, computers are often needed to process the frequently large amounts of data and to execute the statistical procedures. Some reference is made in this text to computer applications, and the final chapter lists some available programs. However, to deal adequately with this expanding area a companion textbook is being prepared dealing specifically with the topic. The idea is that while the two books are very much independent they may also be used together for a combined course in statistics and computing. This is made possible since both contain points of common reference and, as far as possible, similar exercises.

This book, and the companion text on computing in geography, had their origins in a scheme for a programme of related texts, on geographical analysis and methods, which was planned by Peter Toyne (now Deputy Director of North East London Polytechnic) and a group of geographers some of whom were at the University of Exeter. Since then the scheme for the books, and the locations of some of the authors, have changed; but we are glad to acknowledge Peter Toyne's role in originating this textbook and thank him for his continuing advice and comments during the preparation of the manuscript.

Our gratitude must also be extended to Terry Bacon of the Drawing Office of Exeter University's Geography Department, who prepared the diagrams for this book with such care and patience. Finally, this would not be complete without an expression of our thanks to our wives, Margaret and Marion, and our children for their forebearance over the many disrupted evenings and weekends suffered so that the book could be completed. We are grateful to them all.

Dr Gareth Shaw
Department of Geography
University of Exeter

Dr Dennis Wheeler
Department of Geography
Sunderland Polytechnic

Acknowledgements

The statistical tables in the appendices of this book have been prepared from published items, and the authors are indebted to the following organizations and individuals for allowing us to use their material. The table of z values is reproduced by permission of William Heinemann Ltd. from *Statistics Made Simple* by H. T. Hayslett. The Biometrika Trustees gave permission for us to use the tables of chi-square statistics initially published in *Biometrika*, volume 32, and the tables of critical F values from *Biometrika Tables for Statisticians*, volume 1, by E. S. Pearson and H. O. Hartley. The tables of Kolmogorov–Smirnov and Kruskall–Wallis statistics were adapted from original meterial in the *Journal of the American Statistical Association*, with whose permission they are reproduced. The tables of Mann–Whitney statistics are reproduced by permission of the Institute of Mathematical Statistics from original material in the *Annals of Mathematical Statistics*. The tables of critical Spearman and Pearson product-moment correlation coefficients are reproduced by permission of George Allen and Unwin from *Statistical Tables* by H. R. Neave. The authors would also like to thank J. Koerts and A. P. J. Abrahmase of the Erasmus Universiteit Rotterdam for permission to publish the tables of Durbin–Watson statistics which originally appeared in their text *On the Theory and Application of the General Linear Model*, published by the University of Rotterdam Press. We are grateful to the Literary Executor of the late Sir Ronald A. Fisher FRS, to Dr Frank Yates FRS and to Longman Group Ltd., London, for permission to reprint what appears here as Appendix II, from their book *Statistical Tables for Biological, Agricultural and Medical Research* (6th edition 1974).

Chapter 1

Introduction

With the acceptance and incorporation of statistical methods in geographical analysis, all geography students are now exposed to some form of instruction in quantitative techniques. However, for many students this is often a painful and sometimes unrewarding experience. Problems arise initially due to the fact that some students lack any comprehensive background in numerical analysis. This is often compounded by the way that many statistical textbooks present material, by choosing the most complicated mathematical notations. Further problems centre around the disillusionment that many undergraduates show towards statistical methods. This usually arises because most courses and textbooks never go beyond presenting univariate and bivariate statistics, when many of the problems tackled by students in their project work are of a multivariate nature.

These difficulties are not new and are probably well recognized by most people involved in the teaching of quantitative techniques. However, few serious attempts have been made to solve these problems. Some organizations, such as the Quantitative Methods Study Group of the IBG, have attempted to fill a need in the teaching of statistical methods, through the publication of a series of booklets (Catmog, 1975). Unfortunately, as Gregory (1983) points out these have suffered from two major problems: first, the unstructured order in which themes have been published; and second, the very different levels of complexity. Indeed, many are far too difficult to be of use as a teaching set. There are, therefore, only a limited number of publications that bridge the gap between current undergraduate textbooks and the purely statistical texts dealing with multivariate methods.

This book attempts to tackle these problems by presenting a wide range of statistical procedures. The authors trust that the topics covered are varied enough to satisfy the needs of most undergraduate and many postgraduate students.

It must be confessed at the outset that the selection is no more than that — a selection — and several topics have of necessity been left out. In any book of reasonable length such exclusions are unavoidable, and to readers who feel that their favoured, and doubtless indispensible, technique has been relegated we can only apoligize. Our selection was based on two considerations. First we looked at the methods found most often in published research material and balanced this, secondly, with items that we have found useful to our own researches. We hope that the final mixture is a practical one in which our often, sometimes bitter, experiences bear fruit.

1

The layout of the text is progressive from the simpler towards the more advanced topics, and the reader will also be aware at the same time of four general areas of methodology: univariate, bivariate, multivariate and spatial techniques. It must be stressed, however, that topics, chapters, sections, and the methods themselves are not treated in isolation. We have tried to emphasize the links and common ground between, for example, trend surface analysis and multiple and nonlinear regression; or the common strategies of hypothesis testing and the use of probability distributions. Through such links it is possible to allow those topics and methods not included to be approached with greater confidence and not necessarily as 'new' and unfamiliar items.

The question of confidence brings us to the matter of mathematics, for in the authors' experiences it is the almost traditional fear of mathematics that underlies many of the difficulties of teaching statistical methods to geography students; students who very often arrive at their degree courses having had a training in the arts or social sciences and less often in the physical sciences. As a consequence of this, the 'typical' student may not have pursued any serious mathematics after the age of 15 or 16. It is a common experience to start a lecture with a room full of students who know exactly what is meant by the term 'an average' and have a more or less clear notion of how to obtain it from a given set of data. Yet express that concept by using the equation $\bar{X} = \Sigma X_i/n$ and confusion will prevail. Unfortunately, such expressions and many far worse than this cannot be avoided, although we have kept the purely mathematical aspects to a minimum. As a result we have included an introductory chapter that discusses the mathematical procedures that crop up most frequently. As with the statistical procedures themselves this chapter is far from exhaustive but should act as a simple guide through the treacherous waters that lie ahead of the not-fully-numerate student. All equations have been kept as simple as possible and we have tried to avoid the use of frequent subscripts. We have also chosen not to go into the mathematical details of the more advanced methods, details that would tax even a mathematics student. Nevertheless, it must be emphasized that very many of the methods discussed involve only simple mathematics. What makes them appear so difficult is the unusual, probably new, notation used by statisticians and the repetitive nature of the calculations that require many observations to be processed. In the vast majority of cases discussed, the reader will need to be familiar with no more advanced arithmetic than addition, subtraction, multiplication, division, powers and logarithms. As an additional aid we have annotated many of the equations in adjacent keys to explain quickly and clearly how each component of the equation works. In this way what might appear to be an unwieldy expression can be interpreted as a construction of data codes linked by various operational instructions such as summations, divisions or powers.

Inevitably there will be some areas of greater interest than others to readers. There may be questions raised in the reader's mind that we may not have answered fully (we hope not too many). To cope with both of these demands we have provided references for further reading. One set of references is a standard list of items that have been cited in the course of the text; but we have supplemented these with shorter and annotated bibliographies or 'recommended reading'. These selections were made bearing in mind the undergraduate reader who wishes to see specific applications of the methods discussed in each chapter or who wishes more detail on those methods but, once again, does not wish to be exposed to an over-abundance of mathematical theory. Hence some cherished, though demanding, texts and papers have not succeeded to these pages where equally informative but simpler material is at hand. The selections have also, whenever possible, been drawn from the more widely available texts and journals. We cannot, and have not tried to, match the comprehensiveness of carefully assembled bibliographies — such as Greer-Wooton (1972) who lists a startling number of papers itemized under the statistical techniques used in them — although we recommend this publication to anyone who requires a more exhaustive list of material.

Worked examples form a major part of the discussion and are based on 'real' data drawn from original research or published secondary sources. Because of this they may not always be perfect for the problem (or vice versa), but that is a difficulty to be faced by all geographers engaged in research and it is pointless to pretend, even at this stage, that it is otherwise. Similarly, most of the exercises are based on real data. They are designed to be carried out using no more than good pocket calculators. With regard to the latter, those with even simple statistical options, such as means, standard deviations and sums of squares, will be found to be the most useful. No exercises are given for the multivariate methods where the calculations would be too arduous without the aid of a computer. Thus, the opening and middle sections of this book contain exercises and the reader is strongly advised to attempt at least some for each chapter. By doing so he will reinforce his understanding, as it is very easy to read a section and believe its contents to be clearly understood. It is only when those points are put into practice that hitherto unimagined difficulties will arise, and be solved.

Computer programs and manuals

Chapter 18 presents a summary table to indicate where computational assistance for particular methods can be found. Attention is specifically directed to those packages that are widely available — SPSS, BMDP and Minitab — and the appropriate manuals.

REFERENCES

Catmog (1975–). *Concepts and Techniques in Modern Geography*, Geo-Abstracts, Norwich.

Greer-Wooten, B. (1972). *A Bibliography of Statistical Applications in Geography*, Assoc. Am. Geogrs Tech. Paper 9, Washington, DC.

Gregory, S. (1983). 'Quantitative geography: the British experience and the role of the Institute', *Trans. Inst. Brit. Georgrs* (N.S.) **8**, 1, 80–89.

RECOMMENDED READING

Burton, I. (1963). 'The quantitative revolution and theoretical geography', *Can. Geog.* **7.** 151–162. Gives an initial view of the so called 'quantitative revolution' in geography.

Gregory, S. (1976). 'On geographical myths and statistical fables', *Trans. Inst. Brit. Geogrs* (N.S.) **1,** 4, 385–400. Presents a brief review of the development and teaching of statistical methods in Britain.

Wrigley, N. (1981). 'Quantitative methods: a view on the wider scene', *Prog. Hum. Geog.* **5.** 548–561. Sets the role of techniques in a wider and current perspective.

Chapter 2

The Mathematical Basis of Statistics

2.1 INTRODUCTION

The aim here is not to educate the reader in all or even many of the branches of mathematics, but rather to familiarize him with the arithmetic tasks required for the following chapters. Those who have studied mathematics to 'A' level (in England) or to High School (in America) or equivalent grades may feel confident in omitting this chapter. Many undergraduate geographers, however, have not had the benefits of such training, having come through with a so-called 'arts' training, and it is to these geographers that this chapter is directed.

Such students are rarely at ease when handling quantitative data, the successful analysis of which is not so much a question of mathematical ability as of confidence and clear thinking. All too often students allow themselves to be intimidated by the vocabulary and algebraic shorthand of statistical analysis. In reality many equations used by statisticians are surprisingly simple and usually involve the irksome repetition of basic arithmetic procedures (addition, subtraction, multiplication etc.) rather than challenging mathematical enterprises. This of course, is why computers are such an invaluable aid. Perhaps it is the abundant use of Greek symbols or the wealth of subscripts that prompts such fears. But whatever the precise cause this chapter demonstrates by the simple application of elementary arithmetic, that such fears are groundless.

2.2 ARITHMETIC METHODS

The first need is to make clear the order in which the tasks set by various equations are to be carried out, as this is often misunderstood, with serious consequences. It will be recalled that arithmetic must follow certain prescribed orders in which multiplication and division precede addition and subtraction. Thus in the sequence $2.5 + 17.2 \times 3.1$ it is the multiplication of 17.2 by 3.1 that must be executed first, with 2.5 then added to that product. This sequence gives an answer of 55.82. Had the addition been carried out first the answer (incorrect) would have been 61.07. By the same convention divisions also precede additions and subtractions, so that $4.7 \div 2.1 - 2.0 = 2.238 - 2.0 = 0.238$, and not $4.7 \div 0.1 = 47.0$. Within strings of additions and subtractions alone or multiplications and divisions alone the order of execution is immaterial, hence;

5

$$2.1 + 4.7 - 3.7 + 1.4 = 4.5$$
$$3.2 \times 1.1 \times 7.0 \div 2.0 = 12.32$$

irrespective of the order of execution (but remember that in, for example, $17 - 2 - 1$ the answer is 14, and not 16 as the last two items are $-2-1$ and not $2-1$.

Powers and indices take precedence over all the above four operations and must be carried out first for the items to which they relate; for example:

$$2 \times 4^3 = 2 \times 64 = 128 \text{ and not } 2 \times 4^3 = 8^3 = 512$$
$$4 + 7^2 = 4 + 49 = 53 \text{ and not } 4 + 7^2 = 11^2 = 121$$

There are, however, many occasions when statisticians wish to override these orderings. This can be done by organizing the expressions within brackets. All items within brackets must be evaluated before those outside them. A simple example that is not uncommon in statistical analysis is the need to square the difference between two quantities. It would be incorrect to write:

$$7.2 - 2.4^2 = 7.2 - 5.76 = 1.44$$

as the item 2.4^2 would then be evaluated before the subtraction. In this case the correct expression for squaring the difference would be:

$$(7.2 - 2.4)^2 = 4.8^2 = 23.04$$

which would require that the bracketed subtraction be performed before the squaring. Where several pairs of brackets are present the innermost are evaluated first; for example:

$$[2(3.2 + 4.0)]^2 = [2(7.2)]^2 = 14.4^2 = 207.36$$

The four executable symbols of $+$, $-$, $\times$ and $\div$ together with indices or powers provide the basis for the greater part of all statistical operations. The four symbols are straightforward and present no problems. The use of indices is more complex and requires further elaboration. At the most elementary level the use of an index implies repetitive multiplication, so that:

$$2^2 = 2 \times 2 = 4$$
$$4.2^4 = 4.2 \times 4.2 \times 4.2 \times 4.2 = 311.17$$

Indices provide a convenient mathematical shorthand by which 2.1^7 is far easier to express than its longhand counterpart, and less subject to misreading and error. This shorthand facility is often used by computers and pocket calculators when dealing with very large or very small numbers. Both forms of machine, notably the latter, have limited abilities to display very long numbers; 8—digit displays are common in calculators. Rather than abandon numbers greater than 99,999,999 these machines use a 'floating point' format. For example, 22,180 × 27,560 gives 611,280,800, which might be presented as 6.1128 08 or 6.1128 E08. This should be interpreted, in both cases, as 6.1128×10^8 in which the use of indices combines with the multiplication symbol to provide a most useful shorthand notation which avoids long strings of digits. Even very powerful computers often prefer to express long numbers in this form to avoid confusing layouts in the output. There is, nevertheless, a price to be paid for this utility which can be seen if the terms in 6.1128×10^8 are fully expanded; this gives 611,280,000 which is not the same as 611,280,800. Hence some rounding occurs and trailing digits are lost so far as the user is concerned, even though the machine may use the complete number in subsequent calculations. Such errors because they are part of very large numbers, are relatively small; but they can lead to marginal, though perplexing, differences between calculations carried out on different calculators. Such errors need not cause any concern but readers should be alive to their possibility. Even powerful computers are not immune to rounding errors and differing results.

The floating point convention is equally applicable to very small numbers with many leading zeros following the decimal point, but their explanation requires a discussion of negative indices. It is quite possible to evaluate expressions such as 2^{-2} or 3.6^{-3}, but it must be remembered that although such quantities contain negative indices they are not themselves negative numbers unless preceeded by a minus sign. Negative indices are best dealt with by using the reciprocals of the numbers involved, i.e. by taking the quantities 'below the line'. Thus:

$$2^{-2} = \frac{1}{2^2} = \frac{1}{4} = 0.25$$

$$3.6^{-3} = \frac{1}{3.6^3} = \frac{1}{46.66} = 0.0214$$

This makes use of the fact that the sign of the index changes when it is moved below the line. We can also note that when numbers larger than 1.0 are raised to negative powers they become fractions while remaining positive. Conversely, numbers less than 1.0 when raised to negative powers are again positive but become larger; for example:

$$0.5^{-2} = \frac{1}{0.5^2} = \frac{1}{0.25} = 4.0$$

It should also be noted that only the term(s) directly raised by the power move in these operations, and in general algebraic terms:

$$2X^{-2} = \frac{2}{X^2}$$

Returning to the theme of floating point formats we can now appreciate that a display reading 7.4 −03 or 7.4E−03 will represent 7.4×10^{-3}, which is another way of writing 0.0074 being derived from $7.4 \times 1/10^3$.

This brings us to fractional indices. Integer (whole number) indices should present few problems, but fractional terms are less easy to deal with. How should one set about evaluating $4^{2.5}$, $7^{1.5}$ or $16^{\frac{1}{2}}$? From the purely practical point of view most good pocket calculators will perform the task if you employ the correct sequence of keys; or, as will be seen later, the task can be performed with the use of logarithms. By either method $4^{2.5}$ becomes 32.0, $7^{1.5}$ becomes 18.520 and $16^{\frac{1}{2}}$ is 4.0. Negative fractional indices may also be encountered and similarly dealt with. We do not intend to demonstrate how to deal 'longhand' with such expressions since calculators deal with them most efficiently, but attention is drawn to Table 2.1 where the contrasting results of positive and negative indices are demonstrated. Note that for numbers greater than 1.0 positive indices cannot yield results less than 1.0 and negative indices cannot produce results greater than 1.0.

Table 2.1 Numerical equivalents of exponential terms

Exponential term	Result	Exponential term	Result
9^1	9.0	9^{-1}	0.1111
$9^{0.8}$	5.799	$9^{-0.8}$	0.1724
$9^{0.5}$	3.0	$9^{-0.5}$	0.3333
$9^{0.2}$	1.552	$9^{-0.2}$	0.6444
9^0	1.0	$9^{-0.1}$	0.8027

There are three 'laws' of operation with indices:

Law 1: Powers may be added for multiplication of a common base:

$$X^m \times X^n = X^{m+n} \quad \text{e.g. } 3^3 \times 3^2 = 3^5 = 243$$

Law 2: Powers may be subtracted for division with a common base:

$$X^m/X^n = X^{m-n} \quad \text{e.g. } 3^3/3^2 = 3^{3-2} = 3^1 = 3$$

Law 3: Powers may be multiplied when they appear in sequence:

$$(X^m)^n = X^{mn} \quad \text{e.g. } (3^3)^2 = 3^6 = 729$$

Finally, two methods of representing the square root of a number can be identified. Most readers will be familiar with the $\sqrt{X}$ convention, but fractional indices can also be employed to convey the same meaning. Both $X^{0.5}$ and $X^{\frac{1}{2}}$ indicate that X is to be square-rooted and will be used from time to time in the following chapters.

2.3 LOGARITHMS

Any number on the 'arithmetic' scale which is greater than zero can be expressed by its counterpart on the logarithmic scale. The advantages of such re-expression are many and varied, but at this stage only the principles of logarithms (or 'logs') shall be discussed.

The reader will need to be familiar with both of the main logarithmic systems — natural and common logs — the latter being the simpler and better known. Common logs are sometimes referred to as 'logs to the base 10'. Table 2.2 lists some of their more obvious characteristics and indicates how the logarithmic and arithmetic scales are related.

Table 2.2 Logarithmic equivalents of numbers greater than 1.0

Number	Logarithm	Number	Logarithm	Number	Logarithm	Number	Logarithm
1.0	0	3.0	0.4771	5.0	0.6989	7.0	0.8451
10.0	1.0	30.0	1.4771	50.0	1.6989	70.0	1.8451
100.0	2.0	300.0	2.4771	500.0	2.6989	700.0	2.8451
1000.0	3.0	3,000.0	3.4771	5,000.0	3.6989	7,000.0	3.8451

The common log of a number is the power to which 10 must be raised in order to obtain the number. Thus:

$$100 = 10^2 \quad \text{so that} \quad \log_{10}100 = 2.0$$
$$50 = 10^{1.6989} \quad \text{so that} \quad \log_{10}50 = 1.6989$$

Table 2.3 Logarithmic equivalents of numbers less than 1.0

Number	Logarithm	Number	Logarithm	Number	Logarithm
0.1	−1.0	0.3	−0.5229	0.5	−0.301
0.01	−2.0	0.03	−1.5229	0.05	−1.301
0.001	−3.0	0.003	−2.5229	0.005	−2.301

No log system permits logs of negative numbers or of zero, but there are numbers — fractions less than 1.0 — whose logs are negative. Table 2.3 presents the structure of this part of the logarithmic scale. From the table we see that there can be no log for zero; smaller and smaller fractions, when expressed in log form, simply become larger and larger negative numbers towards negative infinity. The mathematical relationship between fractions and logs is again determined by the power to which 10 must be raised in order to produce that number. For example:

$$0.1 = 10^{-1} \quad \text{so that} \quad \log_{10}0.1 = -1.0$$
$$0.5 = 10^{-0.301} \quad \text{so that} \quad \log_{10}0.5 = -0.301$$

Students familiar with logarithmic tables will notice that the values given in Table 2.3 do not correspond with those that they might use, though those in Table 2.2

do tally. This difference arises because, with logarithmic tables, the 'bar' system is used for dealing with numbers of less than 1.0. In this sytem the log of 0.3, for example, would be $\bar{1}.4771$ and not -0.5229. In the same manner the log of 0.03 would be $\bar{2}.4771$. It must be remembered that computers and calculators are unable to work with this convention and use, as we shall see, the system employed for Table 2.3. The relationship between the two systems is made clear if the 'bar' system notation is written in full. The complete notation for $\bar{1}.4771$ is $-1+0.4771$, which equates to -0.5229.

It has already been suggested that logarithms can be used to solve problems with exponents. They can also provide a means of solving long multiplications. In both of these cases the laws governing operations with logs are well exemplified.

First law of logarithms

When multiplying two numbers one may instead add their respective logarithms, the antilogarithm ('antilog') of the result being the correct answer. To take a simple example, the product of 28.47×39.23 can be found by this means:

$$\log_{10}28.47 = 1.4544 \qquad \text{and} \log_{10}39.23 = 1.5936$$
$$\log_{10}(28.47 \times 39.23) = 1.4544 + 1.5936 = 3.048$$

The antilog of 3.048 gives the correct answer as 1116.86.

Divisions can be performed in the same fashion but by subtracting the logarithms; thus:

$$\log_{10}(28.47 \div 39.23) = 1.4544 - 1.5936 = -0.1392$$

the antilog of which gives the answer of 0.7258.

The general algebraic expressions of these laws can be written as:

$$a \times b = \text{antilog}(\log_{10}a + \log_{10}b) \qquad (2.1)$$
$$a \div b = \text{antilog}(\log_{10}a - \log_{10}b) \qquad (2.2)$$

Second law of logarithms

When we are dealing with exponents the position is similar. Take, for example, $3.67^{1.89}$. Many pocket calculators do not have the capacity to perform this operation directly, but the second law of logarithms can be called upon. This states that the value of a number raised to some power can be found by multiplying the log of that number by the exponent, and 'antilogging' the result. Thus $3.67^{1.89}$ becomes:

$$(\log_{10}3.67) \times 1.89 = 0.5647 \times 1.89 = 1.0672$$

the antilog of which is 11.674; hence $3.67^{1.89} = 11.674$. Notice that on this occasion the exponent term is not logged, so that negative exponents can also be dealt with in this manner; $3.67^{-1.89}$ becomes:

$$0.5647 \times (-1.89) = -1.0672$$

the antilog of which is 0.0857. The algebraic expression of this law might be given as:

$$a^b = \text{antilog}(\log_{10}a \times b) \tag{2.3}$$

The examples used to demonstrate these laws have been extremely simple — so simple in fact that it would be absurd to use them in this way. Nevertheless, the principles are important and are critical to making sense of many subsequent points, and of the chapter dealing with nonlinear regression in particular.

Another point should be noted concerning antilogs. The antilog of a number is the value obtained by raising 10 to the power of the log; as should be clear from the opening paragraph of this section. To emphasize this principle, reference back to the example used to illustrate the second law of logarithms shows that $10^{1.0672} = 11.674$.

Natural logarithms

Attention can now be turned to the second of the principal log systems, natural logs. Common logs are based on the number 10, natural logs are based on the number 2.71828 . . . which is designated by the symbol e. As a result the natural log of 2.7183 is 1.0. The square of e has a natural log of 2.0, and so on up the scale. Table 2.4 indicates the structure of the natural log scale.

Table 2.4 Some natural logarithms

Number	Number as an exponent of e	Natural log
1	e^0	0
2.7183	e^1	1
5.0	$e^{1.609}$	1.609
7.3891	e^2	2.0
10.0	$e^{2.303}$	2.303
20.0855	e^3	3.0
1.6487	$e^{0.5}$	0.5

All the rules that apply to common logs apply equally to natural logs. For example $3.67^{1.89}$ would now become:

$$(\log_e 3.67) \times 1.89 = 1.3002 \times 1.89 = 2.4574$$

the natural antilog of which is 11.674. As before the process of antilogging is, in reality, raising the log base, in this case e and not 10, to the power specified by the log:

$$e^{2.4574} = 11.674$$

To avoid any confusion we shall adopt the usual convention of denoting common logs simply by 'log' and natural logs by 'ln'. It is, however, easy to convert from one system to the other and there are two simple equations which allow this. For any quantity x:

$$\log (x) = \ln (x) \times 0.4343 \tag{2.4}$$

$$\ln (x) = \log (x) \times 2.303 \tag{2.5}$$

In fact any number can be used as a log base, although the above two are used almost exclusively. One exception is found in the study of soils and sediments in which particle sizes are usually expressed not in millimetres but in phi (ϕ) units. The phi scale uses the *negative* log of the grain size in millimetres, but the log base is 2.0. Thus as $64 = 2^6$, it follows that $\log_2 64 = 6$. By the same reasoning $\log_2 4 = 2$. A full account of this scale will be found in Briggs (1977), but its general characteristics and relation to the millimetre scale can be seen in Table 2.5.

Table 2.5 Phi-scale for sediment particle sizes

Millimetres	$\log_2$	Phi units
64.0	6.0	−6.0
32.0	5.0	−5.0
16.0	4.0	−4.0
8.0	3.0	−3.0
4.0	2.0	−2.0
2.0	1.0	−1.0
1.0	0.0	0.0
0.5	−1.0	1.0
0.25	−2.0	2.0
0.125	−3.0	3.0

2.4 STATISTICAL METHODS

Thus far attention has focused on arithmetic and mathematical methods. There is however, a notation specific to statistical methods that employs a wealth of subscripts and Greek symbols. As a result the appearance of statistical equations may be quite different from those of purely mathematical ones. At first sight they may appear to be very confusing, but in reality they are often merely a shorthand for what are very simple arithmetic tasks.

The most characteristic and most commonly encountered of these new instructional symbols is the Greek letter Σ (capital sigma). This does not have a numerical quantity associated with it, but informs the operator that he must add together the values of the indicated 'variable'. The variable might, for example, be temperature and be denoted by the symbol X; the instruction to add all the temperature readings would be given by ΣX. Similarly ΣY prompts the addition of the observations for variable Y.

Often a more comprehensive form of this expression is used. If a data set contained a temperature figure for each day of one year the statistician might write the addition instruction as:

$$\sum_{i=1}^{365} X_i$$

in which the subscript i indicates the sequence number between 1 and 365 and the summation takes place for X_1 to X_{365}. Table 2.6 shows how i, X and Σ are connected.

Table 2.6 The index notation in statistical analysis

Sequence number (i)	Numerical value	
1	$X_i = X_1 = 10.2$	
2	$X_i = X_2 = 11.3$	
3	$X_i = X_3 = 9.7$	$\sum_{i=1}^{5} X_i = 50.6$
4	$X_i = X_4 = 7.5$	
5	$X_i = X_5 = 12.1$	

The total number of observations within a data set are often not known in advance and the algebraic letter n (for sample size) can be used to give:

$$\sum_{i=1}^{n} X_i$$

It is a widely adopted convention in statistical analysis to refer to the number of observations as n.

This conventional use of subscripts is most adaptable. The instruction for adding together all the observations between the tenth and twentieth observations in a sequence could be written as:

$$\sum_{i=10}^{20} X_i$$

One subscript only is needed when observations or numbers are arranged in one row (or vector). If, however, they are arranged in a matrix of rows and columns double subscripts are necessary to identify the appropriate location. Thus, the observations for row 2 in column 3 would be $X_{2,3}$. The point is made pictorially in Figure 2.1. Hence in addition to items such as X_i the reader may also come across its two-dimensional matrix equivalent of X_{ij}. To avoid a possibly confusing welter of subscripts in much of the following text these are left out of equations unless some particular point needs to be made. This leads also to the use of double summation signs when addition must take place for all rows and all columns; such as:

$$\sum_{i=1}^{r} \sum_{j=1}^{k} X_{ij}$$

for a matrix of numbers composed of r rows and k columns.

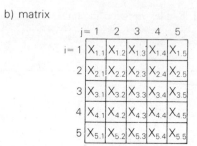

Figure 2.1 Vector and matrix subscripts

The need to add sets of numbers is commonplace in statistical analysis and the sigma sign is a very simple form of shorthand. For example, the instruction for finding the arithmetic mean of a set of data is given by:

$$\bar{X} = \frac{\Sigma X}{n} \tag{2.6}$$

where n is the number of observations and $\bar{X}$ the mean of those figures. The 'bar' symbol, in statistics, alway denotes the arithmetic mean, so that $\bar{Y}$ is the arithmetic mean of variable Y, $\bar{Z}$ the arithmetic mean of variable Z and so on. This 'bar' must note be confused with that used in logs.

Because Σ is an instruction, much like + or −, it is important to establish where it fits in to the sequence of executions. The use of brackets frequently places Σ in its correct position, but some general rules may be noted. Thus, in equation 2.6 the division line underlying both Σ and X indicates that summation precedes division. But exponents take precedence over summations in cases such as ΣX^2, which requires that all the X's be squared before being summed. Table 2.7 demonstrates some similar conventions and how brackets can be used to overcome them.

A final example, which occurs in many guises, is the need to square a set of differences, sum them and divide by the number of observations. Most often this appears in the form of sequential differences of individuals from their mean. The equation reads:

$$v = \frac{\Sigma(X_i - \bar{X})^2}{n} \tag{2.7}$$

The sequence to be followed here is: (1) subtract the mean from each individual, i.e. perform what is in the brackets first; (2) square each of those differences, remembering that exponents take precedence over summations; (3) sum the squares; and finally (4) divide by n. Had we wished to square the sum instead, the equation would have read:

$$v = \frac{[\Sigma(X_i - \bar{X})]^2}{n} \tag{2.8}$$

thereby demanding that first the inner and then the outer brackets be evaluated prior to squaring.

It is most important that the differences between equations (2.7) and (2.8) be kept clearly in mind as the results obtained are vastly different; some of the contrasts due to changes in order of execution are illustrated in Table 2.7.

Table 2.7 Conventions for the execution of mathematical statements

Instruction	Sequence of execution	Worked examples[1]
ΣX^2	(1) square all X's (2) sum the squares	$2 \times 2 = 4$ $5 \times 5 = 25$ $7 \times 7 = 49$ $3 \times 3 = 9$ $4 + 25 + 49 + 9 = \mathbf{87}$
$(\Sigma X)^2$	(1) sum all X's (2) square the sum of X's	$2 + 5 + 7 + 3 = 17$ $17 \times 17 = \mathbf{289}$
ΣXY	(1) multiply all X and Y pairs[2] (2) sum all XY products	$2 \times 3 = 6$ $5 \times 9 = 45$ $7 \times 2 = 14$ $3 \times 1 = 3$ $6 + 45 + 14 + 3 = \mathbf{68}$
$3(\Sigma X)$	(1) sum all X's (2) multiply the sum by 3	$2 + 5 + 7 + 3 = 17$ $3 \times 17 = \mathbf{51}$
$3[\Sigma(X-1)]$	(1) subtract 1 from all X's (2) sum all subtractions (3) multiply the sum by 3	$2-1 = 1$ $5-1 = 4$ $7-1 = 6$ $3-1 = 2$ $1 + 4 + 6 + 2 = 13$ $13 \times 3 = \mathbf{39}$
$\Sigma \dfrac{(X - Y)^2}{Y}$	(1) subtract all Y's from X's (2) square the $X-Y$ differences (3) divide all squares by Y's (4) sum all divisions	$2-3 = -1$ $5-9 = -4$ $7-2 = 5$ $3-1 = 2$ $-1^2 = 1$ $-4^2 = 16$ $5^2 = 25$ $2^2 = 4$ $\frac{1}{3} = 0.33$ $\frac{16}{9} = 1.78$ $\frac{25}{2} = 12.5$ $\frac{4}{1} = 4$ $0.33 + 1.78 + 12.5 + 4.0 = \mathbf{18.61}$

1. In the worked examples there are four paired X and Y observations; for X these are 2, 5, 7 and 3; for Y they are 3, 9, 2 and 1.

2. Strictly this instruction should read $\Sigma(XY)$, but it is conventionally written as above and will be frequently encountered in later chapters.

2.5 GRAPHS OF STATISTICAL DATA

Graphs, maps and diagrams can be invaluable aids to the investigative geographer. In addition to raw geographical data, however, mathematical equations can be presented in graphic form. In both cases the ability of a graph to generalize what can be perceived only vaguely from a column of figures or an algebraic equation is indispensible.

Although equations can be used to generate a whole variety of lines and curves, just one case will be used to demonstrate how any equation can be handled so as to produce a line on a sheet of graph paper. Consider a simple equation such as $Y = a + bX^2$. By convention the lower-case a and b are coefficients fixed at the outset of the exercise; in this instance we will set a to 1.0 and b to 2.0. With this information we may now take values for the predictor term X, and substitute them into the equation one at a time in order to estimate corresponding values for the so-called dependent term Y. For example, if X is set to 10 then:

$$Y = 1.0 + 2.0 \times 10^2$$
$$= 1.0 + 2.0 \times 100$$
$$= 1.0 + 200.0$$
$$= 201.0$$

By this means corresponding pairs of X and Y values can be found and plotted on a graph in order to see the form of the curve generated by this equation. Table 2.8 demonstrates the derivation of the Y values from the X, values, and Figure 2.2 is a plot of the results.

Table 2.8 Successive substitutions in the expression $Y = 1.0 + 2X^2$

Value of X	Expression	Estimate of Y
2	$1.0 + 2 \times 2^2$	9
4	$1.0 + 2 \times 4^2$	33
6	$1.0 + 2 \times 6^2$	73
8	$1.0 + 2 \times 8^2$	129
10	$1.0 + 2 \times 10^2$	201
14	$1.0 + 2 \times 14^2$	393
18	$1.0 + 2 \times 18^2$	649

The procedure is quite straightforward and can be applied to any simple equation. With long expressions and complex curves, however, it is possible to call upon computer assistance not only for the actual calculations, but also for the plotting of the results. Work of this nature is most important when we are examining the results of statistical methods such as linear and nonlinear regression and trend-surface analysis.

Finally, as logarithms have already been discussed in this chapter, it is worth pointing out that graphs can also be plotted on what is known as logarithmic graph paper. This can be a most useful exercise but needs to be carried out with

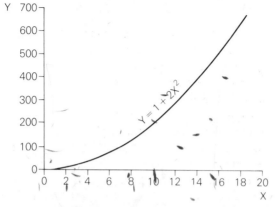

Figure 2.2 Graph of function $Y = 1 + 2X^2$

care. There are two basic forms of logarithmic graph paper: 'double log' or 'log–log', and 'semi-log'. On such paper the spacing of the divisions is performed on a common log scale along either one (semi-log) or both (log–log) axes, the effect being that the distance separating 1 and 10 is the same as that between 10 and 100 (see Figure 2.3). The axes of the graph can be numbered in the usual way, but the scaling is based on cycles of powers of 10. Three-cycle log paper permits the plotting of values within the range, say, 1–10–100–1000 (10^0–10^1–10^2–10^3) or 0.001–0.01–1.0 (10^{-3}–10^{-2}–10^{-1}–10^0). Log paper with between two and five cycles is widely available.

In effect, the use of this form of graph paper permits the plotting of the logs of numbers without those logs having first to be obtained. More importantly, there are a large number of mathematical curves that can be linearized by this method; the consequences of this are discussed more fully in Chapter 12, but by way of an example Figure 2.3 demonstrates what happens to the data from Figure 2.2 when replotted on log–log paper.

2.6 MATHEMATICAL SYMBOLS

Some mathematical symbols may not be familiar to the reader. The quantities e and π are universal numerical constants with values of approximately 2.7128 and 3.1416 respectively, and whenever they appear these values apply. Less well known but also important is the factorial sign, written as an exclamation mark (!). It appears only after integer quantities and carries the instruction to multiply all whole numbers up to and including that indicated. Thus:

$$2! = 1 \times 2 = 2$$
$$4! = 1 \times 2 \times 3 \times 4 = 24$$
$$6! = 1 \times 2 \times 3 \times 4 \times 5 \times 6 = 720$$

Products are taken from 1 upwards. Clearly zero is excluded, although factorial zero is, by convention, taken to be 1.0. Equally clearly it cannot apply to

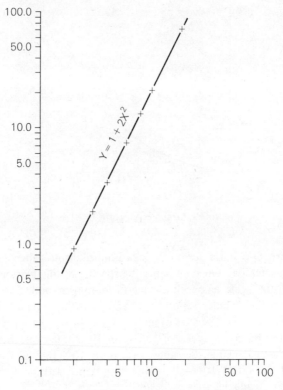

Figure 2.3 Graph of $Y = 1 + 2X^2$ on log–log axes

fractional quantities. Factorials appear in many mathematical tables and it is a standard function of many calculators and computers. Some care is needed when this instruction appears within longer expressions, and the following examples should help to avoid the most obvious pitfalls:

$$5! + 2! = 120 + 2 \text{ (not } 7!)$$
$$5! \times 3! = 120 \times 6 \text{ (not } 8! \text{ or } 15!)$$
$$6! \div 3! = 720 \div 6 \text{ (not } 2!)$$

However, an expression such as $(6 - 3)!$ must be simplified to $3!$.

Another instructional symbol is the modulus sign, a pair of vertical lines, that indicate that the sign of the operation must be ignored. For example:

$$|2 - 4| = 2 \text{ (not } -2)$$
$$|7 - 3 - 10| = 6 \text{ (not } -6)$$

This sign is not a bracket as such but applies to the result of the whole operation between the lines.

Brackets can themselves cause confusion between round and square forms. In this text there is no mathematical distinction between them except that square brackets are used for the outer set and round brackets for the inner; for example:

$$[\Sigma(X - Y)^2]/3$$

In all cases brackets must be paired, with each opening bracket having a closing companion.

Another sign to look out for is the plus or minus sign ($\pm$) which indicates that the numbers either side of this instruction must be both added and subtracted to give two, and not one result. For example:

$$64 \pm 2.5 = 66.5 \text{ or } 61.5$$

Some chapters also use the 'greater than' and 'less than' signs. These are a useful shorthand and appear as follows:

$$100 > 99 \text{ means } 100 \text{ is greater than } 99$$
$$24 < 25 \text{ means } 24 \text{ is less than } 25$$

The 'greater than' and 'less than' signs also appear algebraically as in:

$$X > Y \text{ means } X \text{ is greater than } Y$$
$$a < b \text{ means } a \text{ is less than } b$$

2.7 CONCLUSIONS

Clearly this chapter is not an exhaustive mathematics course. Major areas such as matrix algebra, which is widely used in multivariate statistics, have been omitted as, at this level, it is possible to get by without being familiar with their detailed workings. This also leaves the reader free to concentrate on statistical and interpretational, rather than purely mathematical, issues. Nevertheless, it is hoped that the content of this chapter will make what follows easier to comprehend. The interested reader will find many suitable mathematics texts at all levels should he wish to pursue this topic.

RECOMMENDED READING

In addition to the wealth of published books designed to aid the student mathematician, the authors have found the following items to be particularly interesting or informative.

Briggs, D. (1977). *Sources and Methods in Geography: Sediments*, Butterworths, London. This book contains a valuable discussion and review of the phi scale of sediment size measurement (an application of log-transformations).

Courant, R. (1964). 'Mathematics in the modern world', *Sci.Am.* **211**, 40–49. One of series of papers in this publication that reviews, in a readable manner, the position of mathematics in the twentieth century.

Harwood Clarke, L. (1970). *Modern Mathematics at Ordinary Level*, 2nd edn, Heinemann, London. An excellent book for those seeking a painless introduction to the so-called 'new' mathematics. It contains useful sections on matrix algebra.

Kruglak, H., and Moore, J. T. (1973). *Basic Mathematics,* McGraw-Hill, New York. This is a comprehensive text covering all aspects of the field up to a point where more advanced material can be approached with confidence. The style of presentation is easy to follow and should not offend even the least numerate of geographers.

Monkhouse, F. J., and Wilkinson, H. R. (1971). *Maps and Diagrams,* 3rd edn, Methuen, London. A useful book covering the practicalities of data presentation and graph construction.

Sawyer, W. W. (1955). *Prelude to Mathematics,* Penguin, Harmondsworth.

(1963). *Mathematician's Delight,* Penguin, Harmondsworth.

(1964). 'Algrebra', *Sci.Am.* **211,** 70–78.

All three publications show the more interesting aspects of a subject usually viewed as dull.

EXERCISES

1. Evaluate the following expressions paying attention to the sequences of execution:

$$121.2 + 34.1 \div 6.0 \times 4.82$$
$$46.2 \times 6.3 - 2.84 + 6.71 \div 2.0$$
$$3.6^2 \div 2.74^3 - 2.845$$
$$(3.6 - 2.74)^2$$

2. Express in complete form, the following:

$$7.832 \times 10^2 \quad 5.21 \times 10^4 \quad 2.85 \times 10^7 \quad 3.89 \times 10^{-1} \quad 7.62 \times 10^{-5}$$

3. Evaluate the following expressions using a hand calculator where possible:

$$3.7^{2.5} \quad 8^{3.5} \quad 7.6^{0.5} \quad 81^{0.5} \quad 426^{0.25} \quad 0.45^{\frac{1}{4}}$$

4. Find the common and natural logarithms of the following numbers; either calculators or tables may be used:

$$10.0 \quad 2.7128 \quad 100.0 \quad 124.5 \quad 0.5 \quad 0.1 \quad 6543.0$$

5. Find the common and natural antilogarithms of the following logarithms; either calculators or tables may be used:

$$1.0 \quad 24.0 \quad 6.284 \quad 0.948 \quad 0.55 \quad -0.945 \quad -1.945 \quad -0.217 \quad -2.217$$

6. Using the following sequence of six observations of the variable $X - 7.1, 2.0, 3.5, 19.2, 14.1, 10.6$ — evaluate each of the expressions below:

$$\Sigma X/n \text{ (where } n = 6)$$
$$\Sigma X^2$$
$$(\Sigma X)^2$$
$$\frac{\Sigma(\bar{X} - X)^2}{n}$$

7. For the range of values $X = 0.0$ to $X = 6.0$, plot graphs of the following equations:

$$Y = 3X^2 \qquad Y = 2.5 - 2X^{-2}$$
$$Y = 2.5 + 1.5X^2 \quad Y = X^3$$
$$Y = 1 + 3X^{-2} \quad Y = X^{0.5}$$

8. Using selected values of X up to $X = 100$, plot log–log graphs of the three equations:

$$Y = X^3$$
$$Y = X^{0.5}$$
$$Y = X^{-2}$$

9. Evaluate the following expressions:

$6! + 4!$	$|16 - 4|$
$7!/2!$	$|4 - 16|$
$(4!/2!)^2$	$|18/4 - 16/3|$

Chapter 3

Geographical Sources, Data Collection and Data Handling

3.1 GEOGRAPHICAL SOURCES

The purpose of this chapter is to explore the basic problems associated with data sources in geographical analysis. As we shall see the quality, and in some circumstances the quantity, of our data determines not only which statistical tests should be used, but also the relevance of our results. In this respect it is important that we understand the strengths and weaknesses of our data before we embark on the application of quantitative techniques (see Chapter 4). This section is not, however, intended to be a comprehensive review of geographical sources, but rather an introduction to the major types of data that are available, and a guide to the literature concerned with these sources.

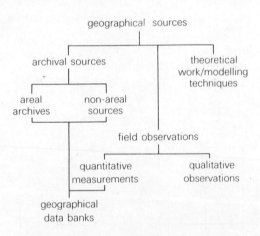

Figure 3.1 Types of geographical sources

The two main types of data source used by geographers are published or 'archival' material and field observations (Figure 3.1). Obviously, the use made of each of these will depend on the type of study being undertaken. For example, Haggett (1965) estimated that about 95 per cent of the articles published in human geography between 1960 and 1965 were based on secondary data sources. We may suppose, considering the nature of physical geography, that such a figure could well be reversed on the side of field observations.

Secondary, or 'archival', sources can be classified in terms of whether they are spatial or non-spatial in form. In terms of the former type, areal data traditionally refers to maps, aerial photographs, and more recently remote sensing information from an increasing variety of satellites. A general review of remote sensing techniques is presented by Townshend (1981), while Dickinson (1979) deals with the collection of urban data by such methods. In both human and physical geography information from satellites, mainly in the form of photographs, is becoming an increasingly important data source.

The changed nature of geography has also transformed the role of maps in many branches of the subject. Thus, traditionally the map was very often the geographer's single means of description, as well as being used to measure the locational properties of objects. While these facets are still important, maps are now just one of a number of data stores for locational information that are available to the geographer. Indeed, some writers believed that the growing interest in quantitative methods would result in the map being replaced by the data matrix as the geographer's primary information system (Smith, 1975). As Figure 3.2 shows, the geographical data matrix in its most general form simply contains information on location, time and conditions. As both Cole and King (1968) and Smith (1975) demonstrate, almost all map-based data can be transformed into an information matrix. The popularity of such approaches stems largely from the use of computers and the organizational advantages of such matrices in the application of computational techniques. Equally, such information matrices can easily be converted back into maps, and in this form used in the testing of statements or hypotheses (Lewis, 1971).

In terms of the use of maps as information stores, a number of texts explore the problems of cartographic design (Monkhouse and Wilkinson, 1971; Robinson, 1952; Board, 1967). The use of maps as actual data sources is less well

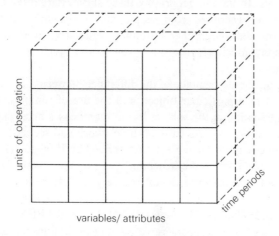

Figure 3.2 Simplified geographical data matrix

documented, although a number of specialist contributions do exist. Thus, Clark (1966) discusses the use of maps as sources in morphometric analysis, while Harley (1980) has reviewed the data contents of Ordnance Survey maps, especially within the context of historical geography.

The non-areal data used by geographers, as we shall see, covers a wide variety of sources, but most share the characteristic of being readily transferred to a map or converted into some type of spatial index. In most cases, however, these information sources have not been compiled for geographical purposes and therefore a number of problems arise in their application. First, many of the units within which the data are collected are unsatisfactory for some types of geographical analysis, usually because such areas vary in size and shape (see Section 10.6 for a discussion of these variations on the results from statistical tests). Second, the release and publication of official data is often only at specific time intervals, such as every ten years for the full British census, which may not be frequent enough for those studies interested in changes over time. Finally, the overall accuracy of these sources is outside our direct control, and as we shall see, many do suffer from problems of inaccuracy.

Geography has traditionally used fieldwork as a major source of information, although field techniques have become increasingly more sophisticated. Thus, much of the early fieldwork merely involved visual observation and description, based on the geographer's powers of perception. Such observations can be recorded by field sketches, written accounts or on field maps. Unfortunately, this type of data collection is very subjective because it depends on the perception of individual geographers. It should be noted, however, that direct field observation is still an important process in the development of ideas, and is often the forerunner to field measurements. It is in the area of field measurement that we can recognize the greatest changes in the processes of data collection. Such measurements can range from the simple enumeration of objects — for example counting numbers of shops in a study of service provision — through to the setting up of large-scale systematic surveys. In human geography these surveys are often associated with collecting views and responses from people, and questionnaire design is now an important part of the geographer's training. Similarly, in many branches of physical geography much emphasis is given to the systematic monitoring and collection of data.

Finally, mention should be made of the difference between the total enumeration and measurement of a particular object, and the use of samples. The increasing amount of quantification in the subject has in many cases necessitated the use of samples in the collection of data from both fieldwork and secondary sources. The traditional prejudices against sample data described by Haggett (1965) are now no longer such a problem, since the acceptance of sampling methods is widespread throughout all branches of the subject (see Section 3.4).

3.2 SOURCES IN HUMAN GEOGRAPHY

As we have already seen, a considerable amount of data in human geography is derived from secondary or 'archival' sources, whose great diversity precludes

Sigrid Denjd

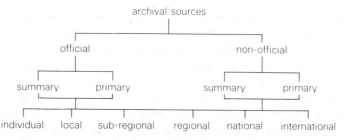

Figure 3.3 Typology of archival sources for geographers

complete coverage. Only the most significant ones are examined here. These sources can be subdivided into a number of different groups, in terms of their origin, purpose and content. Thus, at the most general level we can distinguish between those issued for official purposes, and non-official sources. Furthermore, within each of these broad categories a distinction can be made between primary statistical sources and those which merely summarize other, more detailed, published statistics. As Figure 3.3 shows, each of these categories can be further subdivided according to the geographical scale at which the data is issued and, finally, the topic covered. This simple classification can be used as a framework for our review of 'archival' sources.

In terms of official sources we can start our review by comparing the availability of statistics issued by international bodies with those produced by individual governments. For example, Table 3.1 illustrates the variety of sources available at

Table 3.1 Selected data sources available at an international level

Topics	Worldwide	European Community
General	UN Statistical Yearbook (1948–)	Eurostat (1968–) General statistics (11 issues/year) data on short-term economics trends Basic statistics (1960–)
Population/ social trends	UN Demographic Yearbook (1948–)[1]	Demographic Statistics (1977–)[4]
		Regional Statistics (1971–)
Agriculture	UN Yearbook of Food and Agricultural Statistics (1947–)[2]	Yearbook of Agricultural Statistics (1970–) Agricultural Statistics 6 issues/year, and annual volume, 1959–)
Trade and industry	UN Yearbook of International Statistics (1950–)	Industrial Statistics, quarterly (1959); also a Yearbook (1962–)
	UN Yearbook of National Account Statistics (1957–)[3]	External Trade Statistics (1962–)[5]

) = date of first publication. 1 Contains introduction about previous statistics published by League of Nations, 1922–42. 2. From 1950 onwards publication issued in two volumes, one covering production, and the second, trade and commerce in agriculture. 3. Gives details of accounts by country, e.g. GNP and distribution of national income. 4. First edition in 1977 covers statistics for the period 1960–76, and annually thereafter. 5. *A Users Guide to External Trade Statistics* was published in 1982.

a world scale, produced by the UN and OECD. In most instances these are essentially summary statistics, collated from the official statistics of each individual country. Similar, though more detailed, statistical publications are also available for particular organizations of countries, such as those produced by the European Community (Table 3.1) and EFTA. Such sources are extremely valuable for comparative work, although in the use of summary statistics consideration needs to be given to the type of classification system used. Different individual countries may well have used very different methods of classification and the summary statistics may not resolve such differences, thereby hindering any comparison. An outstanding example of this problem is the definition of 'urban', and the variety of criteria used by national censuses for delimiting urban populations. The attempts to overcome this problem are discussed in a number of United Nations publications, but particular attention is drawn to *Growth of the World's Urban and Rural Population, 1920–2000*, published by the UN Department of Economic and Social Affairs in 1969. Similar types of problem associated with variations in the classification of industrial activities are discussed by Smith (1975) in a comparison of the Standard Industrial Classification used in the USA and the UK.

In the case of the European Community the problems of comparative statistics are being solved by applying standard definitions and in some instances by instituting new surveys for all member states. In this way a comparable set of statistics is becoming available, creating a stronger data base for the geographer.

Table 3.2 Selected official published statistics available for the UK

Topic and source	Frequency of publication	Nationally	Availability County level	Settlement
1. *Population/social*				
Census reports[1]	10-yearly	*	*	*
Housing Statistics (GB)	Quarterly	*		
Local Housing Stastistics (E & W)	Quarterly	*		
Digest of Health Statistics	Annually	*		
2. *Production*				
Agricultural Statistics	Annually	*		
Census of Production	10-yearly	*		
Business Monitor	Periodically	*		
3. *Distribution*				
Census of Distribution (1951–71)	10-yearly	*	*	*
Business Monitor	Periodically	*		
4. *Transport*				
Passenger Transport in GB	Annually	*		
Highway Statistics	Annually	*		

*Indicates availability

In any country the government is the main compiler of statistics, which are published in a variety of forms. This variety can be illustrated in the case of the official statistics published in the UK (Table 3.2). Initially such statistics were usually collected as a by-product of routine administration. However, in advanced economies the collection of statistics becomes more important in the planning of the economy and greater emphasis is placed on the integration of data sources. Despite the recognition of this fact many government statistics are still not available at either a comparable spatial scale or for the same time period. One useful publication that presents data on a variety of topics at a common scale is *Regional Statistics*, which gives information for counties and standard statistical regions. These regions correspond to the Registrar General's Economic Planning regions, although their definitions did change slightly in 1967. Other examples of standard classifications adopted by official sources in the UK are the Standard Industrial Classification and the Registrar General's socio-economic groupings.

A further problem concerns the changing nature of official statistics over time. Unfortunately, for a variety of administrative reasons, the way in which many of these official statistics are collected may have changed, thus making statistical comparisons between one time period and another more difficult. In order to understand the scale and importance of such changes it is worthwhile consulting some of the guides to official statistics. A number of these exist, the most notable being the Central Statistical Office's *Guide to Official Statistics* (1980) and Edwards' *Sources of Economic and Business Statistics* (1972).

In working with official statistics the geographer is often tempted into believing that such data are always totally accurate; but in fact all sources contain some degree of measurement error. Often in official sources this may take the form of under-enumeration, as is the case with the Censuses of Distribution. The first of these censuses in 1951 covered 91 per cent of retail establishments, while in 1961 this figure fell to 88 per cent. What we would need to consider is whether such variations were random or biased, and only related to certain types of shops. Sometimes these measurement errors are due to a lack of cooperation between organizations or people, resulting in less than a total response to government/official surveys. In other cases such errors arise owing to a mishandling of data, which is often more difficult to detect.

Our survey of national statistics has used as one example the data available in the UK. If, however, we are interested in obtaining detailed official sources for other countries we can conveniently use statistical yearbooks, which are published by most countries. One of the most detailed and comprehensive of these is that issued by the USA, which contains not only a wealth of statistics, but also a section on their collection.

Given the large amount of official data that are available it is hardly surprising that human geographers use non-official sources in a limited way. Perhaps two of the most commonly used of these non-official sources are town plans and directories. The former include a variety of fire insurance plans giving details of property ownership in urban areas and, since 1967, Goad plans of British shopping centres (Rowley and Shepherd, 1976). These indicate street numbers, the name of the occupant and the type of retail activity at a scale of 1:1056. Both types are useful sources for data on land-use changes in city centres. However, of greater use are directories that are available in a wide variety, ranging from town to international directories (Bull and Bull, 1978). The importance of directories is that many allow the accurate location of data and give information about individuals and organizations which is not often available in aggregated official statistics. However, problems of inaccuracy often hinder their use, and measurement errors are sometimes extremely difficult to determine (Shaw, 1982).

From the geographer's point of view one of the major problems associated with most published statistics is that the spatial units used for compiling them are seldom meaningful, being often irregular in shape and size, and in some instances changing over time. These problems are well illustrated by the enumeration districts used in the British census of population, which are of variable shapes and, depending on population movements, change from one census to another. Furthermore, many different area divisions have been used in the British censuses, the areal base for the 1971 census being the most complex. This provided statistics for local government areas as defined both before and after the 1974 reorganization, and in total some twenty-two different types of areal units were used (Denham, 1980).

There are two possible methods of solving the problems of unsuitable areal units. The first of these is to compile the official data in statistically sensible and consistent spatial units. To some extent this was attempted in the UK with the 1971 census, which made data available on a 100 m square grid basis. These grid squares were based on the National Grid system and had the advantages of being spatially regular and permanent over time when compared with enumeration districts (Table 3.3). However, they also have their disadvantages in that they do not correspond with any physical features, and in some instances grid lines may be drawn through the middle of a dwelling.

The second method of overcoming these problems is for the geographer to collect his own data from field surveys, and it is to this source that we now turn our

Table 3.3 Advantages and disadvantages of grid squares and enumeration districts

Strengths and weaknesses	Enumeration districts	Grid squares
Ready availability of maps showing boundaries	Yes	Yes
Permanence of boundaries and spatial regularity of areal base	No	Yes
Correspondence with physical features of administrative areas	Yes	No
Correspondence with physical features or administrative areas	Yes	No
Areas recognized in planning	Yes	No
Limited variability of population in areal base	Yes	No

attention. One of the most significant ways of collecting information in human geography is through the use of questionnaires. These require careful planning and preparation and their effectiveness is largely determined by their design. In geography questionnaires are used to measure such things as behaviour, perceptions and attitudes, and characteristics. In most instances a questionnaire may attempt to collect information on all three areas, as for example in work on consumer studies where data are required on patterns of shopping behaviour and attitudes of various consumer types. Obviously, each one will require different approaches in terms of the structure and design of the overall questionnaire.

Table 3.4 Stages in the questionaire design process

1. *Initial decisions*
   (a) Decide what information is required
   (b) Who are respondents?
   (c) What method of survey approach will be used?

2. *Content of questionnaire*
   (a) Which questions are essential?
   (b) Is the question sufficient to generate the required information?
   (c) Are there any factors that might bias or negate the response to the question?

3. *Phrasing and format of questions*
   (a) Is the question phrased correctly or could it mislead respondents?
   (b) Can the question be asked best as an open-ended, multiple-choice or dichotomous question?

4. *Layout of the questionnaire*
   (a) Are the questions organized in a logical way?
   (b) Is the questionnaire designed in a manner to avoid confusion?

5. *Decision to carry out pilot questionnaire*

Only a brief guide is given here to the techniques of questionnaire design, and for more detailed information the reader is directed to more specialist texts (Berdie and Anderson, 1974; Payne, 1951; Oppenheim, 1966). The design of questionnaires can be summarized under five major headings as illustrated in Table 3.4. In the first instance the decisions concern the type of information required, the population under consideration and the method of reaching the respondents. These problems relate not only to the construction of the questionnaire, but also to questions of data collection and survey design which are outlined in Section 3.4. In terms of the phrasing of individual questions one of the major decisions is whether or not open-ended questions should be used. These leave the respondent free to offer any reply that seems appropriate and the answers are not pre-consolidated by a set of response categories. However, these questions suffer from interviewer effects, whereby interviewers vary in their objectivity, and the responses are also extremely difficult to handle in an objective manner and often preclude the use of statistical techniques (Dohrenwend, 1965). A further critical area is the layout of the questionnaire, which is a frequent source of error. In general the initial questions should be simple and interesting in order to gain the attention of the respondent, whilst overall one topic should follow on logically from the next. Finally, the questionnaire should be kept as short as possible, although in instances where respondents were asked to recall past events longer questions appear to be most effective (Laurent, 1972).

A critical problem in the design of questionnaires is measurement error, since differently worded questions on the same topic often produce very different response rates. In one study by Noelle-Newman (1970), for example, a group of housewives were questioned on the same topic using slightly different questions. Thus, in answer to the question 'would you like to have a job if this were possible?', 81 per cent stated that they would like a job, compared with only 32 per cent who made the same reply to the question'would you prefer to have a job, or do you prefer to do just your housework?'. The second version of the question is more explicit and produced a quite different response. Many of these problems associated with questionnaires often only come to light during their application, and it is for this reason that they should be tested before the full survey is undertaken.

The collection of primary data using field surveys is one important method of solving some of the problems associated with archival sources. However, secondary sources are still used extensively because they are so readily available, whereas primary data are often very costly to collect. In some circumstances this type of accessibility can also be given to data collected in the field, through the creation of data banks. A large number of these now exist, some of the earliest experiments having been pioneered in Sweden during the mid 1950s, using census data. In the UK one of the most significant developments has been the SSRC Survey Archive based at Essex University, which is concerned with the dissemination of computer based files containing social and economic data (Tanenbaum, 1980). Furthermore, a growing number of geographical data banks are being initiated, particularly in the field of industrial geography. Some of these are based on questionnaire data collected by surveys of individual firms, and information is therefore available from the data banks at an establishment level (Shaw and Williams1982). The growing importance of such geographical data banks has been assisted not only by the awareness of the need to collect primary data systematically at suitable spatial scales, but also by the use of computer techniques.

3.3 SOURCES IN PHYSICAL GEOGRAPHY

Data sources provide problems for physical geographers that are not necessarily identical to those encountered by their colleagues in the 'human' fields. The differences tell us a good deal about the characters of the two sides of the subject. Human geographers, as we have seen, often express their data within a broadly spatial framework, whereas physical geographers will often examine variations through time at given points in space. Discharge and solute variations at gauging stations or rainfall and temperature variations at given weather stations are good examples of this aspect of the geographer's work in which the spatial element may be subordinate to the temporal. This distinction in methods of data collection and sampling design need not necessarily suggest any fundamental conflict in geographical philosophy. In essence the physical geographer's greater dependence on first-hand field data often requires that he confine himself within stricter spatial limits than do those who rely principally on published material collected by, for example, national governments with substantial organizational and financial means at their disposal. Both logistical and financial considerations can require that the physical geographer gather his data within very clearly defined and limited areas. In some measure we attempt to compensate lack of spatial extensiveness by collecting material over long periods of time. It is, for instance, far more practical to construct one river gauging station and to monitor behaviour over a long period (or intensively over a shorter length of time) rather than to install several such stations throughout a drainage basin. Cost alone would present a significant impediment to the latter course. The same arguments apply to the many notable individual efforts in meteorological studies, and the published works of Manley (1942) testify to the acute practical difficulties in obtaining data from isolated areas. Nevertheless, the many national meteorological organizations have been instrumental in establishing networks of individual observers conforming to standard observational practices.

The question of standardizing observational conditions raises some important issues for published sources in physical geography, for while they assume a less important role than in human geography, they are far from negligible, and gain notable significance in hydrological and meteorological studies. There is also a substantial body of unpublished archived data open to the researcher. Paramount in both the published and archived categories are meteorological data, in some cases extending back over 100 years, held by the national organizations. Even such a wealth of information is not without its problems, however. The observational networks were, and still are, subject to change as stations open and closed, a problem that is more acute with regard to amateur stations rather than those run by official bodies who can replace staff and retain a continuity of record. Hence the data may be drawn from a different network of stations at different times. At any given time the stations are point locations whose representativeness of the local conditions may not be easy to establish, because rainfall and, more notably, temperature are subject to small-scale variations due to topography and exposure. Nevertheless organizations such as the British Meteorological Office have gone to considerable lengths to ensure that while any station is on its network it conforms to standards for exposure and observational procedures (Meteorological Office, 1982). There is also a system of regular station inspections to ensure that these standards are maintained. Such consistency cannot, however, be guaranteed for observations taken before these practices became general, nor can it be assumed for stations not on any official network. But this form of data need not be rejected too hastily, and some workers, through patient enterprise, have been able to standardize information from diverse sources. Manley's Central England temperature reconstructions (Manley, 1974) are a good example of this approach.

National weather networks usually possess a wealth of data which, covering a wide range of daily observations, are archived and published only in summary form, often on a monthly basis. Great Britain is particularly fortunate in this respect with its publication *British Rainfall* (Meteorological Office) which first appeared in 1862 and was published annually until 1968. Initially 500 rain gauge stations were noted, but before ceasing publication this annual report included information from over 5000 sites. The monthly rainfall for the year was given for each station, but other items included snow surveys, chronologies of heavy rainfalls, potential evapotranspiration estimates and notes on thunderstorms. Many of these themes were elaborated by explanatory and analytical notes. The original data from which the summaries were abstracted are archived and available from the Meteorological Office, Bracknell, England. A charge is levied for all such material, but organizations are sympathetic to genuine student requests and strive to keep costs to a minimum. Nevertheless it is always advisable to enquire beforehand concerning costs.

Similar meteorological information is widely available in other countries. In the USA, for example, information can be obtained from both the National Center for Atmospheric Research, Boulder, Colorado or the National Climatic Center, Asheville, North Carolina.

Another subject for which archived and published data are available is hydrology. The USA, especially, has an extensive river gauging network run by the US Geological Survey and maintained to a high level of efficiency. Furthermore, though the spread of stations is necessarily thin in such a large country, this body of data is more regionally applicable than its meteorological counterpart as it represents events occuring within the basin above the sampling point. Once again the data are generally published only in summary form, though the Water Supply Papers of the US Geological Survey contain more detailed material. In England the only published material appears in *Surface Water: United Kingdom* (Water Data Unit, 1978) which has replaced the *Surface Water Yearbook* series. This publication dates back to 1938 when data for 28 stations were included, since when, with only one break in 1952, regular summaries of discharge data have appeared with currently 1205 stations on the network. However, only annual mean figures are now available through these publications, though monthly data were given at one time. Fortunately, the original daily readings are held centrally at the Water Data Unit, Reading, England, where there is an organized system of data retrieval for enquirers. The storage is now fully computerized and can supply not only daily, or monthly, discharge but also run-off equivalents, rainfall data and even annual hydrographs. But, unlike some American stations, there is no organized system for data collection on dissolved or suspended sediment loads. Many of the data collecting agencies (the Regional Water Authorities) do have *ad hoc* sampling schemes for water quality, but this information is not published.

Other countries, particularly in Europe, can offer similar published data. A very good example is Spain, where the Centro de Estudios Hidrograficos has published daily discharge data for its many gauging stations. The publications are based on 10 regional groups with data extending back to 1965. In addition to the daily discharges with monthly and annual summaries, the publications include data for evaporation rates and meteorological conditions. There are also separate volumes dealing with suspended sediment and water quality estimates, though the number of stations engaged in such comprehensive studies are fewer. At the genuinely global scale there are some discharge data available on the world's major rivers (UNESCO, 1969), though inevitably such data have limited application.

The difficulty with all such published materials, and a problem common to all branches of geography, is that the data have been collected with aims that may not coincide with those of a specific geographical project. Daily hydrological data, for example, will often conceal flood hydrographs of a few hours' duration. Similarly, patterns of diurnal temperature variation are lost when only one reading is taken each day. Even terms such as 'daily' and 'yearly' need to be interpreted with care. In meteorology the British rain day extends over the 24-hour period starting at 09.00 GMT, while the sunshine day covers the normal midnight-to-midnight span. In hydrology the year runs from October to September. Users of published data must always be alive to such idiosyncracies.

In brief, although published sources for meteorologists and hydrologists do exist, many workers prefer, or are compelled, to collect their own data; and for almost all other forms of detailed and accurate physical geographical data the researcher is thrown back on his own resources. Soil, landform, karst and glacial studies are obvious members of this category. It is not the purpose of this book to instruct in either field or laboratory methods, and the reader is referred to texts such as that edited by Goudie (1981) or to the booklets published by the British Geomorphological Research Group that deal with specific techniques. Nevertheless, familiarity with a range of field methods is indispensible. Equally important is the need to establish a sensible and workable sampling framework in either a spatial or temporal context. These issues are discussed in more detail in the following section; but it is worth indicating at this stage that physical geographers frequently need to compromise on what is wholly desirable in the face of practical difficulties. Accessibility to desired sites can be obstructed by land ownership problems, for example, while the less inhabited areas often present physiographic obstacles or problems arising from their remoteness.

Nevertheless, physical geographers are not without some published and documentary sources in areas beyond meteorology and hydrology. Accurate topographic maps such as those currently produced by many European governments and in the USA can be a direct source of data themselves. Though never widely adopted, map-based studies have a long history in geomorphology (Gannett, 1901; Wooldridge, 1928; Strahler, 1950, 1952, 1956; Chorley, 1958; Brown, 1951) and should not be too readily dismissed. Their obvious application lies in morphometric studies, and with recent advances in map preparation by photogrammetric methods contour accuracy can be standardized regardless of elevation — whereas previously the density of surveyed, as opposed to interpolated, contours decreased dramatically with elevation. Corresponding developments in the field of automatic digitizers permit map coordinates of contour intersections along transects and areas enclosed by contours to be accurately measured for transmission to computer systems. In this manner plan views of meandering streams, height variations along selected transects and hypsometric curves can all be prepared. Caution is, however, required when map-based data are gathered. Line transects taken across contour maps pose special difficulties in that, for both hillsides and river profiles, relative error terms are only reduced by ensuring that large numbers of contours are crossed by the transect lines. Hence, hillslope profiles based on, say, 25 contour crossing will be more usefully representative than one based on only 4 crossings. This is because, in effect, the number of points used to construct the profile is determined by the number of contour crossings. This problem is reviewed more fully in Strahler (1956) and Wheeler (1979). Hypsometric (height–area) studies pose fewer problems in this respect, but there are a large number of measures available, not all of which are generally applicable. These difficulties are reviewed exhaustively in Clark (1966).

In particular instances geographers have been able to call upon some apparently unusual sources to complement their more traditional forms of evidence. Most notably, the reconstruction of past climates can be aided by historical documents of startling diversity. A good review of such methods can be found in Lamb (1982); and an example of a specific application is the work of Moodie and Catchpole (1975), who employ documentary evidence to interpret environmental changes in the Hudson Bay area between 1714 and 1871. For studies of more recent events sequential editions of topgraphic maps can be used to establish, for example, channel pattern changes and, moving further back in time, ships' logs, estate accounts and farm records all help to piece together a picture of the prevailing environment. As Bell (1970) has demonstrated this process can be extended to any form of documentary record including that of the early Egyptian societies who kept records of flood levels of the River Nile. However, this type of data must always be used with extreme caution as its objectivity and accuracy are at best questionable and its interpretation hazardous. These are difficulties familiar to all historical geographers, which brings us full circle to a point where the distinction between human and physical geography, in terms of data sources at least, is obscured.

3.4 RESEARCH DESIGN AND DATA COLLECTION

Research design is used here as a term to encompass a number of different organizational stages in the gathering of information. Once we have defined our subject of investigation the next step is to collect the necessary data with which to test our ideas. The problem posed is one of research design, which requires decisions on what information to generate, the methods of data collection, the coverage of the data, and the way in which the data are to be analysed. In ideal circumstances these stages in the design of any research project can be viewed as a sequential series of logical steps as outlined in Table 3.5. As we have shown in the preceding sections the major methods of data collection are associated with the use of either 'archival' sources or primary fieldwork. Measurement techniques will vary considerably, within each of these major categories, but both have in common the problem of how much of the 'population' to measure.

Table 3.5 Stages of research design

---

1. Define the problem and the type of information required

2. Select the method of data collection—determine whether secondary or primary data are required, or some combination

3. Select the technique of measurement—for example, decide whether to use questionnaires, and of what type

4. Decide whether to measure total population or sample; if the latter, select sample (see Table 3.6)

5. Determine the appropriate means of analysing the data (see Chapter 4)

Geographers have become increasingly involved with this problem of coverage. As Haggett (1965) suggests, there are two ways in which it may be overcome: first, in a direct fashion by increasing the amount of data available through the accumulation of information, which may be achieved by building up a computer data bank of, say, a number of different surveys; and second, by the use of sampling methods, to which we now turn our attention.

There are a number of factors that need to be considered in the application and use of samples, as illustrated in Table 3.6. The initial step is to specify the population and the individuals contained within it. To the geographer the population in this sense is composed of objects, events or numbers; but as Harvey (1969) points out, geographers have spent very little time analysing the nature and mode of definition of the populations in which they are interested. To some extent the conceptual problems surrounding the definition of geographical populations are not of direct importance here, although as we shall see such considerations do impinge on the interpretation of statistical tests (see Section 11.2). For the present we are merely concerned with the practical issues of defining such populations. For instance, before we could draw a 5 per cent sample of consumers in a particular city, we would need to define the total number of consumers within that area.

Table 3.6 Stages in the sampling process

Stages	Processes
1. Define the population	Defined in terms of (a) units, (b) elements, (c) area, (d) time period
2. Define sampling frame	How the elements of the population can be described
3. Specify sampling unit	Identify units for sampling, e.g. city streets, households
4. Determine sampling method	Method by which units are to be sampled, e.g. probability versus non-probability schemes
5. Determine size of sample	The number of units to be selected
6. Specify sampling plan and method of collecting data	The operational procedures necessary for selecting sample data

From this point we can proceed to the second stage and develop a sampling frame, which locates the individuals within the population. Typical sampling frames list all the objects within the population, and in geography some commonly used ones are electoral registers, street directories and maps. In many geographical problems, however, the construction of a sampling frame is very difficult. For example, sampling mobile populations such as car traffic or river water, where sampling needs simultaneously to be over space and time, poses considerable problems (Harvey, 1969). A further common circumstance is when the total population is not known and is impossible to determine accurately, as for example in some Third World countries with out-of-date or non-existent population census data. In an attempt to ease such problems Krumbein and Graybill (1965) distinguish between a 'target population', the members of which are not all available for sampling; and a 'sampled population', which is available for sampling. This type of sampling, frequently used in geography, is exemplified by Davies and Bennison's (1978) study of consumer perception using an available sample of students to construct mental maps of Newcastle's central shopping streets. In these situations the use of 'sampled populations' precludes probabilistic inferences (see Chapter 6), although conclusions may be extended to 'target population' on the basis of informed judgement.

The third step in the sampling process is to specify the sampling unit, the selection of which depends on the nature of the topic, the sampling frame and the overall design of the project. Thus, in a study of residential mobility our basic sampling unit would ideally be individual households, although data constraints may force us to work with aggregated census information at the enumeration district level. In geography, spatial sampling from maps is obviously of importance, and a number of geometrical sampling units may be used, including points, lines or transects, and quadrats. Each has its advantages and disadvantages which are well reviewed by Haggett (1965), Berry and Baker (1968) and Harvey (1969). For example, quadrat sampling units are usually simple and cheap to use, particularly for the study of point patterns, whereas point sampling of point patterns is less efficient (Greig-Smith, 1964).

A critical, but often misunderstood, stage in the sampling process is to choose the method by which the sample units are to be selected. This is a problem that has received considerable attention among geographers, with discussions ranging from

Table 3.7 Types of non-probability sampling schemes

Types	Characteristics
Convenience	May just select first group of units from population, i.e. first 200 consumers to interview in street
Purposive	Sampling units are selected subjectively by research worker, on basis of background knowledge
Quota	Selection of sample that is as close as possible a replica of the population

the practical aspects of sampling design through to more general reviews (Haggett, 1965). A wide range of sampling methods exist, but an essential distinction can be made between purposive or non-probability sampling and probability sampling.

In the case of non-probability sampling, sample units are selected for economy or convenience, while at the same time representing the characteristics of the population. The reliability of these types of samples depends to a large degree on the skill and knowledge of the research worker. As Table 3.7 shows there are three main types of non-probability sampling schemes, ranging from the simple convenience sample, where the main criterion for selection is ease of collection, through to quota samples, which are selected to represent, as closely as possible, a replica of the population. For example, in a study of consumer behaviour we may select quota samples based on such controls as age, income and geographical location. These methods are used fairly often in geographical field surveys as an alternative to probability sampling schemes. They do, however, suffer from two organizational problems: first, interviewers may fail to secure a representative sample; and second, the method makes the strict control of fieldwork rather difficult, especially if a large number of controls are used. Thus, each set of controls produces a separate cell in a quota sample. For example, if the selection of the sample was controlled by six age brackets, four income ranges and four geographical areas there would be $6 \times 4 \times 4$ or 96 cells in the sample, each of which would require survey information (Kish, 1965).

A particularly common type of purposive sampling used by geographers has been that of the case study. As Blaut (1959) points out this approach amounts to the selection of a 'typical individual' such as a typical city, river or farm. As Harvey (1969) illustrates, however, the problem with the case study approach is its lack of generality and the fact that, statistically, the inferences made from the typical example are impossible to control with respect to the total population.

Given the problems that exist with most types of purposive sampling methods it is not surprising that greatest emphasis is given to probability sampling. A number

Table 3.8 Types of probability sampling designs

Design types	Characteristics
Simple random	Assign to each population element a unique number; select sample units by use of random number of tables (time-consuming for large surveys)
Systematic	Determine the sampling interval, e.g. every fifth individual; select the first sample unit randomly, and select remaining units according to interval
Strafitied proportionate	Determine strata; select from each stratum a random sample of the size dictated by analytical considerations
Cluster or area sampling	Determine the number of levels of cluster; from each level of cluster select randomly, or stratify sample (ideal for large-scale surveys)

of detailed texts exist on the application of probability sampling methods (Kish, 1965; Yates, 1960; Som, 1973), and only the basic facets are examined here.

A probability sample is one in which the sampling units are selected by chance and each unit has a known chance of appearing in the sample. The problem facing the geographer is which of the numerous types of sampling design to select (Tables 3.7 and 3.8).

A simple random sample is a sample of x units selected in such a way that every unit has an equal chance of being chosen. Such a sample could be taken by assigning to each unit in the population a number, use then being made of random number tables (Appendix XI) to select the sample, as is demonstrated by the following example. A simple random sample of 5 students needs to be chosen from a group of 50 students to take part in a study of environmental perception. We must (i) list the students in some form of order, (ii) label the first student 00, the second 01 and so on, (iii) enter the table of random numbers at any line, for example line 109, and read across using the last two digits of each group. Thus, we would select the following students: 09, 12, 38, 16 and 48. It can be seen in this simple example that we must ignore out-of-range labels.

In a stratified random sample, units do not usually have equal changes of being selected, since some strata in the population may be deliberately over-represented in the sample. This method is frequently used by geographers, some of the early applications having been associated with the study of land use (Wood, 1955). It is particularly useful when the environment under study is of an extremely variable nature, as in surveys of coastal sedimentary environments with varying proportions of dunes and intertidal beaches (Cole and King, 1968) or in studies of residential mobility in cities with a diverse range of housing tenure systems. The method of obtaining a stratified random sample is outlined for a simple example in Table 3.9. Such stratified samples have two advantages: first, they make it possible to sample in proportion to the characteristics of the population; and second, by doing so, they increase the level of sampling precision.

A final major type of sampling scheme is multi-stage sampling, in which the sample is selected in stages (Table 3.8). For example, if we were undertaking a

Table 3.9 Example of how to select a stratified random sample

The problem is to select a 10 per cent stratified sample of firms from a survey of 100 businesses of different sizes, namely:

Size of firm (number of employees)	Number in stratum	Sample size
Less than 10	40	4
10–50	30	3
50–100	20	2
Over 100	10	1

In this example, we would then proceed to draw a simple random sample from each of the four different strata.

survey of national trends in consumer behaviour on 5000 households we would in the first stage draw a random sample of 5 districts (stage two), and in the final stage a random sample of 100 households in each district. More commonly such multi-stage designs would be stratified. In our example the strata may be defined in terms of household size and characteristics. This type of sampling scheme is adopted for two main reasons. First, sampling frames may not be available for all units in the population. For example, in a survey of agricultural production in an under-developed country, the sequence may be: stage one, select villages; stage 2, prepare a list of fields within selected villages and sample these fields; stage 3, list individual plots within selected fields and take a sample of these plots. In this way considerable time is saved by not constructing a sampling frame for all of the field plots. Second, multi-stage sample schemes are used even when suitable sampling frames exist, in order to save time and cost in large-scale surveys where samples may be geographically dispersed over a wide area.

The next crucial step in the sampling process is to determine the size of the sample, which depends on the degree of certainty required compared with the resources that are available. Haggett (1963) has demonstrated how the accuracy of a sample increases with sample size, but also that such a relationship is not simple or linear. Indeed, the question of how large a sample to take is related to the concepts of a sampling distribution and the notions of probability discussed in Chapter 6. Here it is sufficient to note a few central points related to the practical aspects of sampling. First, it should be stressed that the form of the relationship between sample size and accuracy has been calculated for most probability sampling schemes. Thus, for a simple random sample the relationship is that the sampling error is proportional to the square root of the number of observations (Berry, 1962). Second, because such relationships are known, it is possible to calculate the required sample size by specifying the allowable error, the confidence level and the coefficient of variation (see Chapters 5 and 6). Tull and Hawkins (1980) illustrate how to construct nomographs that enable sample sizes to be determined quickly for simple random sampling schemes, while Som (1973) discusses the same problem for multi-stage samples.

Unfortunately, very little has been written about determining the size of non-probability samples, despite their widespread use. There is no available basis for determining sampling error, although Stephan and McCarthy (1958) carried out some early work on this problem with regard to quota samples. Generally, sample size is determined purely by the cost and effort involved in collecting information, relative to the available resources.

The final two stages of the sampling process are both concerned with the operational and organizational aspects of data collection (Table 3.6), the details of the procedures varying according to the type of study being carried out. For example, in a questionnaire survey one problem will be to decide how the questionnaires are to be administered, with a choice between personal interviews, postal surveys or telephone surveys (Tull and Hawkins, 1980). Whatever the study, though, some operational procedures or instructions need to be drawn up about how the necessary fieldwork/data collection is going to be carried out.

REFERENCES

Bell, B. (1970). 'The oldest records of the Nile floods', *Geog. Rev.* **136**, 569–573.

Berdie, D. R., and Anderson, J. F. (1974). *Questionnaires: Design and Use*, Pub. New York.

Berry, B. J. L. (1962). *Sampling, Coding and Storing Flood Plain Data*, United States Department of Agriculture.

Berry, B. J. L., and Baker, A. (1968). 'Geographic sampling' in B. J. L. Berry and D. F. Marble (eds) *Spatial Analysis: A Reader in Statistical Geography*, Prentice-Hall, Englewood Cliffs.

Blaut, J. M. (1959). 'Micro-geographic sampling: A quantitative approach to regional agricultural geography', *Econ. Geog* **36**, 254–259.

Board, C. (1966). 'Maps as models' in R. J. Chorley and P. Haggett (eds) *Models in Geography*, Methuen, London.

Brown, E. H. (1957). 'Physiography of Wales', *Geogrl. J.* **123**, 208–230.

Bull, C. J., and Bull, P. J. (1978). 'Regional directories as a potential source for the study of intra-urban manufacturing industry', *Geog.* **63**, 198–204.

Central Statistical Office (1980). *Guide to Office Statistics*, HMSO, London.

Chorley, R. J. (1958). 'Aspects of the morphometry of a polycyclic drainage basin', *Geogrl. J.* **124**, 370–380.

Clark, J. I. (1966). 'Morphometry from maps' in G. H. Dury (ed.) *Essays in Geomorphology*, London.

Cole, J. P., and King, C.A.M. (1968). *Quantitative Geography*, Wiley, London.

Davies, R. L., and Bennison, D. J. (1978). 'Retailing in city centres: the characters of shopping streets', *Tijds. voor Econ. en Soc. Geografie* **69**, 270–285.

Denham, C. (1980). 'The geography of the census: 1971 and 1981', *Population trends* **19**, 6–12.

Dickinson, G. C. (1979). *Maps and Air Photographs*, 2nd edn, Arnold, London.

Dohrenwend, B. S. (1965). 'Some effects of open and closed questions', *Human Organisation*, **24**, 175–184.

Edwards, B. (1972). *Sources of Economic and Business statistics*, Heineman, London.

Gannett, H. (1901) *Profiles of rivers of the United States*, U.S. Geol. Surv. Water Supp. and Irr. Paper 44, Washington D.C.

Goudie, A. (1981). *Geomorphological Techniques*, Allen and Unwin, London.

Greig-Smith, P. (1964). *Quantitative Plant Ecology*, 2nd edn, Butterworths, London.

Haggett, P. (1963). 'Regional and local components in land-use sampling: a case study from the Brazilian Triangulo', *Erdkunde* **17**, 108–114.

——— (1965). *Locational Analysis in Human Geography*, Arnold, London.

Harley, J. B. (1980). *The O.S. and land-use mapping*, Hist. Geog. Research Series 2, Geo-Books, Norwich.

Harvey, D. (1969). *Explanation in Geography*, Arnold, London.

Kish, L. (1967). *Survey Sampling*, Wiley, New York.

Krumbein, W. C., and Graybill, F. A. (1965). *An introduction to Statistical Models in Geology*, McGraw-Hill, New York.

Lamb, H. H. (1982). *Climate, History and the Modern World*, Methuen, London.

Laurent, A. (1972). 'Effects of question length on reporting behaviour in the survey interview', *J. Am. Stats Ass.*, 298–305.

Lewis, P. (1971). *Maps and Statistics*, Methuen, London.

Manley, G. (1942). 'Meteorological observations on Dun Fell, a mountain station in Northern England', *Quart. J. Royal. Met. Soc.* **68**, 151–166. (1974). 'Central England temperatures: monthly means 1659–1973' *Quart. J. Royal Met. Soc.* **100**, 389–405.

Meteorological Office (1974). *British Rainfall 1968*, HMSO, London. (1982). *Observer's Handbook*, HMSO, London.

Monkhouse, F. J., and Wilkinson, H. R. (1971). *Maps and Diagrams*, 3rd edn, Methuen, London.
Moodie, D. W., and Catchpole, A. J. W. (1975). *Environmental Data from Historical Documents by Content Analysis: Freeze-up and Break-up of Estuaries on Hudson Bay* 1714–1871, Manitoba Geographical Studies 5, Univ. of Manitoba, Winnipeg.
Noelle-Newman, E. (1970). 'Wanted: rules for wording structured questionnaires', *Public Opinion Quart.* **34**, 200.
Oppenheim, A. N. (1966). *Questionnaire Design and Attitude Measurement*, Heinemann, London.
Payne, S. L. (1951). *The Art of Asking Questions*, Princeton University Press.
Robinson, A. K. (1952). *Elements of Cartography*, Wiley, New York.
Rowley, G. and Shepherd, P. (1976). 'A source of elementary spatial data for town centre research', *Area* **8**, 201–208.
Shaw, G. (1982). *British Directories and Sources in Historical Geography*, Hist. Geog. Research Series 8, Geo-Books, Norwich.
Shaw, G., and Williams, A. M. (1981). 'Creating a data bank for Cornish industrial estates' in G. Shaw and A. M. Williams (eds) *Industrial change in Cornwall*, South West Papers in Geography 1, Plymouth.
Smith, D. (1975). *Patterns in Human Geography*, Penguin, Harmondsworth.
Som, R. K. (1973). *A Manual of Sampling Techniques*, Heinemann, London.
Stephan, F. & McCarthy, P. J. (1958). *Sampling Opinions – An analysis of survey procedures*. Wiley, New York.
Strahler, A. H. (1950). 'Equilibrium theory of erosional slopes approached by frequency distribution analysis', *Am. J. Sci.* **248**, 673–695.
(1952). 'Hypsometric (area–altitude) analysis of erosional topography', *Bull Geol. Soc. Am.* **63**, 1117–1142.
(1956). 'Quantitative slope analysis', *Bull. Geol. Soc. Am.* **68**, 571–596.
Tanenbaum, E. (1980). 'Secondary analysis, data banks and geography', *Area* **12**, 33–35.
Townshend, J. R. G. (ed.) (1981). *Terrain Analysis and Remote Sensing*, George Allen & Unwin, London.
Tull, D. S., and Hawkins, D. I. (1980). *Market Research*, Macmillan, New York.
United Nations Educational Scientific and Cultural Organisation (1969). *Discharge of Selected Rivers of the World* Vols I–III, Paris.
Water Data Unit (1978). *Surface Water: United Kingdom* 1971–73, HMSO, London.
Wheeler, D. A. (1979). 'Studies of river longitudinal profiles from contoured-maps', *Area* **11**, 321–326.
Wood, W. F. (1955). 'Use of stratified random samples in land use study', *Anns. Assoc. Am. Geogrs.* **48**, 350–367.
Wooldridge, S. W. (1928). 'The 200 foot platform in the London Basin', *Proc. Geol. Assoc.* **39**, 1–26.
Yates, F. (1960). *Sampling Methods for Censuses and Surveys*, Griffin, London.

RECOMMENDED READING

Hakim, C. (1982). *Secondary Analysis in Social Research*, George Allen & Unwin. A good and comprehensive guide to official British sources, EEC surveys, the SSRC archives and a variety of non-official surveys.
Rhind, D. (ed.) (1983). *A Census User's Handbook*, Methuen, London. A comprehensive and advanced collection of papers on the use, analysis and mapping of census data.
Short, J. R. (1980). *Urban Data Sources*, Butterworths, London. A handy, introductory text covering data sources and hypothesis testing in most aspects of urban geography.

Chapter 4

Measurement and Statistical Tests

4.1 DATA CHARACTERISTICS AND SCALES OF MEASUREMENT

Deciding on which particular statistical test to use is an initial problem in most statistical analysis, with the choice being partly conditioned by the quality of the data and the procedures by which they are measured. In this context measurement is concerned with the assignment of values to particular objects or events. However, it should be recognized at the outset that measurement exists in a variety of forms owing to the diversity of data used by physical and human geographers. With this fact in mind geographers, along with other scientists, have found it extremely useful to identify different scales along which data can be measured. These scales simply reflect the set of all values for observations, together with the rules for assigning these values to such observations. In his original work on this topic, Stevens (1946) classified four common scales of measurement, each of which identifies and uses different properties.

Table 4.1 Scales of measurement

Scale	Characteristics	Distribution
Nominal	Determination of equality; data can be placed into classes	Discrete
Ordinal	Determination of greater or less; data can be ranked	Discrete
Interval	Determination of equality of intervals or differences	Continuous
Ratio	Determination of equality of ratios; measurements have a true zero	Continuous

These four scales are summarized in Table 4.1 and are presented in ascending order of strength, with the lowest being that of the nominal scale. As one progresses through these scales the data must satisfy more rigorous requirements. For example, the *nominal* scale involves only the classification or naming of observations, and numbers if used at all can be arbitrarily assigned. Stevens originally distinguished two types of nominal assignments, but for our purposes we need only be aware of the fact that at this level of measurement we are very limited in our choice of statistical tests. The simple classification of consumers by sex or occupation provides an example of the nominal scale, as also does the grouping of settlements by the provision of services that they offer. In contrast, the *ordinal* or *rank* scale of measurement involves the ranking of one observation against another. Thus, we may say that one city is larger than another and proceed to place a number of cities in a rank ordered list, even though we may not have any precise information on the actual populations of each place. Alternatively, we may only be in a position to place our cities into a system of broadly defined ranked classes, such as towns with over a million people, those with populations between 500,000 and one million and so on. In both instances we have measured our data along the ordinal scale. The availability of this type of measurement is important for many of the studies concerning perception and the cognitive components of behaviour, when individuals may be asked to rank particular features, for example, in terms of their perceived attractiveness (Downs, 1970).

The *interval* scale permits us not only to sort and rank observations (as for the nominal and ordinal scales) but also to establish the magnitude of the differences separating each observation. However, the data need not possess an absolute zero; thus, the temperature of river water at 20°C is not twice as warm as water at 10°C, since 0°C is merely an arbitrary baseline and is not the coldest possible temperature or absolute zero. Finally we can identify the *ratio* scale, which is the highest level of measurement; and unlike the interval scale it has a known and absolute origin. If we return to our earlier example of city size we could, under the ratio scale, and assuming that we had the relevant data, say by how much one city was larger than another. Most geographers frequently use data measured on the ratio scale, although for the purposes of many statistical tests data on the interval and ratio scales are treated in the same way.

Two further points are worth consideration at this stage. First, all four scales of measurement involve placing observations along some type of quantitative yardstick, although as we have seen significant differences exist between each one. In addition, a further distinction can be made between the nominal and ordinal scales, and the interval/ratio ones, as in the former the measurements can be termed discrete since observations can be placed only at certain points. This is not, however, the case with interval and ratio measurements, where observations can occupy any position along the measurement continuum and hence are termed continous scales. As we shall see this small fact is often very helpful in deciding on the appropriate statistical test.

The second point is that the type of scale depends on the basic methods of measurement, which in turn is related to the characteristics of the data being measured. In some areas of geographical work it proves difficult to assign precise numerical values to particular features, and in such cases the data severely limit the types of statistical analysis that can be undertaken. In other circumstances it is the constraints placed on the methods of data collection that inhibit measurement and determine the scale along which observations are measured (see Chapter 3). This can be illustrated by returning to our examples of city size relationships. Thus, in countries such as the USA and the UK, we can obtain an accurate measure of population for each urban area from census publications, and we therefore have data that can be assigned to the ratio scale. However, in some of the less developed countries such accurate census data may not exist, and settlements may have to be classified in a more general way using broad groups, thus restricting measurements to the ordinal scale.

It is worth while returning to our opening remark to stress again that each type of measurement scale is amenable to certain forms of statistical tests. Therefore, to recognize the scale of measurement at which a set of data are quantified will go some way in helping us to select the appropriate test.

4.2 AN INVENTORY OF STATISTICAL TESTS

You may at some time get confused over which particular statistical test to use in your work. This tends to be a common problem in geographical studies, partly because of the enormous variety of data we use and also because of the many different tests that are available. Clearly, from what we have already said about scales of measurement some guidance exists. However, this is only part of the picture since the selection of a statistical test is also governed by the tests that are available and what they achieve, relative to the geographer's particular requirements.

To aid in this selection process it is possible to classify most statistical tests into a number of fairly broad groups. In this sense we may talk about univariate, bivariate and multivariate statistics; or secondly, we could distinguish between parametric (classical) and non-parametric tests; or finally between descriptive and inferential statistics. These are three major classifications that reveal quite different characteristics about statistical tests. As you may have already observed this book is organized in major sections under the headings of univariate, bivariate and multivariate statistics. A cursory glance at the introductions to each of these sections will give you a good indication of the differences between each group. In very simple terms this breakdown is related to the number of variables under investigation, and the relationship with the scales of measurement concept is fully explored in Chapter 9 and summarized by Table 9.2.

The major distinguishing feature between parametric and non-parametric statistical tests is concerned with certain assumptions about the background populations from which samples are drawn (have a close look at Chapter 8). Thus, parametric tests often require that the population from which samples are drawn be approximately normally distributed (this term and its implications are discussed in Section 7.4). In contrast, non-parametric statistics impose no such requirements and are therefore commonly referred to as distribution-free tests. These properties closely relate back to our ideas about scales of measurement (Table 4.1). For example, parametric tests can only really be applied to data measured on the interval and ratio scales, whereas non-parametric statistics are more flexible and can be used with all four scales of measurement.

4.3 NON-PARAMETRIC TESTS

The development of non-parametric statistics is fairly recent and much of the initial work dates from the early 1940s. Their general acceptance owes much to the work of Siegal, while their widespread use in geography is primarily related to the rise of social and behavioural approaches to the subject. The growth of these themes has in turn very often necessitated using questionnaries and collecting large amounts of data measured on the nominal and ordinal scales.

Table 4.2 Advantages of non-parametric tests as outlined by Siegal (1956)

1. Probability statements obtained from most non-parametric tests are exact probabilities, regardless of the shape of the probability distribution

2. Non-parametric tests can deal with very small samples

3. Non-parametric statistics can be used on data from a variety of measurement scales

4. There are suitable non-parametric tests for analysing samples drawn from several different populations

5. Non-parametric tests are usually easier to apply than parametric ones

In his eagerness to promote non-parametric tests Siegal (1956) stressed a number of their advantages, the five main ones of which are listed in Table 4.2. However, more recently some statisticians have questioned the validity of these statements. Thus, Anderson (1961) counters the argument that all parametric tests require measurements at least on the interval scale and illustrates how one particular parametric statistic can be used to test data on the ordinal scale of measurement. However, it should be stressed that such tests are unusual and that for most cases parametric statistics relate only to interval and ratio data. Similarly, Norcliffe (1977) refutes Siegal's statement about the ease with which non-parametric statistics can be calculated relative to parametric tests, his line of reasoning being that the widespread availability of computers has largely resolved such problems of manipulation. Indeed, this is quite clearly illustrated in the companion volume referred to in the Preface, both in respect of program development and the availability of a wide range of package programs. One further criticism that is worth mentioning is that by Donaldson (1968), who has found that important parametric statistics, such as the F test, can be used with some types of non-normal data (Table 4.2).

These, and other criticisms (Norcliffe, 1977), are worth keeping in mind but do little to detract from the value of non-parametric tests, especially for the handling of nominal and ordinal data sets. If anything such views merely serve to place Siegal's original comments into a more modern perspective, given the recent advances in statistical and computational knowledge.

4.4 DESCRIPTIVE AND INFERENTIAL STATISTICS

The third type of division we have recognized is that between descriptive and inferential statistics, which distingushes those approaches that allow us to describe in numerical terms an event or data set, and those which enable us to infer relationships between variables.

Descriptive statistics, which are discussed in the following chapter, are the simplest but most effective way of summarizing and presenting data. In geographical studies they take a variety of forms, ranging from the use of basic graphs and frequency tables through to a whole collection of spatial statistics that are presented in Chapter 16. In contrast, inferential statistics are concerned with mathematical probabilities and are characteristic of scientific investigation, which involves a search for principles that have a degree of generality. In this type of study the findings are very often applied to an environment larger than the actual cases or sample that were statistically examined. The making of such generalizations from a sample to a population is termed statistical inference.

Inferential statistics allow us to make probabilistic statements about the following:

(1) hypothesis testing — whether a particular supposition is true or false;
(2) the relationships between variables;
(3) the characteristics of the population from which a sample is drawn.

All these topics are discussed in detail in Chapters 8 and 9.

In practice there are two general types of statistical inference, which we can term 'estimation' and hypothesis testing. The first starts without any stated assumptions about the value of the parameter, but seeks to estimate what the value of that parameter is. For example, we may estimate the population mean from the observation of a sample mean. In hypothesis testing we state our hypothesis before we collect the sample data, which are then used to check the validity of our statement.

Despite the importance of inferential statistics in geography many problems still exist in the application of such techniques. In particular, much debate has arisen over the relative merits of both deductive and inductive forms of reasoning in relation to statistical inquiry (see Section 9.3). More specific difficulties have been outlined by Gudgin and Thornes (1974), who have drawn attention to the fact that many geographical data sets do not fulfil the underlying assumptions of these inferential statistics. Furthermore, they stress that many geographers are often far too conservative in their attitudes toward significance levels, most of which they argue are used in merely an advisory capacity to indicate the likelihood of a randomly generated result. Under such circumstances they claim that the test procedures are far less critical. Quite obviously the acceptance or rejection of these points of view will depend on the particular use that is being made of the results in question.

4.5 THE POWER AND EFFICIENCY OF A TEST

A final factor that influences the selection of a test is its relative power. In statistical terms, the power of a test is associated with its ability to state correctly whether a hypothesis is true or false. In this respect the choice may be between a parametric test, which are generally more powerful, or a non-parametric test (within this context reference can also be made to type I and type II errors, as discussed in Section 9.4). The power of a test can also be influenced by the size of the samples that are being considered, and statisticians often use the term power–efficiency ratio which refers to the increase in sample size that is required to increase the power of a test. Thus, a non-parametric test of low power–efficiency requires a larger sample to achieve the same level of power as a parametric test with a relatively higher power-efficiency ratio. The implication of what we are saying is that tests with high power–efficiency tend to require data measured on the interval or ratio scales.

The power–efficiency of a test is important both in respect of the inferences made about the data, and also in terms of the sample size, which in turn may also have a significant bearing on the design of the sampling framework as outlined in Chapter 3. For example, if data collection is very costly then it may be within the interests of the researcher to employ, if possible, a high power test which can operate effectively with a smaller sample than can a non-parametric test. However, if this strategy is followed then greater emphasis will be placed on the scale of measurement, with the major drawback being to impose much stricter assumptions about the population. Usually in most geographical problems this type of choice does not exist, since very often data are restricted in their quality to a particular scale of measurement.

4.6 SUMMARY

This short chapter has attempted to draw attention to those factors that can influence the selection of a statistical test. From our discussion it is evident that such a decision will depend on the scale at which the data are measured, the number of variables under examination and the power–efficiency level of specific tests. If we understand the role of these factors then the selection of a statistical test becomes a rational decision rather than a lucky guess.

REFERENCES

Anderson, N. M. (1961). 'Scales and statistics: parametric and non-parametric', *Psychological Bull.* **58**, 305–316.

Donaldson, T. S. (1968). 'Robustness of the F-test to errors of both kinds and the correlation between the numerator and denominator of the F-ratio', *J. Am. Stats. Ass.* **63**, 660–676.

Downs, R. (1970). 'The cognitive structure of an urban shopping centre', *Environ. & Behaviour* **2**, 13–39.

Gudgin, G., and Thornes, J. B. (1974). 'Probability in geographic research', *The Statistician* **123**, 157–178.

Norcliffe, G. B. (1977). *Inferential Statistics for Geographers*, Hutchinson, London.

Siegal, S. (1956). *Non-parametric statistics for the Behavioural Sciences*, McGraw Hill, New York.

Stevens, S. S. (1946). 'On the theory of scales of measurement', *Science* **103**, 677–680.

RECOMMENDED READING

Hodge, G. (1963). 'The use and mis-use of measurement scales in city planning', *J. Am. Inst. Planners* **29**, 112–121.

Krumbein, W. C. (1958). 'Measurement and error in regional stratigraphic analysis', *J. Sedimentary Petrology* **28**, 175–85. This and the previous paper provide good examples of the implications and consequence of scales of measurement on statistical analysis.

Labovitz, S. (1967). 'Some observations on measurement and statistics', *Social Forces* **46**, 151–60.

Siegal, S. (1956). *Non-parametric statistics for the Behavioural Sciences*, McGraw-Hill, New York. This remains the most comprehensive book on non parametric statistics.

Chapter 5

Data Reduction and Descriptive Statistics

5.1 DATA REDUCTION

In many aspects of geography we are very often faced with the task of collecting and descibing large amounts of data before we can reach any firm conclusions about particular problems. To illustrate some of the potential difficulties let us take a specific case, and consider the information that may be collected from a survey of consumers in order to assess their patterns of shopping behaviour. In this example, taken from a study carried out in central Exeter, some 2000 consumers were interviewed. The total amount of data that needs to be handled equals the number of respondents multiplied by the number of questions on the questionnaire, which in this instance gives us 36,000 bits of data, i.e. 2000 × 18 (the number of questions). Even from a relatively small study such as this we can be left with a surprisingly large amount of information.

To yield useful results these data must first be summarized and reduced to understandable proportions. This process of data reduction involves a number of steps, some of which, such as data editing, coding and the generation of new variables, have already been explained in Chapter 3. Now we can extend our work to include the tabulation of frequency tables and graphs, and the calculation of descriptive statistics. The latter can either provide measures of the mid-point of a distribution or give a general indication of the amount of variation in the data, known as measures of dispersion. In the remainder of this chapter we shall explore the ways in which these different approaches can be put to use.

5.2 TABLES AND GRAPHS

One of our initial tasks in organizing or describing a set of data is usually to count how often each value occurs. This information may then be used to construct either a frequency table or a graph, and by returning to our example of the consumer survey we can illustrate the simple but effective nature of such techniques. Thus, Table 5.1 shows the breakdown of consumers in terms of their methods of travel (a discrete variable) into Exeter's central shopping area, and therefore conveniently describes the travel characteristics of our sample of 2000 shoppers. Similarly, Table 5.2 illustrates the case for a continuous variable, giving the number of consumers making weekly shopping trips from different distances from the city centre. Frequency tables are a common method of summarizing nominal data, although they are equally important and useful for data measured on both the interval and ratio scales. When large numbers of observations are involved then the production of frequency tables is both laborious and time-consuming. It is for such reasons that geographers often rely on computer programs to carry out these tasks, with probably one of the most widely used, especially by human geographers, being that of SPSS which produces frequency tables extremely quickly.

Table 5.1 Breakdown of consumers in terms
of methods of travel into central Exeter
(1979)

	%
Foot	21.7
Public transport	25.0
Motorcycles	2.5
Cars	50.8
	100.0 ($N = 2000$)

Table 5.2 Distances travelled to shop in
central Exerter by consumers living outside
city (1979)

Distance (miles)	Consumers
2 – 4.9	32
5 – 7.9	37
8 – 10.9	44
11 – 13.9	62
14 – 16.9	60
17 – 19.9	38
	273

One important point to consider in the construction of frequency tables is the selection of class limits and class intervals when dealing with interval and ratio data. Class limits should be assigned to avoid any ambiguity and to ensure that there is only one possible place for each item. Thus, if a continuous variable is being measured, for example distance, then the notation used in Table 5.2 is conventional, since it leaves no gaps and each consumer belongs quitely clearly to a particular class. In instances when discrete units are used, as for example if we were examining the number of employees per factory, then the class limits will obviously reflect that no factory employs say, 10.7 workers. Class intervals, the difference between the upper and lower limits of a class, may sometimes vary in size throughout the table in order to preserve the required detail. In relation to this, a central question is how many class intervals should be used? There are two ways of tackling this problem. First, there exist some rules that relate class intervals to the number of observations; thus Huntsberger (1961) suggests that the number of classes equals $1 \times 3.3\log N$ (where N is the number of observations). Alternatively, Croxton and Cowden (1968) give more general guidelines and state that most frequency tables should have no less than six and no more than sixteen classes. A glance at the literature makes it clear that no specific rules exist and that in most instances the number of class intervals can only be decided after a close inspection of the data.

In a great many situations geographers prefer to present information in the form of graphs; especially important are frequency graphs, either as bar charts (used for discrete variables), histograms or ogives. In addition, a variety of graphs exist that relate to more specialist problems, some of which will be encountered in the section on spatial analysis (Chapter 16). However, for the moment we will concentrate on the use of histograms, in which the height of each column is proportional to the frequency it represents (Figure 5.1). The graphs are constructed using the same principles as for frequency tables, with the exception that open-ended class intervals should be avoided. In some instances more information may be revealed by plotting a cumulative frequency curve (ogive)

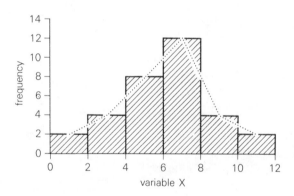

Figure 5.1 Simple histogram and frequency polygon (dashed line)

where the slope of the curve is proportional to the frequency density in a particular range. Thus, the curve changes most sharply where the frequency density is the greatest. Such techniques can be used to represent any type of numerical distribution, although for discrete variables it is usual to plot the data in a step-like fashion. Where the graph illustrates the cumulative frequency of items of 'more than' a given value then the curve falls from the point that corresponds to the overall total, as in Figure 5.2(b). Alternatively, if the graph plots the cumulative frequency of items 'less than' a given magnitude, its curve will rise from zero to the maximum point of the total, presenting the typical sigmoid or ogive type of curve (Figure 5.2(a)).

52

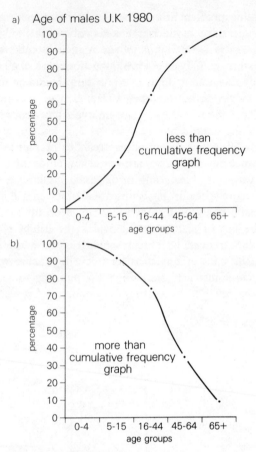

a) Age of males U.K. 1980

less than
cumulative frequency
graph

b) more than
cumulative frequency
graph

Figure 5.2 Cumulative frequency curves or
ogives

5.3 MEASURES OF CENTRAL TENDENCY

A measure of central tendency or average provides a single number to describe
the general magnitude of all observations in a data set. In our work we are likely
to come across five types of statistical average; the mode, the median, the
arithmetic mean, and the less commonly used geometric and harmonic means.
While these statistics have some properties in common they each give a different
measure of central tendency, and are used in varying circumstances.

5.4 THE MODE

The mode is any value occurring most frequently in a set of observations. For
example, the mode of:

10, 9, 8, 12, 12, 7, 12, 12

is 12, as this value occurs four times, and the remainder only once. In terms of a frequency distribution, the mode is the value at which the graph reaches its maximum height. The main advantage of the mode is its usefulness in describing data measured on the nominal scale; thus in our shopping study we could say the car is the most frequently used form of transport, and statistically speaking this is the mode. This measure does, however, have a number of drawbacks, one being that we may have a data set that has no single frequently occurring value. It may, for example, have two equally dominant values (bimodal), or no two values may dominate in which case the distribution is multimodal. Under such circumstances the mode ceases to be a useful measure of central tendency.

Locating the mode in the case of ungrouped data is simple — we just find the most frequently occurring value. However, for grouped data this is established by selecting the modal class, although some caution needs to be exercised over the selection of class intervals since these limits may determine the modal class. In statistics we would say that under these conditions the mode is unstable, as its value can be changed for the same data set merely by changing class intervals. A method does exist to reduce such instability, and this involves using equation 5.1:

$$\text{Mode} = L + \frac{D_1}{D_1 + D_2} \times i \qquad (5.1)$$

Equation 5.1

$D_1 = $ difference between modal frequency and frequency of next lower class
$D_2 = $ difference between modal frequency of next higher class interval
$L \quad = $ lower limit of modal class
$i \quad = $ the class interval

5.5 THE MEDIAN

The median is the mid-point of a data set, where half the values lie above the median and half below. To locate the median the data need first to be ranked. In the case of fairly large data sets of N observations, after placing the values in rank order the median can be calculated by $N/2$. Thus, if we had 430 observations, the median would be, after ranking the data, 430/2 or the 215th number in the ordered list.

A great deal of our work may, however, involve large amounts of grouped data, in which case we need to calculate the median in a slightly different way. It is first necessary to derive cumulative frequencies, as illustrated in Table 5.3; we then apply equation 5.2:

$$\text{Median} = L + \left(\frac{N/2 - cf}{f} \times i \right) \qquad (5.2)$$

> ## Equation 5.2
>
> N = total frequency
> L = lower limit of the class interval in which the median is located
> cf = cumulative frequency
> f = frequency of median class interval
> i = class interval

Table 5.3 Calculation of median for grouped data: population by age, USA (1980)

age	f(millions)	Cumulative f
under 5	16	16
5 – 14	35	51
15 – 24	42	93
25 – 34	37	130
35 – 44	26	156
45 – 54	23	179
55 – 64	22	201
65 plus	25	226
	226	

We must first locate the lower limit L of the median class interval. We start by dividing N by 2: in our example $226/2 = 113$. To locate the class interval we count from the lower end to the 113th case, which is somewhere within the fourth cumulative frequency grouping; it is in the class interval 25 – 34, where $L = 25$. To find the precise position of the median we need also to derive the cumulative frequency of values up to the median class interval, in this case the value before 130, which is 93 (cf). Finally, we need the class width i, which in Table 5.3 is 10. We now have all the necessary information to use equation 5.2; thus:

$$\text{Median} = 25 + \left(\frac{113 - 93}{37} \times 10 \right)$$
$$= 25 + 5.4$$
$$= 30.4 \quad \text{years}$$

Incidentally, the equation also works on percentage frequencies, with the proviso that in equation 5.2 f would be changed to equal the percentage frequency of the median class and cf the percentage cumulative frequency. As the median measures the mid-point of a set of values it is a useful and frequently used statistic that can be applied to ordinal, interval and ratio type data.

5.6 THE ARITHMETIC MEAN

The mean is derived by summing all the observed values and dividing by the number of observations, as shown by equation 5.3:

$$\bar{X} = \frac{\Sigma X}{N} \qquad (5.3)$$

Equation 5.3

$\bar{X}$ = mean

ΣX = sum of all values

N = number of observations

The mean reflects the magnitude of every individual value and any data set can only have one mean. Unlike with the median the observations need not be ordered. For these reasons the mean is an important and commonly used measure. Moreover it can also be manipulated algebraically since the means of sub-groups can in some cases be combined to produce a total mean.

The importance of the mean goes beyond its immediate measure of central tendency, since it posses a number of significant properties. First, the sum of deviations from the mean is always zero — in other words the amounts by which high values exceed the mean are equalled by those which fall below. Second, the sum of the squares of the deviations from the mean is at a minimum; the importance of this will become clear later in this chapter.

As with the mode and the median, when we are calculating the mean of grouped data from a frequency table an alternative approach is required, as outlined in equation 5.4.:

$$\text{Mean } (\bar{X}) = \frac{\Sigma m_j f_j}{\Sigma f_j} \qquad (5.4)$$

Equation 5.4

see Table 5.4 for definitions

Table 5.4 Calculation of mean for grouped data: population sizes of Dorset villages (1979)

Settlement size	Mid-point m	Frequency f	mf
0 – 49	25	53	1325
50 – 99	75	67	5025
100 – 149	125	29	3625
150 – 199	175	36	6300
200 – 249	225	16	3600
		$\Sigma f = 201$	$\Sigma mf = 19875$

$$\therefore \bar{X} = \frac{19875}{201}$$

The mean population is therefore 99 people

Sometimes it may be necessary to weight the means of sub-groups before they are combined together to give a mean of the whole data set. If this is the case then some significance or importance will need to be attached to each sub-group mean. In calculating such weighted means it is important to remember that the sum of the values needs to be divided by the sum of the weights, as illustrated by equation 5.5:

$$\text{Weighted mean} = \frac{\Sigma WX}{\Sigma W} \tag{5.5}$$

Equation 5.5

ΣWX = sum of weighted values

ΣW = sum of weights

5.7 THE GEOMETRIC AND HARMONIC MEANS

Two less used averages are the geometric and harmonic means. The first becomes useful when we are calculating the mean of data that may increase in a geometric fashion; for example it could be used to find the size of a population at the mid-point between census years. In such a case we are making the assumption that the population of a particular area is growing geometrically. Thus, to find the geometric mean of the population statistics 3000, 9000, and 27,000 we would multiply the values together and then find the cube root of the product. In practice the best method is to use equation 5.6:

$$\text{Geometric mean antilog} = \left(\frac{\Sigma \log X}{N} \right) \tag{5.6}$$

Equation 5.6

$\Sigma \log X$ = sum of logarithms of all values

N = number of observations

The final stage is to look up the result of this calculation in a table of antilogs.

In contrast, the harmonic mean given in equation 5.7 may be used in circumstances concerned with rates of movement. For example, we may use this mean to calculate average travel speeds of commuters making their journeys to work by car and from a city centre. However, owing to different traffic conditions average speeds may vary significantly on each stage of the journey; hence the need to use the harmonic mean to calculate the overall average speed:

$$\text{Harmonic mean} = \frac{N}{\Sigma(1/X)} \tag{5.7}$$

Equation 5.7

N = number of observations

$\Sigma(1/X)$ = sum of reciprocals of values

5.8 CRITERIA FOR SELECTION OF AVERAGE

From our previous discussion it should be clear that the different measures of central tendency are not directly comparable and that each have very different characteristics. Thus we could not compare the mean of one set of data with the median or mode of another, the simple reason being that each one measures a different aspect of central location. For example, the mode measures the highest frequency, the mean the centrality of values and the median the middle position of the ranked data. The selection of a particular type of average will therefore depend on a number of criteria, and in particular:

(1) the individual properties of the averages;
(2) the type of question we are asking and the points we want to illustrate;
(3) the characteristics of the data and its pattern of distribution.

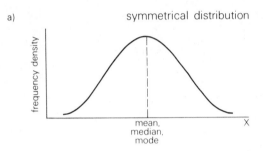

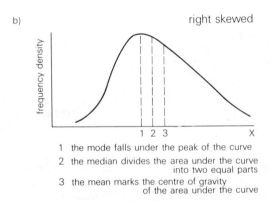

1 the mode falls under the peak of the curve
2 the median divides the area under the curve into two equal parts
3 the mean marks the centre of gravity of the area under the curve

Figure 5.3 A normal distribution and a skewed

Some of these basic points can be illustrated with reference to the effects of different types of distribution on the mean, median and mode. In Figure 5.3(a) the three averages all occupy the same position on the graph of a symmetrical distribution. If the distribution is not symmetrical but skewed, the three types of average will be located in very different positions, with the median always lying between the mode and the mean (Figure 5.3(b)). The mean is influenced by each individual value and will therefore be pulled in the direction of extreme values; thus in a distribution skewed to the right the mean will be positioned to the right of the mode and the median. In unimodal distributions the mode is not influenced by extreme values as in the case of a skewed pattern.

Finally, one important warning concerns making comparisions of the same type of averages, for example the comparison of two means. It may well be that the means of two distributions are the same, but that the two distribution curves are very different in character, as would be the case in Figure 5.4. This raises further questions about the ways in which we can describe and compare data sets by examining not only central tendencies but also by measuring deviations about such locations.

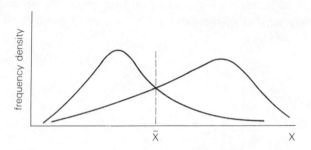

Figure 5.4 Different distributions with the same mean

5.9 MEASURES OF DISPERSION

In many circumstances it may be important that we can describe the variation of data about their average value. Or equally we may want to indicate how well the average represents individual values in a distribution. In either case we want some measure of variation or dispersion about a statistical average. Such dispersion can be measured in terms of the range, the standard deviation, and the coefficient of variation. In addition, we can measure the skewness and kurtosis or peakedness of a distribution. As with the different types of averages we have already encountered, these statistics all offer very different ways of measuring dispersion.

5.10 RANGE AND INTERQUARTILE RANGE

The range is simply the difference between the highest and lowest values in a distribution. Thus if the mean gauged discharge of a river over a 6-month period (measured at monthly intervals) was 20, 40, 50, 25, 80, 70, (m^3/s), then the range would be 80–20=60. While the range is a simple measure to calculate, it has the disadvantage of being sensitive to extreme values, and disregards the pattern of variation between the two extremes.

To some extent such disadvantages can be overcome by ignoring extreme values — taking out observations that fall within the top and bottom quarter of a distribution — and considering the remaining data. The simplest way to apply this measure is to rank the data from the lowest number to the highest. The three quartile values consist of Q_1 (the value below which 25 per cent of the observations occur), Q_2 (the value below which 50 per cent of the observations occur, i.e. the median), and Q_3 (the value below which 75 per cent of the observations occur). These three quartiles are also described as the 25th, 50th and 75th percentiles. The interquartile range is the difference between the 3rd and 1st quartile values ($Q_3 - Q_1$, see figure 5.5) Furthermore, there is no restriction on the construction of such ranges; for example, an interpercentile range could be used simply by taking the 10th and 90th percentiles. However, like the range, these intermediate measures have disadvantages, since the interquartile range leaves out 50 per cent of the data and ignores important features of the overall distribution, as shown in Figure 5.5.

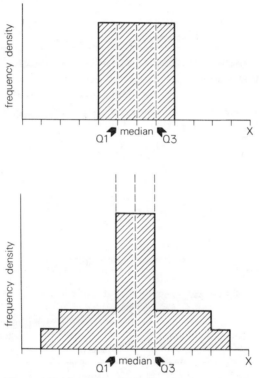

Figure 5.5 Limitations of the interquartile range for measuring the dispersion of frequency distributions

5.11 VARIANCE AND STANDARD DEVIATION

As we have seen, the range does have some problems associated with its use, since it merely represents the limits of variation. It does not reflect the variations of individual items, but only the limits of observed extremes. In contrast, the variance — and its square root, the standard deviation — are measures of dispersion that do take into account every item in a distribution. In statistical terms these measures are the most comprehensive descriptions of dispersion, since they are given in terms of average deviation about a point of central tendency. This will become clear if we examine more closely the properties and uses of these two statistics.

The variance is found by calculating the deviation of each item in a distribution from the mean, squaring these deviations and then deriving the mean square deviation. The symbol employed for the variance is σ^2, as is shown in equation 5.8:

$$\sigma^2_x = \frac{\Sigma(X_i - \bar{X})^2}{N} \qquad (5.8)$$

Equation 5.8
σ^2 = variance
X_i = ith value of X
$\bar{X}$ = mean
N = number of observations

Although the variance adequately measures the extent of dispersion of individual items, it does so in units of squared deviations. To obtain a measure of dispersion in terms of the units of the original data, the square root of the variance is taken, and this is called the standard deviation. It can be calculated, as is shown in Table 5.5, by using equation 5.9:

$$\sigma_x = \sqrt{\left(\frac{\Sigma(X_i - \bar{X})^2}{N} \right)} \qquad (5.9)$$

Equation 5.9
σ_x = standard deviation
X_i = ith value of x
$\bar{X}$ = mean
N = number of observations

Table 5.5 Calculation of standard deviation for rates of car licence holders by county in Wales

County	Car and van licences per 1000 population	Deviations from mean $(X - \bar{X})$	Squared deviations $(X - \bar{X})^2$
Clwyd	270	$270 - 264.5 = 5.5$	30.25
Dyfed	295	$= 30.5$	930.25
Gwent	258	$= -6.5$	42.25
Gwynedd	292	$= 27.5$	756.25
Mid Glamorgan	197	$= -67.5$	4556.25
Powys	302	$= 37.5$	1406.25
S. Glamorgan	257	$= -7.5$	56.25
W Glamorgan	245	$= -19.5$	380.25
	$\bar{X} = 264.5$		8158

$$\sigma_x = \sqrt{\left(\frac{8158}{8} \right)} = 31.93$$

The calculation of the standard deviation for grouped data varies slightly and, as demonstrated in Table 5.6, uses equation 5.10:

$$\sigma_x = \sqrt{\left(\frac{\Sigma X^2}{N} \right)} \qquad (5.10)$$

Equation 5.10

see Table 5.6 for definitions

Table. 5.6 Calculation of standard deviation for grouped data: household incomes, USA (1979)

Income class	Class mark m (thousands of $)	m^2	f (thousands of $)
Under $5000	2.5	6.25	10411
5000 – 9999	7.5	56.25	13006
10 – 14,999	12.5	156.25	12574
15 – 19,999	17.5	306.25	11099
20 – 24,999	22.5	506.25	9783
			$N = 56873$

$$\Sigma X^2 = \frac{N\Sigma m^2 f - (\Sigma mf)^2}{N}$$

$$= \frac{56873 \times 11113056.25 - (695097.5)^2}{56873}$$

$$= 2617627.233$$

$$\sigma^2 = \frac{\Sigma X^2}{N} = 46.0258$$

$$\sigma = \$6.78 \text{ (thousand)}$$

The standard deviation is extremely useful in describing the general characteristics of data and their dispersion about the mean. Indeed, it is often used to categorize the distribution of values when comparisons are being made between, for example, different areas. This is illustrated by Figure 5.6, which shows the distribution of car licences for the UK in terms of standard deviations about the national average. In this way it is possible to examine variations in the spatial pattern of potential car users. The standard deviation is also important within the context of the so-called normal curve, which is discussed in Chapter 7.

Car and van licences/1000 population

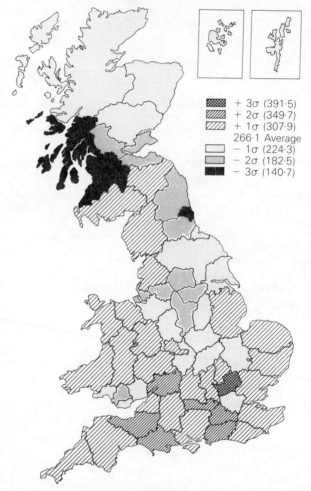

+ 3σ (391·5)
+ 2σ (349·7)
+ 1σ (307·9)
266·1 Average
− 1σ (224·3)
− 2σ (182·5)
− 3σ (140·7)

Figure 5.6 Use of standard deviations as class intervals
for mapping purposes

5.12 COEFFICIENT OF VARIATION

The standard deviation is an absolute measure of dispersion which is often of little
use when we want to compare the variations of one data set with those of another.
Usually under these circumstances we require a relative measure of dispersion;
and this becomes essential whenever the sets of data being compared are of
different orders of magnitude, or are measured in different units.

This can be illustrated by examining the data in Table 5.7, which compares the
rainfall of two very different areas. Despite the differences in magnitude, however,
both data sets have the same standard deviation about their respective means.

Clearly a deviation of 4.63 about a mean of 11.7 is greater than that about a mean of 61.7 In this situation we can use a relative measure of dispersion, such as the coefficient of variation, which is calculated by dividing the standard deviation by the mean. It can be expressed also as a percentage, simply by multiplying by 100. If this is applied to the data given in Table 5.7, then the coefficient of variation for station 1 is 39.5 per cent compared with a figure of 7.5 per cent for station 2. Therefore, in such circumstances this measure can give us a more sensible view of deviations about the means of two or more distributions.

Table 5.7 Application of the coefficient of variation to hypothetical rainfall data for two stations

Station 1	Station 2
6	56
8	58
10	60
12	62
16	66
18	68
$\bar{X} = 11.67$	$\bar{X} = 61.67$
$\sigma = 4.23$	$\sigma = 4.23$

5.13 SKEWNESS AND KURTOSIS

Two other measures of dispersion are skewness and kurtosis, both of which relate to the shape of the frequency distribution. We have already come across skewness in relation to the symmetry of frequency distributions and the locations of the mean, mode and median. It thus measures the extent to which the values in a frequency distribution are concentrated.

We can define skewness statistically in a number of different ways, such as by equation 5.11, which uses the mode, or by equation 5.12 in which the mode is replaced by the median:

$$\text{Skewness} = \frac{\text{mean} - \text{mode}}{\text{standard deviation}} \quad (5.11)$$

$$\text{Skewness} = \frac{3 \, (\text{mean} - \text{median})}{\text{standard deviation}} \quad (5.12)$$

However, one of the most common measures is momental skewness and this is calculated using equation 5.13:

$$\text{Skewness} = \frac{\Sigma(X_i - \bar{X})^3}{N\sigma_x^3} \quad (5.13)$$

> **Equation 5.13**
> X_i = ith value of x
> $\bar{X}$ = mean
> N = number of observations
> σ_x = standard deviation

The value of skewness for a perfectly symmetrical distribution is zero, negative values equal negative skewness and positive values positive skewness.

Skewness is important not only as a test of data normality (see Chapter 7) but also as a valuable descriptive statistic with which to compare frequency distributions. Thus, if we obtain quantitative measures of skewness using for example equation 5.13, we can then make objective comparisions between different histograms.

Kurtosis is the degree of peakedness of a frequency distribution and is usually related to deviation away from a perfectly symmetrical curve (see Chapter 7). For example, a frequency distribution as in Figure 5.7(a) which has a high peak is termed leptokurtic. In contrast, the distribution in Figure 5.7(b) has only a fairly moderate degree of peakedness and is termed mesokurtic; while in Figure 5.7(c) we have a platykurtic distribution characterized by its flat-topped nature.

Once again we can add further statistical meaning to these terms by calculating a coefficient of kurtosis, as defined by equation 5.14:

$$\text{Kurtosis} = \frac{\Sigma(X_i - \bar{X})^4}{N\sigma_x^4} \tag{5.14}$$

> **Equation 5.14**
> X_i = ith value of x
> $\bar{X}$ = mean
> N = number of observations
> σ_x = standard deviation

This equation makes it possible to attach kurtosis values to each of the distributions shown in Figure 5.7. Thus, a symmetrical or normal distribution has a coefficient of kurtosis of 3, a leptokurtic distribution a value greater than 3, and a platkurtic distribution less than 3.

Both skewness and kurtosis, although not particularly widely used in geographic analysis, are potentially valuable statistics. First, they can be applied to highly skewed distributions when the mean and standard deviation may give false measures; second, in such circumstances they can provide us with objective measures with which to compare different distributions; finally, they can also help in determining how far a particular frequency distribution deviates from the normal curve, and are thus important in the application of parametric statistical tests.

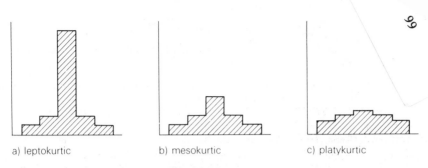

a) leptokurtic   b) mesokurtic   c) platykurtic

Figure 5.7 Types of frequency distributions as described by their peakedness

5.14 CONCLUSIONS

In this section we have examined a wide range of descriptive statistics which can be used in a number of ways. In most types of geographical analysis these statistics would be used, first, to organize large data sets, and second to summarize such data either by measuring central tendency or data dispersion.

It should be stressed that the use of these different descriptive statistics must be exercised with care. They may not provide the same answers when performed on the same data, and some are more appropriate than others — a point already stressed in the text.

Table 5.8 summarizes the position with regard to descriptive statistics and scales of measurement. The relevance of the mean and standard deviation to the normal curve cannot be overstressed; and, as will be shown in Chapters 6 and 7, both measures contain probabilistic implications that are only valid for normally distributed data.

Table 5.8 Descriptive statistics and scales of measurement

	Nominal	Ordinal	Interval/ratio (non-normal)	Interval/ratio (normal)
Central tendency	mode	mode or median	median	mean
Dispersion		Interquartile range	Interquartile range	Standard deviation

Measures of skew and kurtosis are applicable to all interval/ratio scale data. The geometric mean is applicable to data dealing with growth rates The harmonic mean is applicable to data dealing with rates of activity

REFERENCES

Croxton, F. E., and Cowden, D. J. (1968). *Applied General Statistics*, 3rd edn, Pitman, London.
Huntsberger, D. V. (1961). *Elements of Statistical Inference*, Allyn & Bacon, Boston.

RECOMMENDED READING

The following papers provide some good examples of the use of descriptive statistics in geographical analysis.

Armstrong, R. W. (1969). 'Standardized class intervals and rate compilation in statistical maps of mortality', *Annls Ass. Am. Geogrs.* **59**, 382–390.

Evans, I. S. (1977). 'The selection of class intervals', *Trans. Inst. Brit. Geogs* (N.S.) **2**, 98–124.

Gardiner, V. (1973). 'Univariate distributional characteristics of some morphometric variables', *Geografiska Annaler A* 3–4, 147–53.

Gardiner, V., and Gardiner, G. (1978). 'Analysis of frequency distributions', *Catmog* **19**, Geo-Abstracts, Norwich.

Meyer, D. R. (1972). 'Geographical population data: statistical descriptions not statistical inference', *Prof. Geog.* **24**, 26–27.

Scripter, M. W. (1970). 'Nested-means map classes for statistical maps', *Annls Ass. Am. Geogrs.* **60**, 385–393.

Chapter 6

An Introduction to Probability

6.1 INTRODUCTION

Probability can be defined in both the general and the mathematical senses. We might, for example, say 'it is probable that he will understand this book'. From this we could infer that the reader has a greater chance of understanding the text than of being bewildered by it. But this meaning can be interpreted in only the vaguest terms and scientists prefer to use the word in a more rigorous fashion, attaching some numerical value to the probability of an event. This numerical probability can be expressed in either of two ways — on an absolute scale of zero to one, or on a percentage scale of zero to 100. Both are widely used.

It is possible to think of some events to which numerical probability values can be readily attached. It is, for instance, absolutely certain that, at some unspecifiable time, we will all die. Such an absolute certainty has a probability of 1.0, or 100 per cent. Conversely, there is absolutely no chance of a human being lifting, unaided, a 10–ton weight, and in this case the probability is 0.0 or 0 per cent. However, many events are by no means as clear-cut and have probabilities of realization which lie between the two extremes. These events may be thought of as lying at some location along a probability spectrum between certainty and impossibility. Figure 6.1 illustrates the position for some simple, if non-geographical events, some of which occupy central locations with equal probabilities for the realization or non-realization of the event. Other events can be more, or less, likely without being certain or impossible. How can these values be assessed?

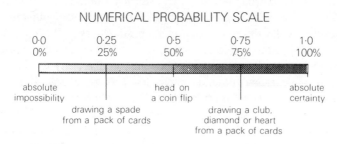

Figure 6.1 Probability scale from 0 to 1 showing the likelihood of some common events

6.2 ASSESSMENT OF PROBABILITY VALUES

The numerical probabilities of inevitable and impossible events can be derived by logical reasoning, as above, but other methods are needed in the less clear-cut cases. The French mathematician Pierre Simon de Laplace (1749–1827) was the first to define and solve the problem algebraically. If the numerical probability of an event x is denoted by $p(x)$, then:

$$p(x) = n/N \qquad (6.1)$$

> n = number of ways in which a particular event can be realized
>
> N = total number of possible outcomes (the sample space).

Consider the problem of drawing one card, at random, from a pack of 52. There are 52 possible outcomes, hence N is 52. Suppose, also, that the specified outcome is a spade card. There are 13 such cards in the deck and, hence, 13 ways in which the specified event can be realized. From equation 6.1 the probability of a spade is:

$$p(\text{spade}) = 13/52 = 0.25$$

Correspondingly, the probability of drawing a card from a suit other than a spade — a heart, club or diamond — is given by:

$$p(\text{non-spade}) = 39/52 = 0.75$$

as there are 39 non-spade cards in the pack. Because the outcome is certain to be either a spade or a non-spade, the total of the two probabilities is 1.0.

Another example is provided by the flip of a coin. In this case the sample space provides only two outcomes, a head or a tail. Each outcome can be realized in only one way. Hence $n = 1$ and $N = 2$, so that:

$$p(\text{head}) = 1/2 = 0.5$$
$$p(\text{tail}) = 1/2 = 0.5$$

The outcome, if we ignore the vanishingly small probability of the coin landing on its edge, must be one of the above and the sum of the two probabilities is 1.0.

We may now introduce two new terms, 'mutual exclusiveness' and 'independence'. Draws of cards and flips of coins produce outcomes that are mutually exclusive. By this we mean that if a head occurs on the flip of a coin then, for that flip, a tail is impossible. If a spade is drawn from a pack of cards then, for that draw, a heart, club or diamond are impossible. This issue, for cards and coins at least, may appear obvious and simple, but it is a valuable illustration of an important concept that will be referred to frequently later.

The other important concept is that of independence — another simple notion at this level where it can be easily illustrated, but fundamental to much of what is to come. Flips of coins and draws of cards fall under a general heading of 'trials'. For these two forms of trials the outcomes are independent. If at some time a trial with a coin yields a head then the next trial is neither more nor less likely to provide a second head. The outcomes are independent and are not influenced by previous results nor do they influence later trials. This applies no matter how many trials we consider, and no matter what our intuitive reactions might indicate. Consider, for example, a sequence of ten trials with a coin. If all ten trials yielded a head, unlikely though such a sequence might be, it in no way affects the probability of the next outcome also being a head, which remains at 0.5. Similarly, when a spade card is drawn from a pack, provided that it is replaced in the pack, the probability of drawing another spade is unaffected.

Here, then, are clear examples of exclusive and independent outcomes. Very many of the probabilistic concepts and statistical analyses which follow assume the events they treat to be equally independent. This is a most important assumption and one that is not always fulfilled by geographical events. When collecting geographical data we should always be aware of this problem. Questionnaire responses, soil survey results, meteorological data and even more disparate phenomena may all provide data sets in which the constituent observations are not independent of one another, and our sampling framework (Chapter 3) should take this into account as it may often be avoided. Chapter 7, in particular, elaborates this issue and the basic arguments put forward here should not be forgotten.

6.3 PROBABILITY ASSESSMENTS AND THE GEOGRAPHER

Laplace's method for estimating probability depends on a complete understanding of the circumstances surrounding each event. In particular this means being able to define both the sample space N and the realizations n on *a priori* grounds. For example, for one throw of a die the sample space consists of the six known faces and the outcome is known to be one of those six possibilities. This thorough understanding is impossible in most geographical studies where the processes and various forms of outcome are not known sufficiently to enable N or n to be accurately defined.

Consider the notoriously unpredictable mid-latitude climates. Our knowledge of the complex interplay of forces involved is inadequate to allow the Laplace model to be applied. Fortunately it is possible to get around this problem by examining the observable consequences of these complex atmospheric processes and to evaluate what can best be termed 'empirical probabilities'. Suppose we wish to know the probability of rain falling in a given area, on a specified day during March. No rigid mathematical model can yield an answer to this problem but, with sufficient records of March rainfall in the selected area, an empirical probability assessment can be attempted. A study of the records of the Durham University meteorological station, England, indicates that, over the past ten years, some form of rain has been recorded on 196 of the 310 days (we are not here concerned with the depths of rainfall). From the two figures above we can derive the probability of rain falling:

$$p(\text{rain}) = 196/310 \quad = 0.63$$

and, logically,

$$p(\text{dry}) = 1 - 0.63 \quad = 0.37$$

The analogies to the Laplace method are clear: the sample space N becomes the total number of available days (310), while the number of realizations is now represented by the days on which rain was recorded. There are, however, some very important contrasts. The results can apply, at best, only to an area around the Durham Observatory whose extent depends on local variability in rainfall patterns. The result also applies only to March. Seasonal variations are barely apparent within the span of one month, and all days may be considered to be equally probable with respect to rain. Over the whole year, however, this important requirement of equal probability will not be met. Winter days, in England, are more likely to experience rain than are summer days. Estimates based on data collected over the whole year would overestimate the probability of rain during the drier summer days and understimate it in winter. This necessarily resticted application of the empirical method contrasts with the universality of conclusions gained by the Laplace method — those for cards and the throwing of dice apply for all events, anywhere and at all times.

But even when equiprobable conditions apply problems may still arise owing to random variations of the sample from which the estimates are made. Returning to the question of coin flipping: a sequence of 10 trials could provide 7 heads and 3 tails. It is a far from impossible result, but would lead to an inferred empirical probability of a head of 0.7. This particular difficulty is best overcome by using long sequences. A run of 500 trials would be far less likely to produce such an erroneous conclusion — long sequences of events tend always to converge on outcome proportions that are close to the true probabilities of those events. Thus for 500 trials with a coin, 350 heads — which also gives an empirical probability of p(head) $= 0.7$, — is far less likely than 7 heads out of 10. Hence, geographers should always strive to obtain the largest samples that are practicable in order to be confident that the data are representative of the conditions.

We have also assumed that the background circumstances controlling rainfall have not changed. For this data set the assumption may be fulfilled. But where social or economic data are used the fluid interplay of forces surrounding their behaviour require a little more thought. Meteorological climates may not change over ten years, but social and economic climates certainly can. Geographical arguments may be substantially weakened by extending conclusions based on one time or space zone to another in which the prevailing forces behave differently.

Lastly we may return to the question of independence. For our probabilistic conclusions to be valid each day's weather must be independent of the previous day's. If it is not so the probability of rain on a given day is not a random event with a fixed probability but a function of the conditions on the day before. Our climate is certainly variable and any one day may be meteorologically independent. But we cannot be certain without some further study. This, however, is the task of later sections.

Despite these restrictions, and provided that they are always borne in mind, this method of probability assessment is useful and is often the only one available. We can now proceed to apply it in a number of more directly useful ways.

6.4 CONDITIONAL AND UNCONDITIONAL PROBABILITY

We have thus far examined only single events — one day's rain or one draw of a card. Frequently, however, we need to examine compound events — events composed of several single incidences of the same type. What, for example, is the probability of three selected days in March all yielding rain? Or of three draws of a card all providing a spade? The probabilities of such compound events remain within the range of 0.0 to 1.0, but are more properly termed conditional probabilities in that they are conditional on a prescribed combination of individual events. It is important to note that these conditional probabilities are found by multiplying the probabilities for each of the individual components. For example, the probability of drawing two spades from a pack (provided that the first was replaced) is expressed by the multiplication law of probabilities:

$$p(2 \text{ spades}) = 0.25 \times 0.25 = 0.0625$$

This theme opens up interesting possibilities when we consider less elementary compound events. Returning to the March rainfall example, rain days might be termed W and dry days D. As we know any day is certain to be either wet of dry; hence:

$$p(W) + p(D) = 0.63 + 0.37 = 1.0$$

When we are considering three days, however, the sample space can no longer be divided into just two unequal parts. We need instead to consider the number of ways in which the two options (W and D) can be permutated in three events. The probability of the three selected days being wet is also found from the multiplication law:

$$p(WWW) = 0.63 \times 0.63 \times 0.63 = 0.25$$

The probability of the first two days being wet and the third dry is found by the same method:

$$p(WWD) = 0.63 \times 0.63 \times 0.37 = 0.147$$

The above specification makes particular reference to the order of events. In this case the sequence is wet – wet – dry. When order is overlooked a different picture emerges. The specification two wet and one dry day can be met in three ways, with the single dry day falling on either the first (DWW), second (WDW) or third (WWD) day of the sequence. Expressed correctly we might say that there are three permutations of the combination two wet and one dry day. The probability of any of these three permutations occuring is found by adding their individual probabilities. The latter is, in each case, 0.147. If we denote the combination of two wet and one dry day by '2W1D', then:

$$p(2W1D) = p(WWD) + p(WDW) + p(DWW) = 0.147 + 0.147 + 0.147 = 0.441$$

The same principles apply to the specification for one wet and two dry days

(1W2D). The probability for the permutation wet–dry–dry, i.e. with the single wet day first in the sequence, is:

$$p(\text{WDD}) = 0.63 \times 0.37 \times 0.37 = 0.086$$

but, for all three permutations:

$$p(1\text{W}2\text{D}) = p(\text{WDD}) + p(\text{DWD}) + p(\text{DDW}) = 0.086 + 0.086 + 0.086 = 0.258$$

Finally, we may consider the probability of all three days being dry. As with all three days being wet, this can only be realized in only one way, and so:

$$p(\text{DDD}) = 0.37 \times 0.37 \times 0.37 = 0.051$$

These four combinations of events cover all the possible outcomes. They are themselves mutually exclusive, so the addition of their individual probabilities (Table 6.1) gives a total of 1.0.

Table 6.1 Combinations and probabilities for three dichotomous events

Combination	Probability
3W	0.258
2W1D	0.441
1D2W	0.259
3D	0.051
	1.000

6.5 HISTOGRAMS AND PROBABILITY

Having dealt, very briefly, with some fundamental concepts in probability, we can now consider their broader application to geographical statistics. There are few more rewarding areas of common ground with which to start than that offered by the histogram. The histogram is generally used as a graphical or descriptive device, but it has other applications. It consists of data classes that may be genuinely discrete, as in the case of the housing categories in Figure 6.2; or the classes may be the arbitrary sub-divisions of a continuous measurement scale such as the measured slopes on a hillside (Figure 6.3). In either case it is the frequency of events within each class that is the focus of attention. Most often these frequencies are represented as absolute numbers, as on the left-hand scales of Figures 6.2 and 6.3. But these are easily converted into relative frequencies, i.e. as ratios of the sample size, and as such are also the empirical probabilities of events within the respective classes.

For example, in Figure 6.2 the sample size was 11,652 residences of which 3612 fell into the category of semi-detached property, giving a proportion of $3612/11652 = 0.31$. This proportion is also the probability that a randomly selected member of the sample will fall into the semi-detached class, and the process can be repeated for all six classes. The same is true for histograms

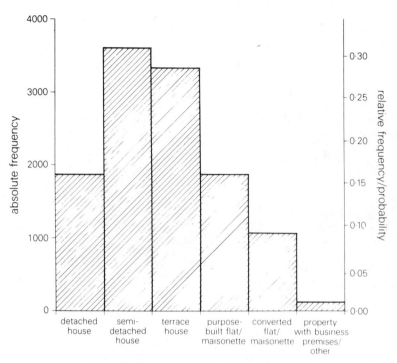

Figure 6.2 Histogram of occupancy types in Great Britain
(Source: Central Statistical Office, 1977)

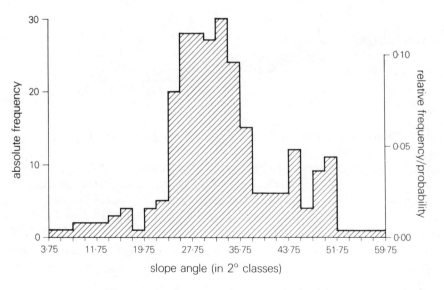

Figure 6.3 Histogram of hillslope frequencies
(Source: Pitty, 1969)

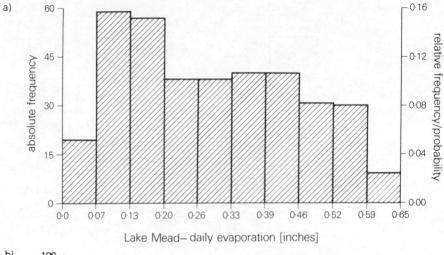

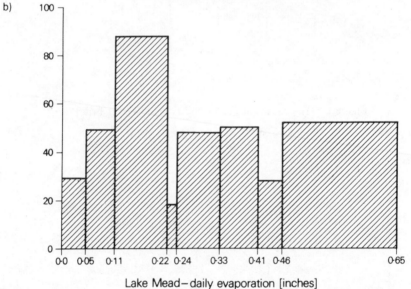

Figure 6.4 Histogram of daily evaporation losses using (a) equal and (b) unequal classes

comprising arbitrary classes along a continuous (interval or ratio) measurement scale. In Figure 6.3 the modal class of $31.75 - 33.75°$ has an absolute frequency of 30 in a sample of 253 observations. This gives a probability $30/253 = 0.12$ of slope measurements within that steepness range. The right-hand scale on Figure 6.3 should be consulted for other class probabilities. These classes represent mutually exclusive events, or ranges of events, covering all observed outcomes. It is consistent with the earlier observation that the individual class probabilities should sum to 1.0.

In reality individuals are free to fall into any of the classes. But the probability values suggest that events are more likely to occur within certain ranges. This variation is displayed by the heights and areas of the histogram's constituent columns. The only prerequisite in this respect is that the class intervals must be regular. If this were not the case classes with wider limits would record disproportionately higher frequencies than those with smaller ranges. The inferred probabilities would be correspondingly distorted. Compare, for example, Figures 6.4(a) and 6.4(b) which are based on data for daily evaporation losses from Lake Mead, Nevada. The data are measured as equivalent water depth, in inches, over the Lake and are taken from the work of Harbeck *et al.* (1958). Despite the use of identical data the two histograms differ considerably and the irregular class intervals in Figure 6.4(b) distort the otherwise orderly pattern in Figure 6.4(a).

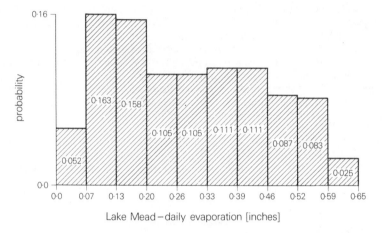

Figure 6.5 Probability bar graph of evaporation losses

The outline of the histogram, then, represents more than a graphical device and demonstrates the partitioning of a sample space whose total area is a notional 1.0 unit. Figure 6.5 treats the evaporation data in just this manner. There is a total probability of 1.0 divided between the classes. Daily evaporation of between 0.13 and 0.20 inches occurs with a probability of 0.158, for example. The probability of water losses between wider limits, say 0.13 and 0.33 inches, is found from the sum of the respective class probabilities; in this case 0.158 + 0.105 + 0.105 = 0.368.

The sequence of changing probabilities along the base scale and from class to class is known as a probability distribution. It is in this sense of changing probabilities and of equating those probabilities with the heights and areas of the histogram columns that we may introduce a fundamental statistical and probabilistic concept — that of the so-called normal distribution to be discussed in the next chapter.

REFERENCES

Harbeck, G. E., Kohler, M.A., and Gordon, E. K. (1958). 'Water loss investigations: Lake Mead studies' US Geol. Surv. Prof. Paper 298, Washington, DC.

Cent al Statistical Office, (1977). *Social Trends* **8,** HMSO, London.
(1978). *General Household Survey,* London.
Perla, R. I., and Martinelli, M. (1975). *Avalanche Handbook,* US Dept. Agric., Agric.
Handbook 489, Washington, DC.
Pitty, A. F. (1969). 'A scheme for hillslope analysis: initial considerations and calculations',
Univ. of Hull Occasional Papers 9, Hull.

RECOMMENDED READING

David, F. N. (1962) 'Games, Gods and Gambling' Griffin, London. An interesting, if non-
geographical, text providing an alternative view of this fascinating subject.
Harvey, D. (1969) 'Explanation in Geography' Arnold, London. This an invaluable book
for geographers, not only for the elements of probability (Chapter 15) but also for all
aspects of theoretical Geography.
Hays, S. (1960) 'An Outline of Statistics' Longman, London.
Kae, M. (1964) 'Probability' *Sci. Am.,* 211(3), 92–104. Another non-geographical item, but
eminently readable and highly informative.
Lipschutz, S. (1965) 'Thoery and Problems of Probability' Schaum's Outline Series, New
York. This is a useful volume for those readers who wish to extend their knowledge of
probability beyond that possible within the limits of this chapter. Plenty of good worked
examples.
Mosteller, F., Rourke, R. E. K. and Thomas, G. B. (1961) 'Probability: a First Course'
Addison-Wesley, Reading, Mass.
Reichmann, W. J. (1964) 'Use and Abuse of Statistics' Penguin, Harmondsworth.

EXERCISES

1. When a dice is thrown six outcomes are equally probable (1, 2, 3, 4, 5 or 6). What are
    the probabilities of the following: (i) a 6 with one throw, (ii) two 6's in two throws,
    (iii) any face other than a 5 or 6?

2. Assuming a complete pack of 52 cards and that cards are replaced after being drawn,
    with what probability will the following events be realized: (i) an ace with one draw,
    (ii) an ace and a king with two draws, (iii) two kings *or* two queens in two draws?

3. According to one report, in 1950 the Puerto Rican population of New York was
    distributed about the city in the following fashion:

Borough	Puerto Ricans	Relative proportion
Bronx	61,924	0.251
Brooklyn	40,299	0.164
Manhattan	138,507	0.562
Queens	4,836	0.020
Richmond	740	0.003
Totals	246,306	1.000

    From the above data what are the probabilities of: (i) a randomly selected Puerto
    Rican living in either Brooklyn or Manhattan, (ii) two randomly selected Puerto
    Ricans coming from the Bronx and Manhattan, (iii) three randomly selected Puerto
    Ricans, all from Queens?

Carl Gauss (1777–1855), a notable German mathematician after whom the distribution is sometimes known. Gauss was the first person to acknowledge that many events accorded to a scheme whereby those close to the mean were relatively common while more extreme cases were proportionally less frequent. Gauss' conclusions were based on his studies of errors in astronomical observations, but it is characteristic of the normal distribution that they apply with equal force to rainfall data and other, widely differing, phenomena. Most importantly, however, Gauss was able to reduce his conclusions to a mathematical form, the results of which we may now examine.

7.2 THE MATHEMATICAL DEFINITION OF THE NORMAL DISTRIBUTION

The pattern of probability changes about the mean of a normal distribution are best described as 'bell-shaped' with the mean at the peak of the bell. Furthermore, the normal distribution is based on continuous data scales (interval and ratio) and not on discrete measurements (nominal and ordinal). Consequently, the distribution should be described by a curve and not a histogram.

One way to construct a normal curve would be to use small class intervals and provide a closer approximation to a continuous line, as in Figure 7.2. To do so, however, is impractical as increasingly large amounts of data would be required to provide sufficient frequencies in each class. However, one of the many notable properties of the normal curve is its mathematical definition which permits a perfectly accurate reconstruction given certain information. The equation (see (7.1) below) is hardly the most inviting but upon closer inspection can be seen to consist of familiar terms though combined in an unfamiliar fashion. In principle it is an equation much like any other and predicts an unknown quantity from a collection of known terms. In this instance the unknown (Y) is the height of the normal curve above the base line at a given point (X) along the measurement scale.

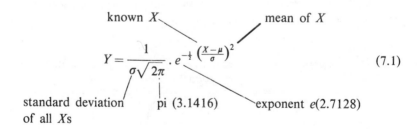

$$Y = \frac{1}{\sigma\sqrt{2\pi}} \cdot e^{-\frac{1}{2}\left(\frac{X-\mu}{\sigma}\right)^2} \tag{7.1}$$

known X mean of X standard deviation of all Xs   pi (3.1416)   exponent e(2.7128)

The equation components fall under three headings. First, the universal constants of pi (3. 1416) and e (2.7128) which never vary. Second, the mean and standard deviation of the variable under study; these vary from case to case. The arithmetic mean and standard deviation are measures of data central tendency and dispersion respectively; they have already been introduced as descriptive devices in Chapter 5. Third, the value of X for which the height Y of the curve is required.

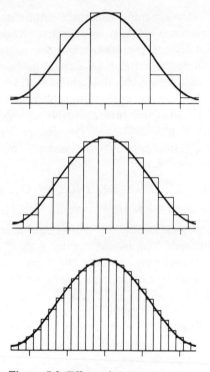

Figure 7.2 Effect of decreasing class
intervals of histogram form

For given μ and σ the substitution of several X values will yield estimates of Y which allow the curve to be constructed. This has been done for the rainfall data depicted in Figure 7.1 and the resulting curve has been superimposed on the histogram. The reader may be put at ease to know that he will not, in the usual course of events, need to use the equation for the normal curve. Nevertheless, it is informative to examine how the curve in Figure 7.1 was generated, and Table 7.1 provides the derivation of some of the points used to construct its outline.

Table 7.1 Calculation of probability density functions (height of the ordinate) on the normal curve

X	$\dfrac{1}{\sigma\sqrt{(2\pi)}}$	$\dfrac{X-\mu}{\sigma}$	$\dfrac{1}{2}\left(\dfrac{X-\mu}{\sigma}\right)^2$	$e^{-\frac{1}{2}\left(\frac{X-\mu}{\sigma}\right)^2}$	Y
650	0.00205	−1.54	1.186	0.305	0.00063
870	0.00205	−0.41	0.084	0.919	0.00188
1000	0.00205	+0.26	0.034	0.966	0.00198
1250	0.00205	+1.54	1.186	0.305	0.00063

Mean (μ) = 950 mm; standard deviation (σ) = 195 mm.

At this point we may note an important distinction between probability histograms and continuous curves. In the former the height of the columns and their respective areas was directly proportional to the probability of that event or range of events (see, for example, Figures 6.2 and 6.3). This is not the case for the normal curve — the height of the curve at any point X is not a direct measure of that event's probability and represents a more abstract quantity known as the probability density function, the discussion of which lies beyond the scope of this text. Because the distribution is continuous and not broken into discrete units the measurement scale (X) is an assemblage of an infinite number of points. Nevertheless the area-probability concept outlined in Section 6.5 remains valid, but because points have no area probability, statements for the normal curve are sensible only when expressed in relation to ranges of events between which an area can be defined (see Figure 7.3). The total area beneath the normal curve possesses a nominal area of 1.0 as it covers all events; and between any two events on that scale a fractional proportion can be established.

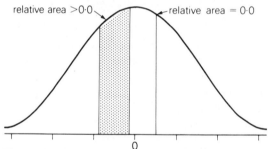

relative area $>$0·0   relative area $=$ 0·0

0

Figure 7.3 The contrast between probabilities at points
and between points on the normal curve

Hence, if required, points on the normal curve can be estimated without great difficulty. In addition, the role of the mean and standard deviation, hitherto used as summary statistics only in Chapter 5, are thrown more sharply into focus as parameters which determine the location (Figure 7.4) and the spread (Figure 7.5) of the normal distribution.

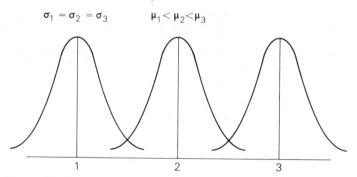

$\sigma_1 = \sigma_2 = \sigma_3$ $\mu_1 < \mu_2 < \mu_3$

1   2   3

Figure 7.4 Frequency distributions with identical standard deviations
but different means

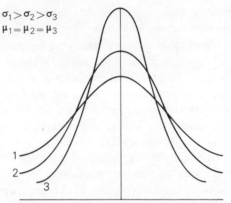

$\sigma_1 > \sigma_2 > \sigma_3$
$\mu_1 = \mu_2 = \mu_3$

Figure 7.5 Frequency distributions with identical means and different standard deviations

✳7.3 THE STANDARDIZED NORMAL CURVE

Individually the form of the normal curve differs widely in response to the characteristics of the data set. This difficulty may be overcome by expressing the 'raw' values as deviations from the mean, a process known as standardization. The units of expression are proportions of the standard deviation and are termed z values. In this way observations numerically greater than the mean become positive z values and those less than the mean become negative z values. Expressed generally (for populations):

$$z = \frac{X - \mu}{\sigma}$$
(7.2)

Equation 7.2
z = required z value
μ = mean of variable X
σ = standard deviation of variable X
X = raw value of X to converted

Hence any raw observation can be readily converted to an equivalent z value. Standardization of the observations also results in standardization of the normal curve to one which has a mean of zero and a variance and standard deviation of 1.0. Some of the data used in Figure 7.1 may now be re-examined. A raw observation of 1300 mm of rainfall, for example, is 1.79 standard deviations greater than the mean, so $z = +1.79$:

$$z_{1300} = \frac{1300 - 950}{195} = 1.79$$

Other observations, may be similarly treated.

It may not, however, have escaped the reader's attention that this standardization process can be equally applied to the equation of the normal curve. The item on the right-hand side of equation 7.2 also appears in the normal equation 7.1, and in the latter it can be replaced by z. Simultaneously, the use of standardized data renders all remaining σ terms equal to 1.0; as a result equation 7.1 can be simplified to:

$$Y = \frac{1}{\sqrt{(2\pi)}}\, e^{-(z/2)^2} \tag{7.3}$$

Equation 7.3
z = z value (see also equations 7.1 and 7.2)
e = the constant 2.7183 ...
π = the constant 3.1416 ...
Y = height of curve at point z

The resulting standardized normal curve retains the characteristic bell shape but possesses the same detailed form irrespective of the nature of the raw data, provided only that the latter is normally distributed. As a result the probabilistic properties interpreted from the standardized curve must also be identical.

7.4 THE NORMAL CURVE AND PROBABILITY

The normal curve contrasts with probability polygons in a number of ways. Its mathematical regularity has been examined and to this we may add that, in theory at least, the curve may extend to infinity in both directions, positive and negative. Thus, events of any magnitude are theoretically possible, both the exceedingly large and the minutely small. But because of the form of the normal curve they have remote probabilities. Refer to Figure 7.1, for example: annual rainfall totals exceeding 1500 mm are possible but highly improbable, and none were recorded during the survey period. In reality the inherent nature of much geographical data imposes real limitations on these idealized properties. Phenomena as varied as rainfall, crop yields, river discharges, incomes, and weight and heights of objects all have very effective 'cut-off' points at zero, i.e. negative values are impossible. Upper limits are far less definite and limits on high annual rainfall totals, for example are hard to identify in any but the vaguest of terms.

But the reader must not be misled into imagining that probability assessments from normal distributions are difficult to obtain. Quite the contrary, they are surprisingly easy to find from the relative area beneath the curve between the limits specified by the researcher. The normal curve, extended to plus and minus infinity, for example, includes all possible events and possesses a notional area of 1.0. The curve is also symmetrical about the mean, and it follows that there are equal probabilities (both 0.5) of events being greater or less than the mean.

Fortunately the probabilities of events within other, less convenient, ranges are not difficult to assess. The problem could be approached by a form of mathematics known as integral calculus. However, because of the consistent form of the standardized normal curve such a recourse is unnecessary. Instead, prepared tables which provide the relative area (probability) between any two z values may be consulted. These z tables, as they are known, appear in many different forms, one of the commoner of which is reproduced in Appendix I. Their use, with practice, is easy and accurate. A few examples should demonstrate their usefulness.

Example 1.

What is the probability of observations with z values of between 0.0 and +0.75?

In effect we need to know the relative area beneath the normal curve between the central point ($z = 0.0$) and +0.75 (see Figure 7.6). Appendice I is arranged to provide the relative area between 0.0 and the required value. The value may be read off directly from the table and appears adjacent to the selected z value. In this example $z = 0.75$ which, from the tables, delimits a relative area, and probability, of 0.2734.

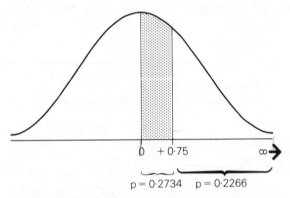

Figure 7.6 Probability of events under the normal distribution between $z = 0.0$ and $z = + 0.75$

Example 2.

What is the probability of observations with z values greater than +0.75?

The required relative area is now that between +0.75 and infinity. But as the tabled figure is that between zero and the relevant z values, some secondary, but simple, calculations are necessary. As the area between $z = 0.0$ and infinity is 0.5 we need only subtract the tabled value from 0.5 in order to establish the required probability:

$$p(z > 0.75) = 0.5 - 0.2734 = 0.2266$$

Example 3

What is the probability of observations with z values between 0.0 and −1.2?

The normal curve is perfectly symmetrical and the table applies equally to negative values of z. In this case the negative sign can, effectively, be ignored as we will abstract the relative area for the limits 0.0 to 1.2. This is found to be 0.3839 (the reader should check this):

$$p(0.0 > z > -1.2) = 0.3839$$

For z values less than -1.2 (Figure 7.7) the probability is found, as in the second example, by simple subtraction:

$$p(z < -1.2) = 0.5 - 0.3839 = 0.1161$$

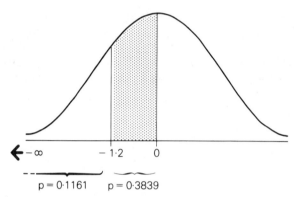

$$p = 0\cdot1161 \qquad p = 0\cdot3839$$

Figure 7.7 Probability of events under the normal distribution below $z = -1.20$

Example 4

What is the probability of events within the range

$$z = +0.3 \text{ to } z = -0.7$$

The same principles apply here, but the problem is best treated in two parts, one dealing with the probability of events within the range 0.0 to 0.3 and the other within the range 0.0 to -0.7 (Figure 7.8). The two probabilities can be taken directly from the table and added:

$$p(z < 0.3) = 0.1179 \text{ and } p(z > -0.7) = 0.2580$$

thus

$$p(0.3 > z > -0.7) = 0.1179 + 0.2580 = 0.3759$$

Example 5

From the annual rainfall information already supplied for Chain Bridge, Herefordshire, what is the probability of rainfall totals of between 800 and 1000 mm?

Again there is no change in principle, but the values need first to be converted to the equivalent z values before the table can be employed. With a mean and standard deviation of 950 and 195 mm respectively:

$$z_{800} = \frac{800 - 950}{195} = -0.77$$

and

$$z_{1000} = \frac{1000 - 950}{195} = 0.26$$

which, in turn, give:

$$p(z > -0.77) = 0.2794 \text{ and } p(z < 0.26) = 0.1026$$
$$p(0.26 > z > -0.77) = 0.2794 + 0.1026 = 0.3820$$

The probability of an annual rainfall total of between 800 and 1000 mm is thus 0.3820.

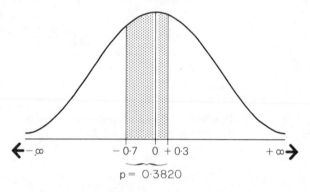

Figure 7.8 Probability of events under the normal distribution between $z = -0.7$ and $z = +0.3$

7.5 FURTHER APPLICATIONS OF THE NORMAL DISTRIBUTION

Any normally distributed variable has predetermined probabilities of events falling within ranges specified by z values. The z tables indicate that there is a 0.6826 probability (0.3313 either side of $z = 0.0$) of an event falling with the range $z = -1.0$ to $z = +1.0$, i.e. within one standard deviation either side of the mean. On the same basis there is a 0.9545 probability of events falling within two standard deviations either side of the mean and a 0.9973 probability of events being within three standard deviations (see Figure 7.9). Thus, in the long run, we would expect 99.73 per cent of all observations to fall within the range $\mu \pm 3\sigma$. These properties hold good for all normally distributed variables; hydraulic, economic, demographic and social data may all conform to this scheme.

It is often convenient, however, to express probabilities within less awkward ranges. For this reason the 95 per cent or 0.95 probability limits are often quoted in preference to the 95.45 per cent limits. Fortunately, it is not difficult to establish

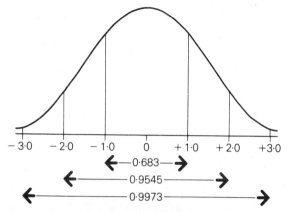

Figure 7.9 Standard deviations and probabilities on the
normal distribution

the z values associated with such probabilities. Once again the published table
can be used, but now in a reverse fashion with the probability, say 0.95, being
known beforehand leaving the z value that delimits this relative area to be found.
Remembering that the table gives information for only half of a perfectly
symmetrical distribution, we must first look for the probability 0.475, i.e. 0.95/2,
in the body of the table from which a corresponding z value of $(\pm)$ 1.96 is read off.
Expressed another way, this means that there is a 0.95 probability of events
falling within the limits $\pm$ 1.96 standard deviations. Correspondingly, events so
extreme as to fall beyond those limits do so with a probability of only 0.05. Figure
7.10 demonstratres how these extreme events may be realized with the 0.05 area
apportioned equally between the two ends of the distribution.

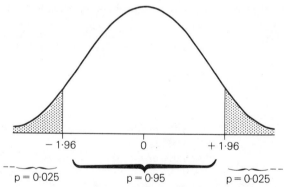

Figure 7.10 Probability of events beyond the 0.95 range

For any normally distributed data sets these z values of $\pm$ 1.96 are readily
transformed into equivalent raw data terms. Consider the rainfall example once
more. By substitution and transposition of equation 7.2 we get, for the upper 0.95
limit:

$$1.96 = \frac{X - 950}{195}$$

Therefore:

$$X = 950 + 1.96 \times 195 = 1.332.2 \text{ mm}$$

and, for the lower limit:

$$X = 950 - 1.96 \times 195 = 567.8 \text{ mm}$$

Hence there is a 0.95 probability of annual rainfall totals at Chain Bridge being between 1332.2 and 567.8 mm. More generally, the required limits can be expressed as:

$$X = \mu \pm z_c\sigma \tag{7.4}$$

Equation 7.4

μ = mean of variable X
z_c = critical z value for selected limits
σ = standard deviation of X
X = required raw observation

But interest may not necessarily lie within the 0.95 limits, and we might be concerned with broader (0.99) or narrower (0.90) ranges of events. The critical z values for these, and other, limits are provided in Table 7.2, which also indicates the complementary probabilities with which events occur beyond those ranges. These extreme probability areas are divided equally between the two ends of the distribution in all instances. Also given are the commonly used alternative expressions for unlikely events. An event with a 0.05 probability, for example, has a 1 in 20 chance of being realized; a 0.01 probability event has a 1 in 100 chance, and so on.

Table 7.2 Extreme probabilities and associated critical z values

Probability of events		Critical z value
within the range	beyond the range	
0.90	0.10 or 1 in 10	± 1.647
0.95	0.05 or 1 in 20	± 1.960
0.99	0.01 or 1 in 100	± 2.586
0.999	0.001 or 1 in 1,000	± 3.290

7.6 WHAT TO DO WITH NON-NORMAL DATA

Despite the wide applicability of the normal distribution there remain many instances when observed distributions are manifestly neither normal nor symmetrical. The most common problem in this respect is provided by data skewness (see Chapter 5) in which there is a long 'tail' of values towards either the high (positive skewness) or low (negative skewness) ends of the distribution. The former is found more often and is frequently associated with a lower 'cut-off' point on the distribution; rainfall and river discharge values cannot, for example, fall below zero. Such impositions prevent the distribution from spreading in that direction. An example is given by the daily discharge data for the Moka River in Mauritius (Figure 7.11).

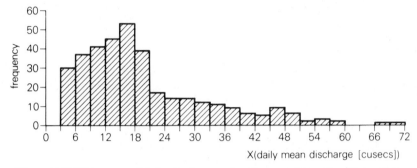

Figure 7.11 Histogram of daily discharge values (source: Mauritius Water Data, 1977)

Fortunately many positively skewed distributions can have a measure of normality imposed upon them by being log-transformed, the results of which are shown in Figure 7.12. In mathematical terms the logs of such data have no lower cut-off value as smaller fractions simply become ever larger negative logarithms (see Chapter 2), and the log of zero is minus infinity. In statistical terms such conversions are valuable because one of the prerequisites for many of the analytical methods that follow is that the data be normally distributed. The use of logged instead of raw data allows those methods to be used. Data behaving and responding in this way is known as log-normal.

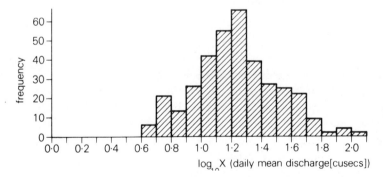

Figure 7.12 Histogram of log-transformed daily discharge values in Figure 7.11

It must be remembered, however, that logging observations will not always result in the successful normalization of the distribution, and neither can the transformation be carried out on data containing zero or negative entries as they have no logarithmic equivalents.

Negatively skewed data are obtained infrequently in geographical studies; but if they are encountered they may be removed by taking squares, or even higher powers, of the raw data. This transformation tends to 'draw out' the tail of the higher values. Taking powers of higher numbers has a proportionally greater effect than on smaller values, e.g. $2^2 = 4$ but $3^2 = 9$.

A word of warning, however: it is often difficult to make geographical sense of results when transformed data have been used. Normality is also now recognized as a less important prerequisite for statistical testing and analysis than had hitherto been supposed. Consequently normalization methods should only be resorted to when their use will not lead to interpretational difficulties at later stages.

7.7 ALTERNATIVES TO THE NORMAL DISTRIBUTION

Thus far only the normal distribution has been examined; but other formal probability distributions exist, many of which are useful to geographers. Two of the more important are the binomial and the Poisson distributions. Both apply to a range of natural events, but their manner of application and their specific data requirements are widely different and extend the geographer's ability to examine and interpret the world about him. Nevertheless, as we will demonstrate, they retain close links with the normal distribution and should not be regarded as wholly independent entities.

Both the binomial and Poisson distribution possess their own mathematical expressions. They can also be summarized by measures of central tendency, dispersion and skewness. These points are best elaborated by examining the distributions in more detail. We shall begin with the binomial.

7.8 THE BINOMINAL DISTRIBUTION

The binomial distribution is concerned, as the title suggest, with events for which there are only two, not many, outcomes, i.e. measurement takes place at the most elementary level of the nominal scale. Such dichotomous variables contrast strongly with the continuous character of interval and ratio data used when defining normal distributions. Coin flipping yields dichotomous data as only two outcomes are possible, a head or a tail. The single 'trial' conditions are clear enough — either a head or a tail will appear with equal probability. But when sequences of trials are involved then various combinations of heads and tails become possible outcomes. If a head is termed a 'success' then, for a sequence of three trials, four outcomes may be realized; either zero, one, two or three successes. The binomial distribution describes the probability of these outcomes and may be used for sequences of any length, from one upwards.

The properties of the binomial distribution were first defined by James Bernoulli in the late seventeenth century and the distribution is ocassionally referred to by his name. Unlike the standardized normal distribution it has no characteristic form and its shape will vary according to the sequence length and probability of a 'success'. Because binomial variables consist of specific and countable events the distribution is discrete and not continuous. In contrast to the rainfall example used to illustrate the normal distribution, fractional values are impossible and the distribution takes this into account. It should also be noted that binomial distributions concern themselves with only one of the two aspects of the dichotomous variable, i.e. either heads or tails but not both simultaneously.

Although the distribution is discrete its form can still be described in precise mathematical terms. The equation that is used to accomplish this provides a series of individual probabilities from which a histogram can be constructed. If the probability of X successes in a sequence of N trials is denoted by $p(X)$, the binomial distribution is defined by:

$$p(X) = \binom{N}{X} (j^X)(1-j)^{N-X} \tag{7.5}$$

Equation 7.5

$\binom{N}{X}$ = the combinatorial expression

j = probability of a 'success'

X = specified number of successes

N = sequence length (number of trials)

Of the quantities indicated in equation 7.5 only the 'combinatorial' expression requires any elaboration. In so far as the binomial distribution is concerned it is a method, using factorials (see Section 2.6), for determining the number of combinations of X items that can be abstracted out of a total of N items. In general terms:

$$\binom{N}{X} = \frac{N!}{X!\,(N-X)!} \tag{7.6}$$

If we wish to know how many different combinations there are of three successes in five trials we would proceed as follows: $N = 5$, $X = 3$ and so:

$$\binom{N}{X} = \binom{5}{3} = \frac{5!}{3!\,(5-3)!} = \frac{120}{6(2)} = 10$$

In this case we obtain the mildly suprising result that there are 10 permutations of 3 successes in a sequence of 5 trials.

These principles may now be included within a full, if simple, example. What is the probability of obtaining 4 heads ($p(4)$) from a sequence of 6 flips of a coin?

Substituting into equation 7.5 we get:

$$p(4) = \binom{6}{4} (0.5)^4 (1 - 0.5)^{6-4}$$

$$= \frac{720}{24(2)} (0.5)^4 (0.5)^2 = 15 \times 0.0625 \times 0.25$$

$$= 0.2344$$

But this probability of 0.2344 provides only a partial picture and the same procedure is employed in order to obtain the probabilities for all successes (heads) from zero to six. Table 7.3 shows how these probabilities are derived and Figure 7.13 is a pictorial representation of the complete distribution.

Table 7.3 Derivation of probabilities on the binomial distribution

X	$\binom{N}{X}$	j^X	$(1-j)^{N-X}$	$p(X)$
0	1	1	0.0156	0.0156
1	6	0.5	0.0313	0.0938
2	15	0.25	0.0625	0.2344
3	20	0.125	0.125	0.3125
4	15	0.0625	0.25	0.2344
5	6	0.0313	0.5	0.0938
6	1	0.0156	1	0.0156

0! and anything raised to the power of 0 are both, by convention, 1.0.

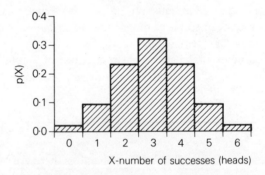

Figure 7.13 Histogram of probabilities on the binomial distribution

7.9 FURTHER PROPERTIES OF THE BINOMIAL DISTRIBUTION

Provided that the probability of a success remains close to 0.5, then the binomial distribution will retain the degree of symmetry indicated in Figure 7.13. Inequalities between the probabilities of success (j) and failure ($1-j$) manifest themselves as distributional skewness. An example is provided by the incidence of wet and dry March days in Durham (see Section 6.3). If a dry, rainless, day is termed a success then j, by empirical observation this time, is 0.37 and the probability of a rainy day 0.63. For any three-day, that is three-trial, sequence the probabilities of 0, 1, 2 and 3 successes are obtained by the methods indicated above. Again we have presented the derivation and results in tabular (Table 7.4) and histogram (Figure 7.14) forms. By comparison with the results of coin flipping the effect of changes in sequence length and success probability can be clearly appreciated.

Table 7.4 Derivation of probabilities on the binomial distribution when the two components of the dichotomous variable are not equally probable

X	$\binom{N}{X}$	j^X	$(1-j)^{N-X}$	$p(X)$
0	1	1	0.2500	0.2500
1	3	0.37	0.3969	0.4406
2	3	0.1369	0.63	0.2587
3	1	0.0507	1	0.0507

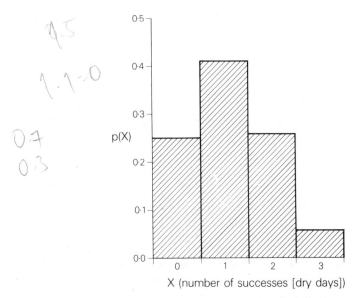

Figure 7.14 Histogram of probabilities of dry days

These observations bring us conveniently to the next point, that of the moments of the binomial distribution. Just as the normal distribution can be summarized in terms of its central tendency (mean), statistical scatter (standard deviation) and symmetry (skewness), so can the binomial. Their meaning and interpretation are much the same, but the derivation differs owing to obvious implications of different measurement scales. The mean of a binomial distribution is best described as the long-term average number of successes in repeated sequences of trial length N. The position is in many ways analogous to that of the normal distribution in which the most probable events are those close to the mean. The derivation of the binomial mean, however, is a simple matter as:

$$\mu = Nj \tag{7.7}$$

where N is the sequence length and j the probability of a success. A sequence of 6 flips of a coin has a most probable outcome, and hence a distribution mean, of 3 successes, i.e. heads, given by:

$$\mu = 6 \times 0.5 = 3$$

The situation, however, is not always so simple. Using the Durham rainfall example we find that the mean of the distribution for three-day sequences, with a dry day as a success, is:

$$\mu = 3 \times 0.37 = 1.11$$

How can we reconcile a fractional mean with a distribution composed of discrete, integer, numbers? To accommodate this seeming paradox it must be recalled that the mean is the theoretical average number of dry days for all three-day sequences. We need to consider only two three-day sequences, one with one dry day and the other with two dry days, to appreciate that a fractional average, in this case 1.5, is quite possible.

The variance, standard deviation and skewness of a binomial distribution are obtained with equal ease though, again, fractional indices may be obtained for the same reasons. The equations are:

$$\text{Variance} = \sigma^2 = Nj(1-j) \tag{7.8}$$

$$\text{Standard deviation} = \sigma = \sqrt{[Nj(1-j)]} \tag{7.9}$$

$$\text{Skewness} = \alpha = \frac{j-(1-j)}{Nj(1-j)} \tag{7.10}$$

The characteristics implicit in these indices are similar to their normal distribution counterparts. It would not, for example, be expected that every N-trial sequence would provide the most probable outcome — some degree of variation would be likely. Six flips of a coin will not always yield three heads. On different occasions 2, 4, 1, 5 or even 0 or 6 heads may be registered. The standard deviation of the binomial distribution measures this dispersive tendency. Skewness, again, is the degree of distributional asymmetry. It is highly responsive to changes in the probability of a success (j) and will vary accordingly. When $j = 0.5$, as in Figure 7.13, skewness is zero but soon increases as j departs from 0.5.

7.10 THE BINOMIAL DISTRIBUTION AND CONTINUOUS SCALE DATA

The binomial distribution is applicable only to dichotomous variables. Coin flipping undeniably falls into this category, as might the division of human populations into male and female. But dichotomous variables may also come from the twofold division of continuous variables, particularly where such variables have within them thresholds either side of which different geographical consequences may be realized. Human behaviour, for example, may change abruptly at the threshold age of 65 when retirement occurs, and social geographers might regard age as a dichotomous variable for certain purposes: above and below 65. Hence, in addition to the obviously dichotomous variables such as wet and dry days, urban and rural locations etc., they can also be synthesized from interval and ratio data.

More substantial links also exist with the continuous normal distribution. Consider the case for large N as it affects the binomial distribution. As the trial sequence increases in length the number of possible outcomes grows proportionally and the columnar histogram begins to resemble the probability curves of Figure 7.2. Provided that the probability of a success (j) does not differ greatly from 0.5, the histogram will tend towards a normal curve with surprising rapidity. As a working rule, if both Nj and $N(1-j)$ are greater than 5.0, then approximation to the normal form can be assumed. Hence, if $j = 0.5$ all sequences longer than 10 will be normal, but for $j = 0.3$ the length must be 16 before normality occurs and must be 50 if $j = 0.1$ Under these circumstances it becomes possible to express outcomes on the binomial distribution in terms of z values by using the moments of the latter and equating them with those of the normal distribution. Thus:

$$z = \frac{\text{difference between individual and mean}}{\text{standard deviation}}$$

$$z = \frac{X - \mu}{\sigma} = \frac{X - Nj}{\sqrt{[Nj(1-j)]}} \qquad (7.11)$$

Equation 7.11
μ = normal mean
σ = normal standard deviation
Nj = binomial mean
$\sqrt{[Nj(1-j)]}$ = binomial standard deviation
X = individual observation

This conversion permits us to express deviations from the binomial mean as z values with all the attendant probabilistic properties.

⚹ 7.11 THE POISSON DISTRIBUTION

This is the last of the three 'fundamental' distributions to be introduced. Named after its French discoverer, it has proved to be a fruitful area of application for geographical and spatial data and it is examined in more detail in the concluding sections of this book dealing with spatial statistics. It may, however, be introduced at this point as it has many points in common with its binomial and normal counterparts.

The Poisson distribution is most commonly applied to problems concerned with discrete events in space or time. The distribution is plotted in histogram, and not curve, form and it shares with the normal and binomial distribution the capacity for a mathematical definition of its form. While the binomial curve can be used without reservation when the probabilities of a 'success' and a 'failure' are not dissimilar, problems soon arise when we are dealing with rare events for which the probabilities of the outcomes are highly disproportionate. It is under these circumstances that the Poisson distribution may be employed. This clearly is the situation when dealing with 'point' events in space or time, when the number of events can be easily counted but it may not be sensible to contemplate the number of points where an event did not occur. For example, a shop location is a point event in space and it is clearly possible to count the number of points where such events have occured within a given area. But because any area is composed of an infinite number of points it is impossible to estimate the number of locations where a shop did not occur. This eliminates any possibility of attaching empirical probabilities to either the presence or absence category and, consequently, of employing the binomial distribution. Stream junctions, factory locations, crime locations and a whole variety of geographical phenomena fall firmly into this category.

As an example we may take the location of grocers shops in Sunderland England (Figure 7.15). The first task is to delimit the study area. In this case the extent of the built-up area was used, but the issue is not always so easily resolved and the choice of spatial limits can affect the final outcome. The selected area can, once delimited, be divided into regular grid squares and the average density of points (in this case shops) per grid square estimated. This quantity, denoted by λ, is vital to the mathematical description of the Poisson distribution, as we can see from its appearance in equation 7.12 below. This equation describes the Poisson distribution, and if it is remembered that each shop has now become a statistical event the distribution's form allows us to estimate the probability p of X events per grid square:

$$p(X) = \frac{\lambda^X}{X!} e^{-\lambda} \qquad (7.12)$$

Equation 7.12

λ = average density of events

e = the constant 2.7183 ...

X = specific number of events

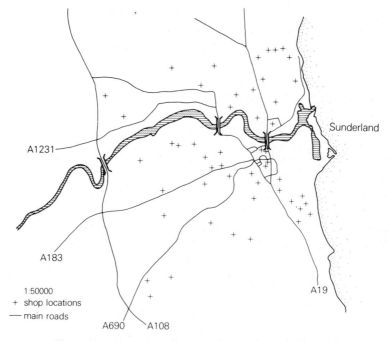

Figure 7.15 Location of grocers in Sunderland, England

The number of events (shops) per grid square may vary between zero (a square with no shops) and a theoretical limit of infinity as each square consists of an infinite number of points. In practice the distributions are evaluated only within sensible limits (in this case 6). But because only integer quantities of X are possible (there can be no fractions of shops in this context), the distribution is discreet and represented in histogram and not curve form. The established probabilities for 0, 1 and 2 etc. events applies equally to all units, in this case grid squares, in the study. area. The importance of λ can now be appreciated as it governs the nature of the distribution; consequently care is needed in defining the study area as it will affect this quantity.

Figure 7.15 shows the scatter of grocers' shops, of which there are 54 over the 135 grid squares hence $2 = 54/135 = 0.4$. With this information any desired $p(x)$ can be estimated; for example, the probability of a square with two shops is:

$$p(2) = \frac{0.4^2}{2!} \times 2.7183^{-0.4}$$

$$= \frac{0.16}{2} \times 0.6703 = 0.0536$$

A more complete impression is obtained by evaluating the distribution up to $X = 6$. The results are shown in Table 7.5 and in Figure 7.16.

The very low probability of 6 events per square, with the prospect of even smaller probabilities for 7 and above, demonstrates why, in this case, no further evaluation is necessary. However, the range zero to infinity covers all possible events in the Poisson distribution. Therefore $p(0) + p(2) \ldots p(\infty) = 1.0$. The sum of probabilities up to $p(6)$ is 0.99998 and it is sufficient to conclude that the probability of 7 or more events per square is only $1 - 0.99998 = 0.00002$.

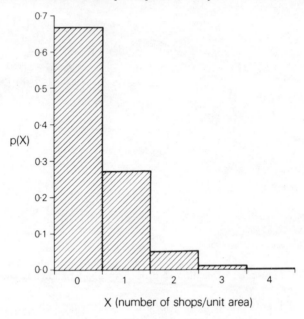

X (number of shops/unit area)

Figure 7.16 Probability histogram for Sunderland shops

Table 7.5 Probabilities on the Poisson distribution for Sunderland shops per grid square

Shops per square X	$\dfrac{\lambda^X}{X!}$	$e^{-\lambda}$	Probability $p(X)$
0	1	0.6703	0.6703
1	0.4	0.6703	0.2681
2	0.08	0.6703	0.0536
3	0.0106	0.6703	0.0072
4	0.00107	0.6703	0.00072
5	0.00009	0.6703	0.00006
6	0.0000007	0.6703	0.0000004

7.12 MOMENTS OF THE POISSON DISTRIBUTION

The Poisson distribution has its own measures of central tendency, dispersion and skewness. Again, they are simply derived but depend almost wholly on the observed density of events. The moments are found as follows:

$$\text{Mean} \ (\mu) = \lambda \qquad (7.13)$$
$$\text{Variance} \ (\sigma^2) = \lambda \qquad (7.14)$$
$$\text{Standard deviation} \ (\sigma) = \sqrt{\lambda} \qquad (7.15)$$
$$\text{Skewness} \ (\alpha_3) = \frac{1}{\sqrt{\lambda}} \qquad (7.16)$$

One of the most remarkable qualities of the Poisson distribution, almost its hallmark, is the equality of variance and mean. The remaining moments are simple transformations of this quantity. The mean represents the most probable long-term average density of events. The above example used points per unit area, but events may also be studied on the basis of points in time, e.g. per year, day or hour. But whatever the basis of measurement it would be unusual to find that every observation conformed to the most probable outcome, and the standard deviation is used to assess the degree of variation. Clearly the latter is very much influenced by the mean. In both cases we are again presented with fractional moments for a discrete distribution. The mean of the Sunderland shops, for example is 0.4 shops per unit space. Obviously we will never achieve 0.4 shops which is, correctly, the overall trend of the data set; some have more shops, others less, and the variance measures the extent of this variation.

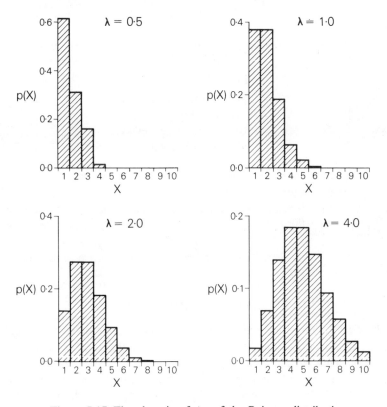

Figure 7.17 The changing form of the Poisson distribution

Finally, the inherent properties of the Poisson distribution dictate that skewness will be a common feature. The zero cut-off point (negative frequencies are impossible) at one end, together with the theoretical extension to infinity at the other, encourages distributional asymmetry. The skewness for the Sunderland shops data is 1.58 and the asymmetric character of the distribution emerges clearly from Figure 7.16.

This question of skewness bring us conveniently to the links between the normal and Poisson distribution. An earlier section drew attention to the links between the binomial and the Poisson distribution through increasing inequalities between the probabilities of the two components of a dichotomous variable. As the density of events (λ) increases a similar link with the normal distribution can be established. The shift of the Poisson mean in response to an increasing λ reduces the influence of the zero cut-off point and produces a more symmetrical distribution. These changing circumstances are illustrated in Figure 7.17, in which the asymmetry of low density events gradually gives way to the greater symmetry of higher density occurrences, eventually approximating to a normal curve. When this happens deviations about the Poisson mean can be expressed by z values. Thus, if the mean and standard deviation of the Poisson distribution are known, equation 7.2 for z can be rewritten as:

$$z = \frac{X - \mu}{\sigma} = \frac{X - \lambda}{\sqrt{\lambda}}$$

normal z values.

$$(7.17)$$

> ### Equation 7.17
>
> μ = normal mean
> σ = normal standard deviation
> λ = Poisson mean
> $\sqrt{\lambda}$ = Poisson standard deviation
> X = individual observation
> (cf. equation 7.11)

In an extreme example of an event with an average density (and therefore a Poisson mean) of 10.8, an observed unit with a density of 6.0 can be converted to an equivalent z value thus:

$$z = \frac{6.0 - 10.8}{3.286} = -1.46$$

7.13 PROBABILITY AND EXPECTATION

The three distributions introduced here may all be clearly defined in quantitative terms, and thus they contrast with many of the distributions encountered in Chapter 6 which lacked this definition of form. The hillslope data in Figure 6.3 appears to demonstrate normal characteristics, but the house tenure data of Figure 6.2 can be readily rearranged to provide a totally different-looking distribution.

In the latter instances observed frequencies were used from which probabilities could be inferred. Now, however, we have travelled full circle and can use the theoretical distributions and their properties to estimate 'expected' frequencies. The case of Sunderland shops offers an example. The probabilities of 0, 1, 2 etc. shops per square are given in Table 7.5. These values can easily be converted into expected frequencies for a perfect Poisson distribution by multiplying the $p(X)$ values by the number of squares (N) used in the study, in this case 135. The result is known as the expectation E of the event X. In general:

$$E(X) = N \times p(X) \tag{7.18}$$

For example, the expected frequency of squares with no shops is the product of the total number of squares and the probability for $X = 0$:

$$E(0) = 135 \times 0.6703 = 90.5$$

In this example the observed and expected numbers of squares with no shops do not differ by much (96 as opposed to 90.5), but such similarities are not always encountered.

But why might observed frequencies not follow these theoretical patterns? To answer this question we need to emphasize a point hitherto overlooked. The normal, binomial and Poisson distributions apply only to random variables consisting of independent observations scattered randomly either side of the mean with no external, non-random forces acting to concentrate events into specific ranges on the distribution or to exclude them from others. In the case of the binomial distribution the probabilities of different outcomes are conditioned solely by the probability of the success event (j) and not by other considerations. Within this condition is that of independence between events, and we assume that one event cannot influence the outcome for any other. Thus because a head results from one flip of a coin it does not mean that the outcome of the next flip is any more or less likely to be another head. Similarly, the Poisson distribution requires events to be randomly located in space, thus also being independent of one another. Many spatial phenomena are not so arranged. The incidence of sufferers from contagious diseases is spatially highly clustered and non-random because of the strong dependence between these events. As a result we cannot assume that all possible points in space are equally available for this event because those closer to existing cases will have a greater probability of registering a new one. It is when these conditions of independence and randomness are not fulfilled that differences between the theoretical and observed distribution arise. Later, in Chapter 9, we shall examine some of the ways in which these differences can be objectively compared.

7.14 SUMMARY

This chapter has attempted to familiarize the reader with three fundamental probability distributions — the normal, binomial and Poisson. Only the barest principles have been examined and many of the unresolved issues are left for later, and more appropriate sections. Some general conclusions can, however, be drawn and have bearing on what will follow.

First, all three distributions are theoretically defined, and 'real world' data may not necessarily conform precisely to the characteristics of these random variable distributions. A degree of difference between observed and theoretical behaviour can, nevertheless, be accepted and might be expected. But when external, non-random influences are active the departure of observed from expected behaviour may be marked. These inconsistencies need not spell disaster for the geographer — they frequently identify profitable areas of further investigation. What influences, for example, might cause Sunderland shops to adopt a non-random scatter — market forces, planning control or aggolmerative tendencies?

Second, the three distributions may all be expressed in clear quantitative terms, though only the standardized normal curve has a consistent and characteristic form. All three also posses measurable degrees of central tendency, statistical scatter and asymmetry. The equations describing both the distributions and their moments are drawn together in Tables 7.6 and 7.7, where they may be compared and noted for future reference.

Table 7.6 Summary of probability distribution equations

Distribution	Mathematical expression	Required parameters
Normal	$Y = \dfrac{1}{\sigma\sqrt{(2\pi)}}\, e^{-\frac{1}{2}\left(\frac{X-\mu}{\sigma}\right)^2}$	σ, μ, X
Standardized normal	$Y = \dfrac{1}{\sqrt{(2\pi)}}\, e^{-(z/2)^2}$	z
Binomial	$p(X) = \dbinom{N}{X} j^X (1-j)^{N-X}$	X, N, j
Poisson	$p(X) = \dfrac{\lambda^X}{X!}\, e^{-\lambda}$	λ, X

Table 7.7 Summary of expressions for describing the probability distributions considered in Chapter 7

Distribution	Mean	Variance	Skewness
Normal	$\dfrac{\Sigma X}{N}$	$\dfrac{\Sigma(X-\mu)^2}{N}$	$\dfrac{\Sigma(X-\mu)^3}{N}$
Binomial	Nj	$Nj(1-j)$	$\dfrac{j-(1-j)}{Nj(1-j)}$
Poisson	λ	λ	$1/\sqrt{\lambda}$

In all cases the standard deviation is the square root of the variance.

RECOMMENDED READING

Blalock, H. M. (1960). *Social Statistics,* McGraw-Hill, New York. One of a range of non-geographical texts that should be consulted by all geographers seeking to broaden their grasp of statistical principles. Chapters 8 and 9 are notably pertinent to this section.

Chorley, R. J., and Kennedy, B. A. (1971). *Physical Geography: a Systems Approach,* Prentice-Hall, London. Chapter 5 presents a very clear portrayal of the role that probability studies can play in geography. Human geographers, too, might like to note this title.

Curry, L. (1952). 'Climatic change as a random series', *Anns Assoc. Am. Geogrs* **52,** 21–31. A thought-provoking paper dealing with our understanding of random processes and how they may account for seemingly organised behaviour.

Dury, G. H. (1977). 'Likely hurricane damage by the year 2000', *Prof. Geogr* **29,** 254–258. The title bears out the very practical nature of this application of probability principles.

Ferguson, A. G. (1977). 'Probability mapping of the 1975 Cholera epidemic in Kisuma District, Keyna', *J. Trop. Geog.* **44,** 23–32. An application of the principles of the Poisson distribution, but in a simple and clearly understandable manner.

Gudgin, G., and Taylor, P. J. (1974). 'Electoral bias and the distribution of party voters', *Trans. Inst. Brit. Geogrs.* **63,** 53–74. A good example of the application of empirical probability distributions and their associated properties.

Granger, O. (1979). 'Increasing variability in California precipitation' *Anns Ass. Am. Geogrs* **69,** 533–543. Empirical probabilities are used here to great effect in analysing and predicting trends from a long series of rainfall observations.

Gumbel, E. J. (1958). 'Statistical theory of floods and droughts', *J. Inst. Water Engrs* **12,** 157–184. This singularly important item of research looks at the probabilities associated with extreme and, therefore, unlikely events.

Manning, H. L. (1956). 'The statistical assessment of rainfall probability', *Proc. Royal Soc. Series B* **144,** 460–480.

Moroney, M. J. (1951). *Facts from Figures,* Penguin, Harmondsworth. This book was written primarily for the business and quality control market, but Chapters 8 and 9 are valuable accounts of the nature of probability distributions.

Roder, H. (1974). 'Application of a procedure for statistical assessment of points on a line', *Prof. Geogrs* **26,** 285–290. A demonstration of different forms of empirical probability distributions.

Trenhaile, A. S. (1975). 'The morphology of a drumlin field', *Anns Ass. Am. Geogrs* **65,** 297–312. Distribution moments are used here to determine and distinguish patterns of glacial behaviour. Other statistics are used which relate to topics dealt with at a later point in this book.

EXERCISES

1. If a given normal distribution has a mean of 510 and a standard deviation of 115 units, convert the following 'raw' observations to equivalent z values:

   612   408   502   346

2. Use the equation for the standardized normal curve (equation 7.3) to estimate the probability density function, i.e. the height of the curve (Y) for the following z values.

   0.0   1.5   −1.0   2.6   −4.2

3. What are the probabilities of events under the normal distribution between the following limits (all expressed as z values): (i) 0.0 and 1.6; (ii) 0.0 and −2.9; (iii) −0.9 and 1.2; (iv) 0.3 and 1.8; (v) −0.6 and infinity?

4. For the distribution parameters in question 1, between which 'raw' values will events occur with a 0.95 probability?

5. The mean orientation of drumlins in the Geulph area of Ontario has been established at 297.18° magnetic. The associated standard deviation is 9.76 (Trenhaile, 1975). What is the probability of a randomly selected drumlin having an orientation of between 290° and 300°? What is the probability of a drumlin being orientated at 272° or less?

6. A 1971 survey of car ownership revealed that 42.9 per cent of British households possessed one car. For a random sample of 6 households define the complete binomial distribution. Take car ownership as a 'success', so that $p(\text{success}) = 0.429$. You will need to establish the probabilities of all successes from 0 up to 6.

7. Calculate the mean and standard deviation of the distribution in question 6 and plot a histogram of the probabilities derived.

8. Repeat questions 6 and 7 with reference to the percentage of households found to possess two or more cars. This percentage in 1971 was 9.1 per cent. Take N to be 6 when constructing the probability distribution and assume a household with two or more cars to be a 'success'.

9. In a study of the hydrology of some British drainage basins stream frequencies were measured and the following two basins were noted:
River Urr, Southern Scotland   1.73 junctions/km^2
River Thames (upper part of basin)   0.33 junctions/km^2
Use these average densities to draw up the respective Poisson distributions as far as $p(X) = 6$. junctions per km^2.

10. Using the data in question 9, construct the two probability histograms and assess the distributions' means and standard deviations.

11. In question 10 how many kilometre squares in a sample of 30 might be expected to contain (i) as few as zero stream junctions, (ii) as many as five? Calculate these two expectations for both drainage basins.

12. From the data in question 5, how many drumlins from a sample of 100 might be expected to have an orientation of between 290.6° and 303.8°?

Chapter 8

Samples and Populations

8.1 INTRODUCTION

The reader should now be aware of the important distinctions between the statistical properties of populations, in the mathematical sense of the word, and samples drawn from them. Earlier chapters introduced some aspects of sampling and offered methods by which they may be obtained, and Chapter 7 considered some of the interpretational difficulties of sample data. We may now consider, in more detail, the methods by which inferences can be drawn from sample data.

The reader's attention is first drawn to the puzzling issue of symbols. A wide variety of symbols has been adopted for statistical use, and a major distinction needs to be drawn between those used to indicate population parameters and those for samples. Discussion has so far been largely within the context of populations. When we introduced normal distributions, for example, it was assumed that the inferences drawn were for population parameters. The mean has, consequently, been denoted by μ and the standard deviation of σ. Latin symbols, however, are preferred when sample properties are summarized. This convention applies to many, though not all, statistics. Those most commonly used are listed in Table 8.1. The most frequent exception to this rule is made when referring to population and sample sizes, usually designated by N and n respectively.

Table 8.1 Symbols used for describing the characteristics of samples and populations

Parameter	Population symbol	Sample symbol
Mean	μ	$\bar{X}$
Standard deviation	σ	s
Variance	σ^2	s^2
Skewness	a_3	a_3
Number of observations	N	n

8.2 ESTIMATES OF RELIABILITY : STANDARD ERRORS

Samples will not always reflect the characteristics of the populations from which they are drawn, and the most well-designed sampling framework does not

105

guarantee to identify the precise nature of the population. Even random sampling is capable of providing an unrepresentative impression of a population. But though our understanding of the study population may be limited — a common difficulty in geographical studies — we do not need to be equally vague about the reliability of derived samples, and simple methods exist to establish their accuracy.

Initial interest is with 'large' samples, i.e. those containing more than 30 observations. In such cases particular conditions exist, the most important of which is described by the central limit theorem. This 'law' is best explained in the following fashion: if all possible samples of size n were taken from a population, then the statistical distribution of those sample means would be normal. Furthermore, but usefully for us, this normality prevails irrespective of the distribution of the population. Hence a population with a skewed distribution would still provide a normal 'sampling distribution of means'. This principle is demonstrated in Figure 8.1 using daily discharge data for the River Wear, County Durham, England.

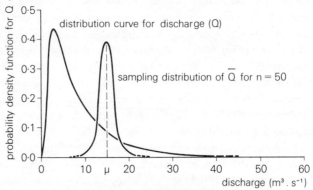

Figure 8.1 Distribution of mean daily discharges and sampling distribution of sample means for $n = 50$

Under all but the most unusual circumstances investigators would take only a very few samples from one population. They might thus be thought to be in a very poor position to examine the sampling distribution of means. But this is far from the case and we may, with as few as one sample, estimate the standard deviation of the sampling distribution. Standard deviations of sampling distributions are generally known as 'standard errors'. Their exact numerical values vary from case to case but, for large samples, possess all those properties and characteristics attributed to the standard deviation of the normal distribution. Furthermore, the standard error of sample means ($\sigma_{\bar{x}}$) is easily evaluated from:

$$\sigma_{\bar{x}} = \frac{s}{\sqrt{n}} \tag{8.1}$$

Equation 8.1
$\sigma_{\bar{x}}$ = standard error of sample means
s = standard deviation of sample
n = sample size

Additionally, if all possible samples of size n were drawn then the mean of these sample means ($\mu_{\bar{X}}$) would equal the population mean μ:

$$\mu = \mu_{\bar{X}} \tag{8.2}$$

It is for this reason that the sampling distribution of the means is usually centred on the population mean, as in Figure 8.1. In effect we are here examining a new random variable, that of sample means, that is normally distributed. As this random variable is normal it is distributed as z, and equation 7.2 can be rewritten to include the new terms:

$$z = \frac{\bar{X} - \mu_{\bar{X}}}{\sigma_{\bar{X}}} \tag{8.3}$$

Equation 8.3
$\bar{X}$ = sample mean
$\mu_{\bar{X}}$ = population mean
$\sigma_{\bar{X}}$ = standard error of sample means

The mathematical identity of these expressions is important because it permits probabilistic assessments concerning sample means using the established properties of the normal curve (see Section 7.3).

8.3 CONFIDENCE LIMITS AND STANDARD ERRORS

It follows from what we know of the normal distribution that the probability of a given sample mean lying within certain limits about the population mean can be found from z tables. For example, there is a 0.683 probability that the sample mean will be within one standard error (standard deviation) of the population mean, a 0.954 probability that it will lie within two standard errors and a 0.997 probability that it will be within three standard errors. The position is identical to that reviewed in the previous chapter for individual observations about the mean. There is, however, one major contrast — here we do not know the value of the population mean. For large, and certainly for infinite, populations it can never be assessed with absolute precision, which is why we often take samples. This failing compels us to readjust our thinking and take the mathematically justifiable alternative of transposing the roles of sample and population means when making probability statements. From this position we argue that there is now a 0.683 probability of the population mean being within one standard error of the sample mean, a 0.954 probability of the population mean being within two standard errors of the sample mean, and so on (Figure 8.2). In this way, and by reference to the magnitude of the standard error, the reliability of sample means can be evaluated from just one sample — even though the population mean is unknown.

Daily mean discharges from the River Wear provide an example of how these procedures may be applied. A random sample of 50 day's observations was drawn giving a sample mean of 14.68 cumecs (cubic metres per second) of discharge and a sample standard deviation of 21.89. How reliable is this estimate? Is the 'real' mean likely to be within 1 or 10 cumecs of this value? The first step in answering these questions is to estimate the standard error ($\sigma_{\bar{x}}$), or spread, of the sampling distribution of means from equation 8.1. Hence:

$$\sigma_{\bar{x}} = \frac{21.89}{\sqrt{50}} = 3.096$$

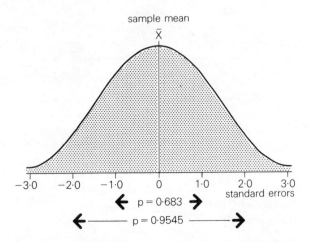

Figure 8.2 Probabilities on the normal distribution defined by standard errors

From this we conclude that there is a 0.682 probability of the population, or 'real', mean lying within 3.096 cumecs of the sample mean and a 0.954 probability of it lying within 6.192 (from 2 × 3.096) cumecs of the sample mean. The 0.682 probability range can be quantified thus:

$$14.68 \pm 1 \times 3.096 = 11.58 \text{ to } 17.78 \text{ cumecs}$$

and the 0.954 probability range by:

$$14.68 \pm 2 \times 3.096 = 8.49 \text{ to } 20.87 \text{ cumecs}$$

In these cases the selected standard errors equated to critical z values (z_c) of 1.0 and 2.0. These values help to define what are termed the 'confidence limits'. The two expressions above define such limits and enclose a band known as the confidence interval. Traditionally these limits are described in percentages rather than absolute values. We speak of the 68.2 per cent and 95.4 per cent confidence limits, for example. But the properties of the normal distribution are well known, and we are not confined to such numerically cumbersome limits but are free to choose values within the range 0 – 100 per cent. In practice, however, the 95 and 99 per cent confidence limits are used, almost to the total exclusion of all others.

Nevertheless, others may be used and the limits are generalized in the expression:

$$\bar{X} \pm z_c\, \sigma_{\bar{X}} \tag{8.4}$$

Equation 8.4

$\bar{X}$ = sample mean of variable X

z_c = critical z value for selected level

$\sigma_{\bar{X}}$ = standard error of sample mean

---

in which the critical z values are determined by the selected confidence level and can be ascertained from the z tables. For the 95 per cent (0.95) and 99 per cent (0.99) levels the critical z values are 1.96 and 2.58 respectively (see also Section 7.5). In the example above, the 95 per cent confidence limits are given by:

$$14.68 \pm 1.96 \times 3.096 = 8.61 \text{ to } 20.75 \text{ cumecs}$$

and the 99 per cent limits by:

$$14.68 \pm 2.58 \times 3.096 = 6.69 \text{ to } 22.67 \text{ cumecs}$$

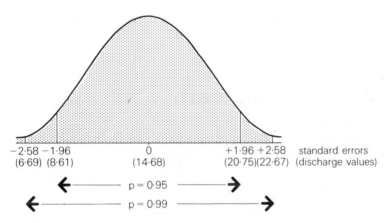

Figure 8.3 Probabilities limits on the normal distribution expressed as z values and equivalent discharges (in brackets)

These results are summarized also in Figure 8.3, in which we see how higher confidence limits can be achieved only at the expense of wider confidence intervals. Hence, we are 99 per cent certain that the population mean lies between 6.69 and 22.67 cumecs, but only 95 per cent certain that it lies between the narrower bounds of 8.61 and 20.75 cumecs.

It may also be clear that the best way to gain greater accuracy in sample estimates is to increase the sample size n. As n increases (see equation 8.1) the standard error, or spread, of the sampling distribution is reduced and the resulting confidence intervals narrowed. The critical z values for given confidence levels are, of course, fixed and independent of the sample size. Here, then, is a substantive justification for the intuitive notion that larger samples may be more representative of the populations from which they are drawn. These changes, expressed as variations in the form of the sampling distribution, are shown in Figure 8.4.

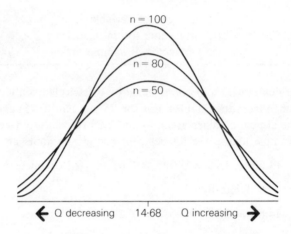

Figure 8.4 Changing shape of the sampling distribution with increasing sample size

8.4 SAMPLES AND STANDARD DEVIATIONS

The mean is frequently the first point of interest in a statistical analysis, but rarely the only one. The standard deviation is also important in providing a clear expression of the behaviour of the study variable. It is of little value to know the mean of a data set if knowledge of variability about this central figure is absent. Fortunately, assessments can also be made of the degree of uncertainty attached to sample estimates of standard deviations. Again, if all possible samples of size n were drawn from a population and the standard deviations of those samples calculated, then their distribution, irrespective of the character of the population, would be normal. This condition is best fulfilled by samples with a minimum size of 100, though this necessity is frequently overlooked. The standard error of this new sampling distribution (σ_s) is given by:

$$\sigma_s = \frac{s}{\sqrt{(2n)}} \tag{8.5}$$

Equation 8.5
s = sample standard deviation
n = sample size
σ_s = standard error of sample standard deviation

As the sampling distribution is again normal, confidence limits about a single estimate may be constructed. The strategy is similar to that for sample means with confidence limits defined by:

$$s \pm z_c \, \sigma_s \tag{8.6}$$

where z_c is determined, as before, from the z tables according to the selected confidence level.

A study of loess and lacustrine deposits by Campbell (1979) showed the variability in the percentage of sand content of the samples (measured as the standard deviation) to be 0.78 per cent. The sample size was 369, but how reliable is this measure of variability? This question can be answered by first of all finding the standard error of the sampling distribution of the standard deviation (σ_s) using equation 8.5. By direct substitution:

$$\sigma_s = \frac{0.78}{\sqrt{(2 \times 369)}} = 0.029$$

This derived standard error can now be substitued into equation 8.6 to give the required confidence limits:

$$0.78 \pm 1.96 \times 0.029$$

$$= 0.78 \pm 0.0568 = 0.723 \text{ to } 0.837$$

By an identical process the 99 per cent confidence limits are found from:

$$0.78 \pm 2.58 \times 0.029 = 0.705 \text{ to } 0.855$$

To summarize these findings, we have 95 per cent confidence that the population standard deviation of percentage sand content lies between 0.723 and 0.837 per cent. We also have 99 per cent confidence that it is between 0.705 and 0.855 per cent. These narrow confidence bands are due, in part, to the large sample size; again testifying to the value of taking large samples whenever possible.

8.5 SAMPLES AND NON-PARAMETRIC DATA

So far only parametric (interval or ratio scale) data have been used. There are, however, frequent occasions when geographers need to sample from nominal scale data. Under such circumstances the background population may follow a binomial distribution. How can these data, being in the form of frequencies or proportions, be treated in sample form?

A random sample of shoppers might be taken to establish the proportion that are from out-of-town locations (possibly and important factor in planning car parking facilities etc.). A sample of 100 might yield a 0.25 (25 per cent) proportion in this category. But how reliable is this figure? At first sight the problem might appear to be intractable, particularly when using just one sample. But, fortunately, the central limit theorem applies equally to proportions and we may proceed much as if parametric data were being used. Thus, if all possible samples of size n (100 in this example) were taken, then the distribution of the resulting proportions would be normal even though the events themselves might be binomially distributed.

Let p be the one sample proportion; then the standard error of the sampling distribution of proportions (σ_p) is given by:

$$\sigma_p = \sqrt{\left[\frac{p(1-p)}{n}\right]} \tag{8.7}$$

This standard error, having all the probabilistic properties of the standard deviation of the normal distribution, can be used to provide confidence limits for p. We need only rearrange the general definitions of confidence limits given in equations 8.4 and 8.6 to give:

$$p \pm z_c \sigma_p \tag{8.8}$$

Proportions and non-parametric data generally are widely used in human geography, particularly when analysing questionnaire surveys. The British Government's decadial census provides a useful example of how the standard error of proportions can be applied. As it takes several years before a complete census is analysed preliminary results for a 10 per cent sample are published to give an impression of trends since the previous census. The 1971 10 per cent sample census report Staffordshire (HMSO) showed that of the 90,100 people constituting the sample of economically active individuals 4047 were unemployed. Our problem is to establish how reliable this latter proportion is. The first step is to convert to a relative proportion p, where:

$$p = \frac{4047}{90,100} = 0.0449$$

This proportion can now be examined using equation 8.7 to find the standard error of the distribution of proportions:

$$\sigma_p = \sqrt{\left[\frac{0.0449(1-0.0449)}{90,100}\right]} = 0.00069$$

With σ_p we can evaluate the 95 per cent (or any other selected) confidence limit. Remembering that the critical values of z for 95 per cent limits are ± 1.96, we use expression 8.8 to find limits of:

$$0.0449 \pm 1.96 \times 0.00069$$
$$= 0.0435 \text{ to } 0.0463$$

In other words we can be 95 per cent confident that the real proportion of unemployed in Staffordshire at the time of the survey lies between 0.0435 and 0.0463 of the economically active population.

8.6 SMALL SAMPLES

Caution is always required when using samples, more so when the sample size, for whatever reason, is less than 30. Such samples are termed 'small samples' and have their own procedures for analysis which acknowledge their peculiarities. When large samples (*n* of over 30) are used the sampling distributions are normal regardless of the population. This is not so for small samples where the sampling distributions can be distorted by the nature of the population's distribution. Hence, if the background population is known, or suspected, to be non-normal, large samples must be used. Should, however, the population be normally distributed then inferences may be drawn using small samples. The sampling distributions, nevertheless, are no longer normal though their behaviour and characteristics are equally well understood and follow a form known as the '*t*-distribution'.

The *t*-distribution is not, in appearances, unlike the normal. It is continuous, extends from plus to minus infinity and is, to varyng degrees, bell-shaped. The greatest contrast with the normal distribution lies in the dependence of its detailed form on sample size. For very small samples the distribution forms an attenuated bell-shape, but as the sample size increases the shape approximates ever more closely to that of a normal curve. Sufficient coincidence is often assumed for sample sizes greater than 30 (Figure 8.5).

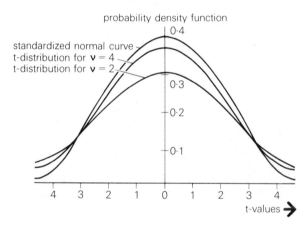

Figure 8.5 Changing shape of the *t*-distribution as *v* (degrees of freedom) become greater

The *t*-distribution was discovered as recently as the 1930s by an industrial statistician, William Gosset, who published his findings under the pseudonym 'Student'; hence the distribution is also known as 'Student's *t*-distribution'. The variability of the *t*-distribution presents no difficulties to the statistician because the form of the curve remains definable regardless of the sample size, and from this mathematical definition *t* tables, much like the normal distribution's *z* tables, can be prepared (see Appendix II). These tables are used to establish the *t* values between which events lie with specified probabilities. In the case of the normal distribution there is a 0.95 probability of events within the range of $z = -1.96$ to $z = +1.96$. We can use *t* tables to evaluate the critical *t* values within which events lie with 0.95, or any other, probability. Their layout requires a little explanation.

Reference to Appendix II indicates that the t values are arranged by rows according to the degrees of freedom (a quantity which, in this case, is one less than the sample size) and by columns according to the probability. The first column gives the plus and minus t values (the distribution is symmetrical) which delimit the 0.9 probability range, the second column the 0.95 range, and so on. The delimiting critical values decrease as the degrees of freedom (v) increase.

Use of t tables is easy but does require some care and practice. A simple example will demonstrate the principles. Suppose we have a sample size of 11 and wish to know the t values between which events occur with a 0.95 probability. The appropriate row of the table, that for 10 degrees of freedom ($v = n - 1$), is first located, then the column, in this case 0.95. The value of 2.23 is read off. Thus between $t = -2.23$ and $t = +2.23$ lies 0.95 of the area of the t distribution, and within that range events will fall with a probability of 0.95. Figures 8.6 and 7.13 can be compared to emphasize the contrasts between the t- and the normal distribution, in which the corresponding t values for the 0.99 probability range are ± 3.17.

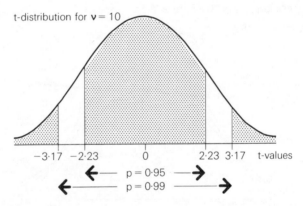

Figure 8.6 The t-distribution for $v = 10$ showing the t values for the 0.95 and 0.99 probability limits

Before progressing further, however, it is necessary to elaborate on the largely unexplained concept of degrees of freedom. The principles can be demonstrated by a simple example. Consider a column of ten numbers which adds up to, say, 17.5. This total can be regarded as fixed for that set of figures. Of the ten constituent numbers nine are free to assume any value whatsoever. But in order for the total of 17.5 to be preserved, the remaining constituent number is predetermined by the other nine. We say that one degree of freedom has been lost. In statistics there are occasions when more than one degree of freedom may be lost, but these will be considered as they arise.

Degrees of freedom are often preferred to sample size as they reduce unwanted bias in parameter estimates. Small sample variances, for example, are usually underestimates of the population parameter. This difficulty can be overcome by dividing the sum of squared deviations from the mean (see Section 5.11) by the sample size minus one ($n - 1$) instead of the complete sample size (n). This provides what is termed the 'best estimate' of population variance, and division by the smaller quantity ensures a higher final value. A useful qualitative expression, and one that will be used frequently in later sections, is:

$$\text{Variance} = \frac{\text{sum of squares}}{\text{degrees of freedom}}$$

This can be represented in algebraic form in many ways depending on the type of variance being measured, but in this case it would be written:

$$\hat{\sigma}^2 = \frac{\Sigma(X - \bar{X})^2}{n - 1} \tag{8.9}$$

Equation 8.9

X = individual observation of X

$\bar{X}$ = sample mean of variable X

n = sample size

$\hat{\sigma}^2$ = best estimate of variance

The corresponding expression for the best estimate of standard deviation ($\hat{\sigma}$) is the square root of equation 8.9:

$$\hat{\sigma} = \sqrt{\left[\frac{\Sigma(X - \bar{X})^2}{n - 1}\right]}$$

There are many applications of the concept of degrees of freedom. Here it is used to provide unbiased variance estimates. In the following chapter we shall see how they are used to avoid bias in an area known as hypothesis testing.

8.7 CONFIDENCE LIMITS AND SMALL SAMPLES

Equipped with the information in the foregoing section we can proceed to apply the principle of confidence limits to small samples. The mean and standard error are estimated as above, but the confidence limits are defined now by:

$$\bar{X} \pm t_c \sigma_{\bar{X}} \tag{8.10}$$

where critical t values (t_c) replace the critical z values of expression 8.4.

Lam (1978) studied soil characteristics in the Hong Kong area. A sample of 20 observations gave a mean soil pH of 4.75. For such a small sample the sampling distribution of the mean follows the t-distribution, not the normal distribution. Confidence limits remain, however, relatively easy to calculate. The standard error of the sampling distribution is estimated as before but using the 'best estimate' for the standard deviation ($\hat{\sigma} = 0.12$) because of the scarcity of data. Thereafter, confidence limits are assessed using expression 8.10. The standard error $\sigma_{\bar{X}}$ of this sampling distribution becomes, from equation 8.1:

$$\sigma_{\bar{X}} = \frac{0.12}{\sqrt{20}} = 0.027$$

The t table now replaces the z table in defining the limits. The degrees of freedom, for a sample size of 20, are 19, giving critical t values (t_c) of ± 2.09 for the 95 per cent limits and ± 2.86 for the 99 per cent limits. When these values are substituted in expression 8.10 we get, for the 95 per cent confidence limits:

$$4.75 \pm 2.09 \times 0.027 = 4.694 \text{ to } 4.806$$

and for the 99 per cent limits:

$$4.75 \pm 2.86 \times 0.027 = 4.673 \text{ to } 4.827$$

Both sets of critical values are greater than their normal counterparts (1.96 and 2.58), reflecting the greater degree of uncertainty in small samples. Nevertheless the probable range of values for the population mean remains narrow and the sample-based estimate is seemingly reliable — we can be 95 per cent confident that the population means lies between 4.694 and 4.816 pH.

8.8 SAMPLE SIZES AND THE SAMPLING FRACTION

The final point to be covered concerns the question of the sample size as a proportion of the population. So far we have assumed the populations to be infinite. As a consequence any sample, no matter how large, can only be an infinite fraction of the total population. But finite populations can, and do, occur and irrespective of sample size we might feel more confident of a sample that represents 50 per cent of the population than one based on only 5 per cent. Can we make allowances for the magnitude of this sampling fraction? The brief answer is yes, though it is not often done. The method employs a correction factor by which the standard error is multiplied. This factor is a function of the ratio of sample to population size, expressed by:

$$\sqrt{\left(\frac{N-n}{N-1} \right)} \tag{8.11}$$

Equation 8.11
$N =$ population size
$n =$ sample size

which, when included with equation 8.1, gives the following expression for the sampling error of the mean:

$$\sigma_{\bar{x}} = \frac{s}{\sqrt{n}} \sqrt{\left(\frac{N-n}{N-1} \right)} \tag{8.12}$$

It may also be used with the standard error of proportions in equation 8.7, giving:

$$\sigma_p = \sqrt{\left(\frac{p(1-p)}{n} \right)} \sqrt{\left(\frac{N-n}{N-1} \right)}$$

Table 8.2 Summary of the formulae in Chapter 8

Description of expression	Expression	Conditions
Standard error of sampling distribution of means ($\sigma_{\bar{X}}$)	$\dfrac{s}{\sqrt{n}}$	Sampling from an infinite population or by replacement[1]
Standard error of sampling distribution of means ($\sigma_{\bar{X}}$)	$\dfrac{s}{\sqrt{n}} \sqrt{\left(\dfrac{N-n}{N-1} \right)}$	Sampling from a finite population[1]
Confidence limits about the sample mean	$\bar{X} \pm z_c \sigma_{\bar{X}}$	n greater than 30
Standard error of sampling distribution of standard deviations (σ_s)	$\dfrac{s}{\sqrt{(2n)}}$	n greater than 100
Confidence limits about sample standard deviation	$s \pm z_c \sigma_s$	n greater than 100
Standard error of proportions (σ_p)	$\sqrt{\left[\dfrac{p(1-p)}{n} \right]}$	n greater than 30 (dichotomous variable)
Confidence limits about sample proportion	$p \pm z_c \sigma_p$	n greater than 30
Best estimate of population variance	$\dfrac{\Sigma(X_i - \bar{X})^2}{n-1}$	n less than 30
Confidence limits about sample mean	$\bar{X} \pm t_c \sigma_{\bar{X}}$	n less than 30
Confidence limits about sample proportion	$\bar{X} \pm t_c \sigma_p$	n less than 30

1 This distribution is normal for n greater than 30 but distributed as t when n is less.

The greater the sampling fraction the smaller the numerical value of the correction factor becomes, and the greater its effect in reducing the value of the standard error; it can never be greater than 1.0.

No rigid rules apply for the application of the sampling fraction correction; but where the correction exceeds 0.9 it may be overlooked, otherwise the researcher may consider its use. For infinite populations the question never arises. However, seemingly finite populations may be rendered 'infinite' if sampling by replacement is undertaken.

As a measure of the importance of this correction we may consider the case of a population of 100 from which a sample of 50 is drawn. Hence $N = 100$ and $n = 50$, and the correction is:

$$\sqrt{\left(\frac{100 - 50}{100 - 1}\right)} = 0.711$$

Thus if the standard error of the sample mean had been established at, say, 12.6 it would be corrected by this factor:

$$12.6 \times 0.711 = 8.96$$

Hence the error range is reduced and results made more reliable when an appreciable proportion of the population is sampled. Nevertheless it must be added that only rarely are geographers in the happy position of being able to use this correction and our statistical populations are usually either infinite or very large.

REFERENCES

Campbell, J. B. (1979). 'Spatial variability of soils', *Ann. Ass. Am. Geogrs* **69**, 544–556.
Lam, K. C. (1978). *'Soil erosion, suspended sediment and solute production in three Hong Kong catchments', J. Trop. Geog.* (Singapore) **47**, 51–62.
Office of Population Censuses and Surveys (1971). *Census 1971 (10% sample)*, HMSO, London.

RECOMMENDED READING

Berry, B. J. L., and Baker, A. M. (1968). 'Geographic sampling' in B. J. L. Berry and D. F. Marble, (eds) *Spatial Analysis*, Prentice-Hall, Eaglewood Cliffs.
Chorley, R. J. (1966). 'The application of statistical methods to geomorphology' in G. H. Dury (ed.) *Essays in Geomorphology*, Heinemann, London. A wide-ranging yet lucid and comprehensive paper which includes a useful section on sampling in geography.
Court, A. (1972). 'All statistical populations are estimated from samples', *Prof. Geogr* **24**, 160–162. A brief but important paper which repays the effort of its reading.
Dixon, C., and Leach, B. (1975). *Sampling Methods for Geographical Research*, Catmog 17, Geobooks, Norwich. An exhaustive treatise, written specifically for geographers yet not too demanding for the student reader.
Harvey D. (1969). *Explanation in Geography*, Arnold, London. Current interest would focus on Chapter 19 of this book, which is a penetrating, if slightly philosophical, study of sampling and data collection.

Krumbein, W. C., and Graybill, F. A. (1965). *Introduction to Statistical Models in Geology*, McGraw-Hill, New York. Chapters 6 and 7 deal specifically with sampling, and although they were written with geologists in mind the material is by no means peripheral to human geographers.

Spiegal, M. R. (1972). *Theory and Problems of Statistics*, Schaum's Outline Series, McGraw-Hill, New York. Chapters 8–11, and the associated worked examples, make a useful general reference.

EXERCISES

1. The average pH value of soil samples taken during a survey of Chalk Downlands in Southern England was 6.95. The sample size was 35 and the standard deviation 0.52. What are the 95 per cent confidence limits for this sample mean?

2. In the same area as question 1 a later, and more exhaustive study using 60 samples, gave a mean of 6.84 and a standard deviation of 0.55. What are the new 95 per cent confidence limits for the sample mean and how do they compare with those found in the first question?

3. Wheat yields from a sample of 25 farms gave a standard deviation of 3.4 cwt/acre. Express the reliability of the estimate as a standard error.

4. A sample survey indicated that of the 400 heads of households interviewed 42 were unemployed. How wide is the 99 per cent confidence interval for this proportion? By transposition of equation 8.7 assess the necessary increase in sample size in order to reduce the standard error to one-half that for $n = 400$.

5. Samples are often unavoidably small in geographical studies. A catchment rainfall study recently undertaken could manage only 12 raingauges. The monthly mean figures (in mm) are given below. Calculate the 'best estimate' standard deviation. By how much does the best estimate differ from the uncorrected value? What are the 95 per cent confidence limits for the sample mean? Remember to use the $t-$ and not the normal distribution.

171.2	149.0	126.5	120.8
124.4	106.5	93.4	96.7
122.1	93.5	96.2	94.2

Chapter 9

Testing Hypotheses — the Univariate Case

9.1 INTRODUCTION

All geographers form general impressions in the course of their work that are the beginnings of ideas and hypotheses. We might observe that some forms of retail function cluster more readily than others, or that some rock types occur on steeper slopes than others. But we want to know if our impressions are correct and, if they are, why.

The first step in answering these questions is that of measurement, and here also is the first of our problems. For while some phenomena can be measured on one or more scales (nominal, ordinal, interval or ratio), others have no agreed forms of measurement. Slopes can be measured by degrees of inclination (interval/ratio scale), rock hardness on the Moh's scale (ordinal scale), and rock type can be allocated by lithological type (nominal scale). But how does one measure aspects of human behaviour such as environmental perception or class attitude? Hence we are not always able to measure the variable in which we are interested or, even worse, may be unsure of its definition. Consequently, a distinction has to be drawn between theoretical and operational definitions or concepts. For example, a geomorphologist might be interested in hillside morphology, hence his theoretical concept is that of hillside shape. In operational terms, however, a more precise and pragmatic definition is required. What exactly is meant by hillside morphology? Is it overall inclination, maximum slope, ruggedness or slope variability? Indeed how is a hillside to be defined and demarcated? Such problems are common to all branches of geography and the solution lies in clearly defining the problem at the outset.

The substitution of operational for theoretical concepts is similar to translating from one language to another, in which care must be taken not to lose sense of the statement. But, with a satisfactory operational definition, speculations and hypotheses can be expressed in a testable form. The term 'test' is used here in a specific context of statistical tests, the study of which consumes the following chapter and large portions of subsequent sections.

9.2 THE NULL HYPOTHESIS

Statistical tests require numerical data collected within a rigorous sampling framework. The data must consist of a sample, or samples, drawn at random from a specified population. This specification implies a clear understanding of the population being sampled, but this understanding is not always possible. Nevertheless the geographer must take every precaution to avoid biased samples.

The hypothesis to be tested is not the general geographical statement but a 'null hypothesis' designated H_0. The null hypothesis is an indispensible part of all statistical tests and, despite what our strongest impressions are concerning the behaviour of the variable, is always one of 'no difference'; for example, that there is no difference between the mean slopes on two geological strata or that there is no difference between the degree of clustering of two retail functions.

The statistical test results in the acceptance or rejection of the null hypothesis. If it is rejected then the alternative hypothesis (H_1) must be accepted. The alternative hypothesis is the logical converse of the null hypothesis and asserts that very real differences exist between the cases under study. Hence we may distinguish three forms of hypothesis:

(1) the research hypothesis or general geographical statement
(2) the null hypothesis of no difference (H_0)
(3) the alternative hypothesis (H_1)

The null and alternative hypotheses are statistical statements only and represent the testable expression of the research hypothesis. They 'exist' only within the statistical test, but the explanation of their acceptance or rejection can form the essence of a geographical project.

9.3 AN INTRODUCTION TO HYPOTHESIS TESTING

The strategy for all statistical tests is broadly similar. The first step is to define the null hypothesis (H_0) which should be made with reference to the measured attributes used in the test. The latter will produce a 'test statistic', the magnitude of which indicates whether H_0 should be accepted or rejected. Each test yields its own statistic; sometimes these are in the form of the now familiar z or t values, but often a test will have its own specific statistic with a known and definable probability distribution. From the latter it is possible to assess the random probability of the test statistic, i.e. is such a figure likely or unlikely by random processes? In general, when the test statistic is improbable by random processes the null hypothesis will be rejected.

At this stage it is sufficient to state the random probability of the test statistic is also the probability of H_0 being correct. Thus low random probabilities such as 0.01 would lead us to reject H_0, but a higher figure of 0.5 would allow us to accept it. But in rejecting H_0 the acceptance of H_1 is mandatory. It is important, before the test begins, to decide on the probability limit beyond which H_0's probability of being correct is unacceptable. This limit defines the so-called rejection region on the probability distribution. Traditionally, though for no specific reason other than numerical convenience, the choice lies between the 0.05 (5 per cent) and 0.01 (1 per cent) limits.

In order to demonstrate the general principles of all statistical tests we can take the example using material already introduced in Chapters 7 and 8. Our problem concerns the need to establish whether a given sample could reasonably be expected to have been drawn from a specified population. In a study of rainfall in a small area of England data were collected from a large number of ground-level recording stations; these data are used to estimate the population parameters. Supplementary data were also available for a rooftop raingauge. The problem is to determine if the latter sample, being drawn from a distinctive location generally regarded as overexposed, was genuinely representative of the population and therefore eligible for inclusion in the main data set. The three hypotheses (the geographical, null and alternative) used in this process are:

(1) The geographical hypothesis. The rooftop site provides rainfall estimates that are unrepresentative of the region owing to overexposure and consequent underestimates.
(2) The null hypothesis. There is no difference between the means of the rooftop sample and ground-level population other than can be attributed to random sampling variations. This hypothesis might be expressed symbolically by $H_0 : \bar{X} = \mu$, where $\bar{X}$ and μ are the sample and population means respectively.
(3) The alternative hypothesis. The difference between the means of the rooftop sample and ground-level population are too great to have arisen by random sampling variations from a common population, or $H_1 : \bar{X} \neq \mu$.

The rejection region for the null hypothesis must also be decided at this stage. This is determined by the choice of probability beyond which H_0 is rejected: a quantity denoted by α and termed the 'significance level'. In this example the 0.05 significance level is chosen and any test statistic (a z value in this case) with a random probability of less will cause H_0 to be rejected. The regions for these remote probabilities occur at either end of the normal distribution (which describes the probability of z values) and are shown in Figure 9.1, where the critical 0.05 region is divided equally between the two ends. We already know from Section 7.5 that the critical z values that delimit these areas are ± 1.96 and a derived z value in excess of either will cause H_0 to be rejected. The corresponding limits for the 0.01 significance level would have been ± 2.58.

The next step is to calculate the test statistic. Chapter 8 has shown how the sampling distribution of sample means is normally distributed about the population mean. By using this distribution we may assess the difference between the two means in probabilistic terms. In the present example the population mean was estimated from the large body of ground-level data, and the sample mean and standard deviation, obtained from the rooftop sample, can be substituted into equation 9.1 to provide the standard error of the sampling distribution of means ($\sigma_{\bar{x}}$). It is then an easy matter to convert the difference between population and sample means into a z value on this distribution (equation 9.2). The probability of values equal to or greater than z can then be found from Appendix I. Thus, with a sample size n of 40 observations, a roof-top site mean $\bar{X}$

of 50.23 mm/month, and a standard deviation s of 28.04 we obtain:

$$\sigma_{\bar{X}} = \frac{s}{\sqrt{n}} \qquad (9.1)$$

$$= \frac{28.04}{\sqrt{40}} = 4.43$$

The difference from the population mean (μ) of 53.19 can now be expressed as a z value.

$$z = \frac{\bar{X} - \mu}{\sigma_{\bar{X}}} \qquad (9.2)$$

$$= \frac{50.23 - 53.19}{4.43} = -0.67$$

A glance at Figure 9.1 alone should be sufficient to confirm that the test statistic of $z = -0.67$ does not fall within the rejection region. Such a small difference between the means is attributed to random sampling variation from a common population, and the null hypothesis of no difference is accepted and the rooftop data is adjudged to be admissable to the general data set.

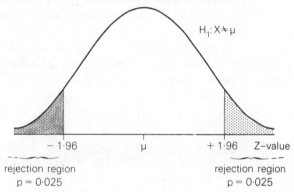

Figure 9.1 Rejection regions on the normal curve for the 0.05 significance level

9.4 ERRORS IN STATISTICAL TESTING

The statistical test does not guarantee an accurate result and there is always the possibility of the null hypothesis being rejected when it is correct or accepted when it is false. In this example it was decided that H_0 that should be rejected if the test statistic's random probability was less than 0.05. But the significance level (α) is also the probability of wrongly rejecting H_0 as it is also the probability of random differences falling within the rejection region. We have simply decided that this probability is too low for an explanation; it is not, however, impossible.

To wrongly reject H_0 is to commit a type I error and the likelihood of this happening can be reduced by lowering the significance level to, say, 0.01. To do so, however, is not a perfect solution because it increases the chances of accepting

H_0 when it is incorrect — this is the type II error. The probability of the latter is termed β and the power of a test is measured by:

$$Power = 1 - \beta$$

The weakness of adhering rigidly to a policy of either 0.05 or 0.01 significance levels should now be clear, and the correct level should be determined not from such a restricted choice but by reference to the consequences of committing either of the two fundamental errors. For most teaching purposes this question can be overlooked, but in industry or commerce it is often necessary to think very carefully about the outcomes of either type I or type II errors and to adopt significance levels in accordance with those consequences.

As an example we can take the case of research with drugs prior to their release for use by humans. If tests are performed on two groups of rats, one of which receive the drug and the other does not, with a view to establishing the possibilities of harmful side effects, the null hypothesis might state that there is no difference between the subsequent ill-health rates of the two groups. Such enquires necessitate that the type II error must be avoided at all costs, i.e. the null hypothesis must not be wrongly accepted. It is far preferable that a type I error, whereby the null hypothesis is wrongly rejected, is committed. In order to accomplish this the significance level would need to be of the order of, say, 0.20. In this way the rejection region is very large and slight differences between ill-health rates become statistically important. In real terms this reduces the risk of exposing the public to a harmful drug; the price to be paid is the modest one of possibly withholding a harmless one.

9.5 ONE– AND TWO–TAILED TESTS

The earlier example is termed a two-tailed test, so called because the rejection region is located at both ends of the test statistic's distribution (Figure 9.1). Because the alternative hypothesis did not specify that the sample mean should be either larger or smaller than the population, merely that it should be different, both eventualities were allowed for by apportioning the rejection equally between the two ends of the normal cuve.

There is often a need to indicate in the alternative hypothesis not just that a difference exists but that the diffeence is in a specified direction. In the rainfall example it would have been sensible to stipulate that the sample mean was both different from and smaller than the population mean as overexposure of raingauges causes underestimation of rainfall totals. In symbolic terms this idea would be expressed by $H_1 : \bar{X} < \mu$ instead of $H_1 : \bar{X} \neq \mu$. As a result the rejection region must now be concentrated at one end of the distribution (hence the term one-tailed). Care must be taken to allocate the region to the correct end of the distribution, and equation 9.2 shows that for $\bar{X}$ to be smaller than μ it must lie at the negative end (Figure 9.2). This reorganization also means that the critical value defining the rejection region has to be adjusted as it now delimits an area in one and not two sections. The new critical values for $\alpha = 0.05$ are given in Table

9.1, but the *z* table can also be used to determine the limits for any significance level. In the present example the 0.05 rejection region, being at one end only of the distribution, lies on the negative side of the *z* value that divides it from the remaining 0.95 of the area. As the distribution is symmetrical the whole positive half accounts for one-half of the area. The critical value deliming the 0.45 and 0.05 regions in the negative half is found directly from the tables (Appensix I) by locating the 0.45 figure (or its nearest value) and reading off the associated *z* values; in this case (−)1.645.

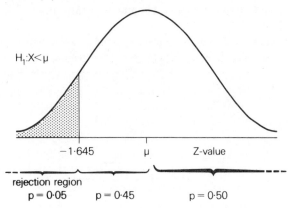

Figure 9.2 One-tailed rejection region when the alternative hypothesis specifies a negative *z* score

Table 9.1 Critical *z* values for the 0.01 and 0.05 rejection regions for one– and two-tailed tests

Tailedness	Critical values	
	0.05 level	0.01 level
One-tailed test	−1.645 or +1.645	−2.33 or +2.33
Two-tailed test	−1.96 and +1.96	−2.58 and +2.58

The changes in the critical values between one- and two-tailed tests have important consequences because it is possible for H_0 to be accepted if the test is two-tailed but rejected if it is one-tailed. This happens with *z* values within the range 1.645 and 1.96 and test statistics of, say, 1.75 could fall outside the two-tailed rejection region but within the one-tailed. Consequently, the phrasing and justification of the alternative hypothesis should be formulated with considerable care.

The more specific tests which follow represent a small number of a very wide range that can be used. The selection has not, however, been made on random grounds but in an attempt to include representatives from each of a number of types of test. The tests have been grouped according to two criteria. First we consider the measurement scale of the data required by the test, be it nominal, ordinal, interval or ratio. In broader terms a distinction is usually made between parametric tests — those requiring interval or ratio scale data — and non-parametric tests needing nominal or ordinal type data. Second, tests can be distinguished by the number of samples on which they operate. In the one-sample case a single sample is compared with a hypothesized population. But there are also tests which allow comparisons to be made directly between two or more samples — the two- and k-sample tests respectively. Table 9.2 expresses this idea in tabular form and shows how this chapter is arranged.

Table 9.2 Categories and examples of statistical tests

	non-parametric	non-parametric	parametric
	nominal	ordinal	interval/ratio
K	X^2 K-sample test section 9·8	Kruskall-Wallis H-test section 9·12	Analysis of Variance section 9·13
2	X Two sample test section 9·8	Mann-Whitney U-test section 9·10	t-test of difference between means section 9·11
1	X^2 One sample test section 9·6	Kolmogorov-Smirnov D-test section 9·7	Z-test t-test section 9·3

no. of samples

scale of measurement

Statistical tests are not only subject to the type I and type II errors. It is also easy to apply the tests to inappropriate data, and although a test statistic may well emerge at the end it is likely to be meaningless unless some thought has been given to the compatibility of the test, the data and problem being studied.

9.6 THE χ^2 TEST : A ONE-SAMPLE TEST FOR NOMINAL DATA

The χ^2 (chi-square) test is widely used in geography, but it is not as powerful as either the parametric tests or even some of its non-parametric counterparts. This weakness is, however, more than compensated for by its simple data requirements, since it needs observations only on the nominal scale, i.e. class frequencies. The test is even more widely applicable when it is recalled that both ordinal and interval/ratio scale data can be easily converted to the nominal form, though such categorization causes a loss of detail.

In its one-sample application the test examines one set of observed categorical frequencies with an hypothesized set. The example chosen uses data for days of snow cover over upland areas of the UK. The test allows regional differences in longevity of snow cover to be detected. But as snow cover is an altitude-dependent variable the data were drawn only from sites with similar altitudes.

The observed frequencies are given in Table 9.3, but the expected frequencies with which they are compared are estimated from the null hypothesis which, in this example, is of no difference between the snow cover of the five regions. Under H_0 the same frequency of days must be expected in each region and the final figure is found by apportioning the total number of snow cover days equally between the five 'cells, thus:

$Sample1$

$$\frac{142 + 178 + 87 + 189 + 87}{5} = \frac{683}{5} = 136.6 \text{ days}$$

Table 9.3 Observed and expected frequencies for the χ^2 test of regional snowfall cover (Source: *British Rainfall*, 1968, HMSO)

	North-West Highlands (Scotland)	Central Highlands (Scotland)	Southern Scotland	Northern England	North Wales	
Observed frequency	142	178	87	189	87	$\Sigma = 683$
Expected frequency	136.6	136.6	136.6	136.6	136.6	
$(O-E)^2/E$	0.213	12.55	18.01	20.10	18.01	

Degrees of freedom $= 4$; $\chi^2 = 68.88$.

The complementary alternative hypothesis is that there are real differences between the observed cell frequencies beyond those that might be expected by random variation within a common population. The 0.05 significance level will be used. Interest focuses on the differences between observed and expected values under H_0. If they are small then H_0 might be correct, but if they are larger H_0 becomes more doubtful.

But how great must the differences be before H_0 can be rejected? The normal distribution is inappropriate in this context of nominal data, but we can consult a non-parametric counterpart based on χ^2 test statistics. These are easily derived using equation 9.3 but, most importantly, they come from a known and definable probability distribution. Thus:

$$\chi^2 = \sum_{i=1}^{k} \left[\frac{(O_i - E_i)^2}{E_i} \right] \tag{9.3}$$

Equation 9.3
$\chi^2 =$ chi-square test statistic
$O =$ observed frequencies
$E =$ expected frequencies
$\Sigma =$ summation sign

The individual divisions are shown in Table 9.3 and sum to 68.88. Clearly the magnitude of the final test statistic varies with the discrepencies between the observed and expected categories, and the random probability of statistics as large as or larger than this can assessed from the χ^2 distribution. This distribution is continuous, like the normal, but dependent also on the sample degrees of freedom found from $v = k - 1$, where k is the number of categories, or cells, used. There are no negative test statistics but the distribution extends towards infinity in the positive direction. Figure 9.3 shows how its character changes with increasing v.

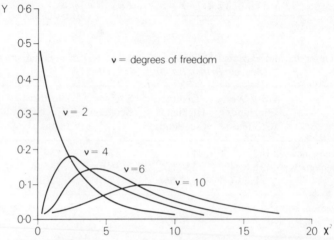

Figure 9.3 Forms of the chi-square distribution with different degrees of freedom v

The rejection region for H_0 is defined using the tabled summary of this distribution (Appendix III). The table is set out by rows and columns designated by degrees of freedom and significance levels respectively. After selecting the significance level and estimating the degrees of freedom, we can read off directly the critical value delimiting the rejection region. In the present example $\alpha = 0.05$ and $v = 5 - 1 = 4$. From the column headed 0.05 and the row for $v = 4$ the critical value of 9.49 is found. In other words there is a 0.95 probability, under H_0 of no difference, that the calculated χ^2 statistic will be less than 9.49 and only a 0.05 probability that, by random variation, it will exceed it.

The test statistic of 68.88 falls well within the defined rejection region (see Figure 9.4) and H_0 is rejected in favour of H_1; the differences between observed and expected frequencies are attributed to non-random variations, perhaps inter-regional climatological contrasts. However, it should not be forgotten that there remains a 0.05 probability that differences giving test statistics greater than 9.49 will arise by random variations. Other points that cannot be overlooked include the test's minimum data requirements. While it is true that non-parametric tests demand fewer preconditions than the parametric equivalents, they are by no means free from such considerations and those for the χ^2 test may be briefly listed:

(1) The data categories must be discrete and unambiguous, i.e. no interchangeability.

(2) The frequencies must be absolute and not percentage or proportional values.

(3) The null hypothesis must not permit zero entries in the expected categories.

(4) There should not be less than five observations in more than 20 per cent of the observed cells.

Lastly it should be noted that the question of 'tailedness' of the alternative hypothesis does not arise in the context of chi-square tests, and because of the manner of its execution the direction of departure is immaterial.

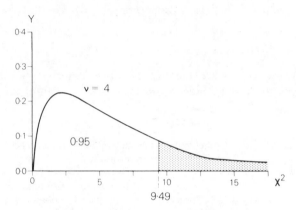

Figure 9.4 Rejection region on the chi-square
distribution for $v = 4$, $\alpha = 0.05$

9.7 THE KOLMOGOROV–SMIRNOV TEST : A ONE-SAMPLE TEST FOR ORDINAL DATA

This is similar to the χ^2 one-sample test in that it compares sample values with a specified theoretical set of numbers. The Kolmogorov–Smirnov test differs in using the cumulative frequencies of the observed and hypothesized observations. Hence ordinal data that can be placed in some order are required for the construction of sensible cumulative frequency curves though, once again, interval/ratio data can easily be reduced to ordinal scale if necessary. The test produces a test statistic denoted by D, given by the largest absolute difference between corresponding points on the two cumulative frequency curves (the observed and hypothesized). The null hypothesis of no difference is rejected when D exceeds the critical value. The distribution of D when subject only to random variations is known and the probability of any D statistic can be assessed from it. From Appendix IV we see that this distribution, as with the t- and chi-square, is dependent on the sample size. If, at the selected significance level, the test statistic exceeds the critical value, then that degree of difference between the two cumulative frequencies would be unlikely to have arisen by chance and suggests real diffeences between them.

A popular application of this technique is to the comparison of observed discrete data with theoretical distributions such as the Poisson. Reverting to the example of the spatial scatter of grocers' shops (see Section 7.11), we can compare the observed pattern of shops with the numbers, per grid square of the town, had they been following a Poisson distribution and, hence, located at random. The observed frequencies consist of the numbers of grid squares containing 0, 1, 2, 3 etc. shops. For example, there are 27 squares with only one shop (Table 9.4). But by the method of expected values (see Section 7.13) the corresponding number for a perfect Poisson distribution can be calculated; these too are given in Table 9.4. The two cumulative frequencies need now to be expressed in relative terms (as a proportion of 1.0). The test statistic is found from:

$$D = \text{maximum} \mid F_0(X) - S_n(X) \mid = 0.041 \qquad (9.4)$$

where $F_0(X)$ and $S_n(X)$ are the expected and observed relative cumulative frequencies and the vertical brackets indicate that the sign of the subtraction is ignored.

Table 9.4 Observed and hypothesized frequencies of Sunderland grocer's shops for the Kolmogorov–Smirnov test

Shops per square	Expected frequency	Observed frequency	Cumulative expected frequency $(F_0(X))$	Cumulative observed frequency $(S_n(X))$	Difference
0	90.5	96	0.670	0.711	0.041
1	36.2	27	0.938	0.911	0.027
2	7.2	9	0.992	0.978	0.012
3	0.97	2	0.999	0.993	0.006
4	0.10	1	1.000	1.000	0.000

In this example the 0.05 significance level is used to test H_0 of no difference between the observed and expected frequencies. The critical value of D is easily determined, as the table in Appendix IV is arranged in columns according to α and rows arranged by sample size n, which is the number of observations used to construct the cumulative frequency curve. When n exceeds 35, however, the table can be replaced by an equation appropriate to that significance level. The sample size in this example is 135 as each grid square used counts as one observation, and the critical D (for $\alpha = 0.05$) is:

$$D = 1.36/\sqrt{n} \qquad (9.5)$$

$$= 1.36/\sqrt{135} = 0.117$$

Test statistics larger than 0.117 lie within the distributions rejection region and have a random probability of less than 0.05. Only if, as in this case, the test statistic is smaller than the critical value is H_0 accepted and the contrasts attributed to random sampling variations. As a consequence the spatial scatter of grocers' shops is assumed to approximate to the Poisson distribution.

In this form the Kolmogorov–Smirnov test is two-tailed and no prediction is made in H_1 concerning the direction of departure of observed from hypothesized cumulative curves. At present there is insufficient theoretical basis for a one-tailed test, and for this reason it remains always two-tailed with the sign of D being ignored.

A final note of warning concerning the Kolmogorov–Smirnov test concerns the manner in which the hypothesized cumulative values are derived. Readers will have noted that the latter are, in the case of tests involving the Poisson distribution, derived from λ (the observed density of points). But the tables of critical values assume that no such connection exists and that the hypothesized distribution is evaluated beforehand. The dependence of the Poisson distribution on λ ensures that this cannot be done. The errors which result from such departures from the correct procedure appear, fortunately, to be small and where the test statistic D indicates clear acceptance or rejection of the null hypothesis the problem can be overlooked. Only in marginal cases might the difficulty arise, and in the absence of any form of correction the best precaution is to pay paticular attention to the choice of significance level. The consequences of commiting either the type I or type II error (see Section 9.4) will probably determine in which direction the rejection region should be moved in order clearly to position the test statistic. This is one of those instances where the strict observance of the conventional 0.05 and 0.01 levels should be dispensed with to provide more latitude for sensible choice.

9.8 THE χ^2 TEST : A TWO-SAMPLE TEST FOR NOMINAL DATA

The basis of this test has been discussed in Section 9.6, but it is equally applicable to the two-sample case where two samples of observations are compared with each other rather than with an hypothesized population. The H_0 remains one of no difference, but now between the two samples with the underlying inference that they were drawn from the same population.

The method of estimating the test statistic is comparable with that for the one-sample case and the same tables of critical values are used. The most apparent difference lies in the method by which the expected frequencies and degrees of freedom are obtained. An example, taking the employment structure for men and women in Batley, England, highlights these differences. Batley has an unusually high proportion of working women (in 1974 they held 46 per cent of all jobs in the town). With this distinctive character planners might well be interested to know the degree to which male and female employment patterns differ.

The two samples are, then, male and female employments categorized under the headings in Table 9.5, and H_0 states that there is no difference between these two samples other than that due to random sampling variations from a common statistical population. H_1 states that the two samples are different and drawn from different populations. The expected values under H_0 are provided by the marginal totals of the rows and columns, so that:

$$\text{Expected cell frequency } E_{ij} = \frac{\text{row total} \times \text{column total}}{\text{grand total}} \quad (9.6)$$

For example, and by reference to Table 9.5, the expected frequency of male employment in the chemicals industry is:

$$\frac{281 \times 6324}{14362} = 123.7$$

Table 9.5 Observed and expected frequencies for the χ^2 test of male and female employment categories in Batley. Yorkshire. (Source: Batley Community Development Report 1974)

Industrial class	Males	Females	Total
Chemicals	173 (123.7)	108 (157.3)	281
Mining, quarrying and agriculture	24 (21.1)	24 (26.9)	48
Food, drink and tobacco	414 (621.3)	997 (789.7)	1411
Textiles	2432 (2763.0)	3844 (3512.0)	6278
Engineering	1357 (893.9)	673 (1136.0)	2030
Service and distribution	1339 (1620.0)	2340 (2059.0)	3679
Construction	405 (194.2)	36 (246.2)	441
Transport and communications	180 (86.3)	16 (109.7)	196
	6276	8038	14362

All such expected frequencies are given in parenthesis in Table 9.5. The equation for the test statistic is similar to the one-sample expression but includes a double summation as the additions take place over rows and columns:

$$\chi^2 = \sum_{i=1}^{r} \sum_{j=1}^{k} \left[\frac{(O_{ij} - E_{ij})^2}{E_{ij}} \right] \quad (9.7)$$

Table 9.6 shows the individual squared cell differences after division by E_{ij}. The two columns each of eight rows are added to give the final test statistic of 1215.86. The degrees of freedom are determined by the product of the number of rows r and columns k, each less one, so that:

$$v = (r - 1) \times (k - 1) \quad (9.8)$$

$$v = (8 - 1) \times (2 - 1) = 7$$

Using then the 0.01 significance level and $v = 7$, the critical value is determined as 18.48. As the test statistic far exceeds this it cannot be explained by random variations between the samples, and H_0 is rejected. Therefore H_1 is accepted and the samples are deemed to indicate different employment structures.

The most useful aspect of this test is that it is directly applicable, with no further changes, to the k-sample case for three or more groups. The expected frequencies remain based on the marginal totals and the greater number of samples requires only that the number of columns k be increased accordingly to match them.

Table 9.6 Derivation of the χ^2 statistic using the data from Table 9.5

Industrial class	$(O-E)^2/E$ males	$(O-E)^2/E$ females
Chemnicals	19.61	15.45
Mining, quarrying and agriculture	0.40	0.31
Food, drink and tobacco	69.17	54.40
Textiles	39.76	31.29
Engineering	239.95	188.70
Service and distribution	48.71	38.34
Construction	228.82	180.05
Transport and communication	101.73	80.03

Total $\chi^2 = 1336.89$

9.9 THE χ^2 TEST USING 2 × 2 CONTINGENCY TABLES

The most frequent use of the two-sample χ^2 test is when each sample is measured on a dichotomous variable giving a two-by-two 'contigency tabie' of frequencies. Where data are in this form equation 9.7 is replaced by 9.9. This has the additional advantage of providing a more accurate result if interval/ratio data are reduced to two classes or ranges of values, as it contains a continuity correction. In all 2 × 2 cases there is one degree of freedom.

An example can be developed from the studies of social attitudes undertaken in rural India by Vlassoff and Vlassoff (1980). A sample of adult males were allocated to one of two groups according to age (25 – 49 and 50 – 59) and questioned concerning their views of family support for the aged. When asked if they considered the problem the younger group gave a greater number of negative responses, but there was a more equable response from the older group (Table 9.7). The differences, however, are not great and we might be justifiably cautious in pressing any arguments based only on visual inspection of these results. The χ^2 2 × 2 test gives firmer ground for argument.

Table 9.7 Data for the χ^2 2 × 2 test (the symbolic notation for use in equation 9.9 is included in parentheses)

Response	Age category 25–49	50–59	Totals
Yes	86 (A)	38 (B)	124
No	119 (C)	39 (D)	158
	205	77	$n = 282$

The test statistic is given by:

$$\chi^2 = \frac{n\left(\,|AD - BC|-\dfrac{n}{2}\,\right)^2}{(A + B)(C + D)(A + C)(B + D)} \tag{9.9}$$

in which the A, B, C and D cell frequencies are as identified in Table 9.7. By substitution:

$$\chi^2 = \frac{282(3354 - 4522 - 141)^2}{124 \times 158 \times 205 \times 77} = 0.96$$

The critical value, using $\alpha = 0.05$, is 3.84 (see Appendix III) and the test statistic falls well short of the rejection region. Consequently H_0 of no difference between the groups (as measured by this response) is accepted. The small differences that do appear are explained, therefore, by random variations and neither group is more conscious of the need for support of the aged.

9.10 THE MANN–WHITNEY U TEST : A TWO-SAMPLE TEST FOR ORDINAL DATA

Where data for two samples can be ranked the Mann-Whitney test is used to examine their differences. The greatest advantage of this test is its ability to use very small samples, and groups of as few as two observations can be employed. Nevertheless, the test remains equally applicable to large data sets.

Our first example takes a small sample application of the test and examines regional agricultural activity in England. There is a long-held view that because of climate and terrain the northern and western regions of England have concentrated on pastural farming whereas the south and east have favoured arable land use. Using data for the proportion of land under pasture the eight regions of England can be studied and ranked from the most to the least pastural (see Table 9.8). Each region may then be allocated to either the generally north and western or southern and eastern halves of the country. The test enables us now to determine if the former group are significantly more pastural than the latter. Thus the Yorkshire, North, West Midlands, North-West and South-West regions form the north/west group (NW), and East Anglia, East Midlands and South-East regions form the south/east group (SE).

Table 9.8 Data for the Mann–Whitney test of degrees of pastural agriculture in North-West and South-East England (Source: *Regional Trends*, 1981, HMSO)

Region	South West	North-West	West Midlands	North	South-East	Yorkshire	East Midlands	East Anglia
Proportion of pasturage	0.662	0.654	0.579	0.507	0.368	0.366	0.338	0.132
Rank	1	2	3	4	5	6	7	8
Group	NW	NW	NW	NW	SE	NW	SE	SE

$n_1 = 3; n_2 = 5; U = 1.$

The H_0 is of no difference in the overall ranks of the two groups, while H_1 states that the differences are too great to be attributed to random sampling variations and the 'object' group has higher ranks because of its genuinely pastural character. Clearly if the latter is correct then the sequenced observations will show some segregation, with members of the object group concentrating at the upper end of the rankings. The test statistic (U) measures the segregation, but as its probability distribution is known it is possible to decide at which point the degree of segregation is sufficient to reject H_0.

For such a small sample the U test statistic is obtained by a simple counting method. The group supposed, under H_1, to possess the higher overall rankings becomes the 'object' group, each member of which is taken in turn and the number of members of the other group which preceed them are counted. The test statistic is the sum of these counts. In this example the north/west areas form the object group, the first four members of which are not preceeded by any representative of the south/east group. Only Yorkshire of the object group is preceeded by any member of the south/east group. This result gives a test statistic of:

$$U = 0 + 0 + 0 + 0 + 1 = 1$$

Clearly the greater the degree of segregation the lower will be U, and this test is unusual in requiring low test statistics in order for H_0 to be rejected. Appendix Va gives, for small samples, the probabilities of specific U values. The distribution of U is discrete because only integer quantities are possible. If the tabled probability is less than the specified significance level, then H_0 is rejected as it falls within the rejection region and is deemed unlikely to have arisen by chance. In order to determine the probability of U attention is directed to the sub-table for n_2 (the larger of the two groups and in this case 5). The required values form the body of the table and are located by reference to n_1 (the size of the smaller group) and the U statistic. For $n_1 = 3$ and $U = 1$ the probability is 0.036. For the 0.05 significance level U falls within the rejection region, its random probability being less than 0.05, and H_0 is rejected. Consequently it is concluded that the northern and western regions are indeed more pastural in their agriculture than the south and east regions.

When used in the manner outlined above the Mann–Whitney test is one-tailed in that H_1 stipulates one group to be ranked higher than the other. The tables are set out in accordance with this. But the test can be made two-tailed if neither group is presupposed to have a higher overall position but merely different rankings. Hence either group may form the object and two U statistics are possible depending on that choice. In the farming example, had the south/east (SE) group been selected the resulting U statistic would have been

$$U = 4 + 5 + 5 = 14$$

as this object group falls so far down the rankings. However, in both one- and two-tailed tests only the smaller of the two possible U statistics can be used, but for two-tailed tests the tabled probabilities must be doubled. If H_1 had been non-directional the probability of $U = 1$ would have been 0.072 and not 0.036, with the result that H_0 would have been accepted; another example of the care needed in phrasing and justifying H_1.

For sample sizes greater than 8 it is often easier to use equation 9.10:

$$U = n_1 n_2 + \frac{n_1(n_1 + 1)}{2} - R_1 \tag{9.10}$$

Equation 9.10
n_1 = size of small group n_2 = size of larger group R_1 = sum of ranks in smaller group U = Mann–Whitney test statistic

When this test is used for groups of equal size two U statistics are possible, and again only the smaller should be taken for the test. There is a simple check to ensure that the smaller value has been obtained. If the two possible U's are denoted by U and U', then:

$$U = n_1 n_2 - U' \tag{9.11}$$

and direct substitution allows the complementary value to be estimated without using the longer equation (9.10).

More comprehensive tables are need for group sizes larger than 8. These are given in Appendix Vd in which the sub-tables are headed now by the significance levels, with the critical U statistics forming the body of the table and being interpreted by the n_1 and n_2 values. But remember that the test statistic U must be *less* than critical U for H_0 to be rejected.

Another variation of the test applies when the size of n_2 exceeds 20, as the distribution of U then approximates to the normal with a mean of μ_u and standard deviation σ_u, so that:

$$\mu_u = \frac{n_1 n_2}{2} \tag{9.12}$$

$$\sigma_u = \sqrt{\left[\frac{n_1 n_2 (n_1 + n_2 + 1)}{12}\right]} \tag{9.13}$$

The test statistic is estimated from equation (9.10) but is then converted to an equivalent z score as follows. In general:

$$z = \frac{\text{difference from mean}}{\text{standard deviation}}$$

hence

$$z = \frac{U - \mu_u}{\sigma_u} \tag{9.14}$$

and from equations 9.12 and 9.13:

$$z = \frac{U - \dfrac{n_1 n_2}{2}}{\sqrt{\left[\dfrac{n_1 n_2 (n_1 + n_2 + 1)}{12} \right]}} \tag{9.15}$$

The critical z values are then established in the usual manner and by reference to the significance levels and tailedness of the test. The probability of z then determines whether H_0 is rejected or accepted.

Data requirements for the tests are few, the most important being that measurement is at least on the ordinal scale. Because of problems of small samples or non-normality of the data, this test often uses interval/ratio data that has been reduced to ranks. The mathematical operation of the test only becomes demanding when large samples are used, and the only further point concerns ties on the ranking scale. These are dealt with by reassigning the tied observations over the average of the ranks involved had a tie not occurred. For example, if two observations tie for rank 3, then they are each awarded 3.5 being the mean of the two involved ranks 3 and 4. The next rank is 5 in the sequence, which now would read ...2, 3.5, 3.5, 5 ....

9.11 THE t-TEST FOR DIFFERENCE BETWEEN MEANS : A TWO-SAMPLE TEST FOR INTERVAL/RATIO DATA

This is another test to determine if two samples could have been drawn from the same population, but using parametric data. Unlike the non-parametric test there are some important prerequisites that the data must fulfil. A general necessity for all parametric tests is that of data normality, though most are tolerant of surprising degrees of data skewness. More specifically, this test exists in two forms; the more accurate requires also that the variances of the two samples are not significantly different. If they are dissimilar, however, a variant form of the t-test can be used.

The two forms of test are in many respects identical and derive their test t statistic from the difference between the sample means divided by the standard error of that difference:

$$t = \frac{\text{difference between means}}{\text{standard error of difference}} = \frac{\bar{X} - \bar{Y}}{\sigma_{\bar{X} - \bar{Y}}} \tag{9.16}$$

where $\bar{X}$ and $\bar{Y}$ are the sample means and $\sigma_{\bar{X} - \bar{Y}}$ the standard error of $\bar{X} - \bar{Y}$. The reader should be familiar with the idea of standard errors for sampling distributions of means, standard deviations and proportions all of which were shown in Chapter 8 to follow the t-distribution. In the same manner there is also a sampling distribution of the difference between the means of pairs of samples from a common population. This, too, is distributed as t.

138

Large differences between the means combined with low variances will yield a high t value, while small differences with a greater spread of observations reduces t. This contrast is displayed visually in Figure 9.5.

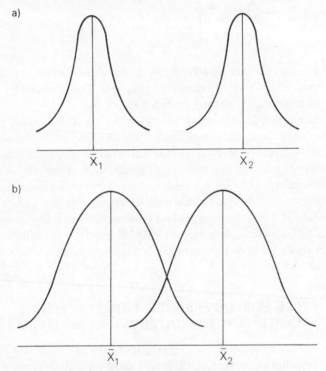

Figure 9.5 Distinct (a), and overlapping (b) groups of data in the t- test of difference between means

As an example we will take annual run-off from two geologically and meteorologically contrasted areas; the Southern Uplands of Scotland and South-East England (see Table 9.9). The former is an area of impermeable Palaeozoic rocks and high rainfall, the latter of permeable Cainozoic strata and lower rainfall. The null hypothesis is of no difference between the two sample means which are, by inference, drawn from a common run-off population; hence $H_0 : \bar{X} = \bar{Y}$. The alternative hypothesis is that the two means differ by a degree too great to be attributed to random sampling variations from a common run-off population and that the mean for the Scottish data ($\bar{X}$ is this case) should be higher than for the English. Thus, $H_1 : \bar{X} > \bar{Y}$ and the test is one-tailed.

The next task is to determine the equality of the sample variances. This is done by estimating the ratio of the greater to the lesser variance to give a variance ratio, known also as an F statistic. The distribution of F is another of those that are well known to statisticians, and the critical values on the F distribution are given in Appendix VI of this volume. These tables are set out with a separate list of critical values for each significance level (only the 0.01 and 0.05 values are used here).

Table 9.9 Run-off data for Southern Scotland and South-East England in mm per year
(Source: Surface Water, United Kingdom, 1971—73, HMSO)

Sample X (S Scotland)	Sample Y (SE England)
1324	401
879	292
583	358
711	91
736	166
1010	222
1089	205
819	237
692	106
821	163
698	184
429	116
627	310
357	270
771	212
430	123
807	190
1323	270
297	460
1597	236
871	421

$\bar{X}$ 803.33 $\bar{Y}$ 239.67

n_x 21 n_Y 21

s_x^2 107950.63 s_y^2 10951.43

The variance estimates are based on n — 1 degrees of freedom

The significance level for the test of variance equality having been selected, the appropriate critical value is found from the degrees of freedom associated with the greater and lesser variances (v_1 and v_2 respectively, where $v = n - 1$). If the observed F ratio exceeds the critical value then the null hypothesis of no difference between the variances is rejected.

In this example the two variances derived from Table 9.9 are 107950.63 for the Scottish sample and 10951.43 for the English sample, giving:

$$F = \frac{107950.63}{10951.43} = 9.86 \text{ with 20 and 20 degrees of freedom}$$

From Appendix VI we find that the critical value at the 0.05 level is 2.05. As the observed F is far greater, H_0 is rejected and variance equality cannot be presumed — the second of the two forms of test must be adopted.

The rejection region for the test can now be defined. Using the 0.05 significance level and remembering that the test is one-tailed, we can consult the table of t values in Appendix II. Each of the samples loses one degree of freedom, but as they are now considered jointly for this test:

$$v = (n_X - 1) + (n_Y - 1) \tag{9.17}$$

Hence $v = 40$ and the critical t value is $+1.68$ as the result must show $\bar{X}$ to be greater than as well as different from $\bar{Y}$.

The formula for the test statistic was loosely defined in equation 9.16, which can now be elaborated to give:

$$t = \frac{\bar{X} - \bar{Y}}{\sqrt{\left(\dfrac{s_X^2}{n_{X-1}} + \dfrac{s_Y^2}{n_{Y-1}} \right)}} \tag{9.18}$$

Equation 9.18

$\bar{X}$ = mean of variable X
$\bar{Y}$ = mean of variable Y
s_X^2 = variance of X
s_Y^2 = variance of Y
n_X = sample size of variable X
n_Y = sample size of variable Y
t = test statistic

One of the advantages of this test is that the quantities needed for its execution have already been provided by the preliminary F test and can be substituted directly; thus:

$$t = \frac{803.33 - 239.67}{\sqrt{\left[\left(\dfrac{107950.63}{21 - 1} \right) + \left(\dfrac{10951.43}{21 - 1} \right) \right]}} = +7.31$$

and, as this value far exceeds the critical t, H_0 is rejected and the samples are concluded to have been drawn from different populations.

Had the requirement of variance equality been met then a different expression would have been used to determine t, in which the standard error of the difference between the means is:

$$\sigma_{\bar{X} - \bar{Y}} = \sqrt{\left(\frac{n_X s_X^2 + n_Y s_Y^2}{n_X + n_Y - 2} \right)} \sqrt{\left(\frac{n_X + n_Y}{n_X n_Y} \right)} \tag{9.19}$$

The first part of equation 9.19 is known as the 'pooled' or combined variance estimate, but it can only be used properly for samples of similar variances. The two methods will give similar results only when the samples are large and the variances similar; hence it is important that the more appropriate of the two be correctly selected at the outset.

9.12 THE KRUSKAL–WALLIS TEST : A k-SAMPLE TEST FOR ORDINAL DATA

Where individuals can be ranked as well as classified into groups the significance of the inter-group contrasts can be examined by the Kruskall–Wallis test. The number of groups may be three or more and the sample sizes vary from three upwards. The test assists in deciding if the groups could have been drawn from a common population. As with its two-sample counterpart, the Mann–Whitney test, this method is useful not only for intrinsically ordinal data, but also for interval and ratio scale data that have been reduced to ranks to avoid problems of normality or small sample size.

The data are first ranked over all observations irrespective of their groups, with the smallest observation as 1 and so on. Ties are dealt with as outlined in Section 9.10. The test proceeds by summing the ranks within each of the k groups to give a set of rank sums (R_i), one for each group. These R values are then included within the equation for the test statistic (H) given by:

$$H = \frac{12}{n_s(n_s + 1)} \sum_{i=1}^{k} \frac{R_i^2}{n_i} - 3(n_s + 1) \tag{9.20}$$

in which n_s is the total number of observations and n_i the number within each group. Where there are more than five observations in each group the test statistic is distributed as chi-square with $k - 1$ degrees of freedom, where k is the number of groups. For samples smaller than five the tables of H statistics in Appendix VII must be used in which the critical values for different group numbers at various significance levels are listed.

In principle, greater inter-group differences provide larger H statistics which, if greater than the critical value, causes H_0 to be rejected. The test is most often used for small groups; but it is equally applicable to larger samples if its distribution-free qualities are needed, as data normality is not a prerequisite. The following example uses data for deserted village incidence in three regions of England, the Sout-East, Central Midlands and North, for which only fragmentary evidence based on county surveys is available. The small group sizes make the Kruskal–Wallis test the only suitable means of deciding whether the occurences have any regional variations. The data for the problem are given in Table 9.10.

The H_0 under test is of no difference between the three samples other than that expected by random sampling variations. H_1 states that there are differences beyond those explicable by random variation. It can be noted at this stage that k-sample tests, by their nature, are 'non-tailed'. The 0.05 significance level will be used. Substitution from Table 9.10 into the equation for H gives:

$$H = \frac{12}{12(12 + 1)} \left(\frac{10^2}{3} + \frac{38^2}{5} + \frac{30^2}{4} \right) - 3(12 + 1) = 3.075$$

Appendix VII shows that the critical H value for $\alpha = 0.05$ and group sizes of 3, 4 and 5 is 5.631. Thus the test statistic lies beyond the rejection region and H_0 is accepted. Deserted village densities do not differ between these three regions.

The question of tied observations has already been mentioned. It may be added that where large numbers of ties are encountered a correction must be imposed on H. This correction (C) is applied by dividing H by C, where:

$$C = 1 - \frac{(T^3 - T)}{n_s^3 - n_s}$$

(9.21)

Equation 9.21

$T =$ number of ties

$n_s =$ number of observations

In most cases C is close to 1.0 but becomes smaller as the proportion of ties increase. The correction should be applied if ties consume 25 per cent of the data.

Finally, it must be remembered that if all groups are greater than 5 then H can be treated as a χ^2 statistic with $k - 1$ degrees of freedom.

Table 9.10 Density of deserted villages in three English regions — the data are prepared for Kruskal–Wallis test (Source Beresford and Hurst, 1971)

South-East			East Midlands			North		
County	Density[1]	Rank	County	Density	Rank	County	Density	Rar
Kent	1.85	3	Northampton	1.46	2	Northumberland	3.16	9
Surrey	1.07	1	Nottingham	3.03	7	W Yorkshire	2.08	5
Hampshire	3.00	6	Oxfordshire	3.94	10	E Yorkshire	6.20	12
			Leicestershire	3.10	8	N Yorkshire	2.06	4
			Warwickshire	5.08	11			
		10			38			30

1. In deserted villages per 100 km^2.

9.13 ONE–WAY ANALYSIS OF VARIANCE : A k-SAMPLE TEST FOR INTERVAL/RATIO DATA

The final test in this chapter is one used to discriminate between k samples measured on the interval/ratio scales. Being a parametric test, the one-way analysis of variance (so called because discrimination is based on one variable only) makes many more demands on both data and researcher than does its non-parametric counterparts. This test is, however, one of the most powerful available.

The raw data needs to be normally distributed and the sample variances should not be grossly dissimilar; the test, nevertheless, is tolerant on both counts. The test assesses the likelihood of the k samples having been drawn from the same population by decomposing the total data variance into within- and between-groups components, i.e. the variance within each group and the variance between the groups about the grand mean. The ratio of these two variances gives an F ratio, the probability of which can be assessed from the tables in Appendix VI in the manner already outlined in Section 9.11. Now, however, the F ratio is obtained by dividing the between- by the within-groups variances. The degrees of freedom remain v_1 for the greater and v_2 for the lesser variance though, as will be shown, the manner of their estimation is different. If the test value of F exceeds the critical value, then H_0 of no difference between the means of the groups is rejected. Naturally for any k samples drawn from a common population some random sampling variation will exist. This F test helps to decide whether the observed differences could have arisen by chance or because the samples come from different populations. But whatever the conclusion it applies equally to all the groups and, by this method, there is no question of concluding that one group is 'more different' than the others.

Interest focuses on the three quantities of total, within-groups and between-groups variances. If the between-groups component far exceeds the within-groups, then high F ratios will result and suggest that most of the variance is accounted for by inter-group variation (Figure 9.6(a)). Less strong contrasts produce lower F ratios with the result that the groups' distributions show far more overlap and similarity (Figure 9.6(b)).

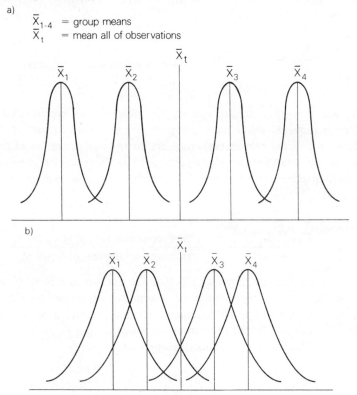

Figure 9.6 Distinct (a), and overlapping (b) groups of data in the F-test analysis of variance

In the following example potential evapotranspiration data have been drawn from four regions of Great Britain; North-West Scotland and North-East, South-East and South-West England. Each sample consists of twenty randomly selected monthly potential evapotranspiration figures expressed in centimetres per month. The H_0 states that there is no difference between the groups; or $H_0 : \bar{X}_1 = \bar{X}_2 = \bar{X}_3 = \bar{X}_4$. H_1 stipulates that there are differences beyond those expected by random sampling from a common population; or $H_1 : \bar{X}_1 \neq \bar{X}_2 \neq \bar{X}_3 \neq \bar{X}_4$. The 0.01 significance level will be used.

Having defined the null and alternative hypotheses and selected a significance level, we may analyse the data to yield the required variances. The general expression for variance is given by:

$$\text{Variance} = \frac{\text{sum of squares}}{\text{degrees of freedom}}$$

This qualitative expression can be adapted for the specific needs of analysis of variance in the following way. The within-groups variance (s_w^2) is found by dividing the sum of squared deviations of all observations about their respective group means by the appropriate degrees of freedom. The latter in this case is $n - k$, where n is the total number of observations and k the number of groups; with the result that:

$$s_w^2 = \frac{\sum\limits_{j=1}^{k} \sum\limits_{i=1}^{n_j} (X_{ij} - \bar{X}_j)^2}{n - k} = \frac{\text{within-group sum of squares}}{\text{within-group degrees of freedom}} \tag{9.22}$$

where $\bar{X}_j$ are the group means and X_{ij} the observations within each group. Summation is repeated within each group to give the required sum of squares.

The between-groups variance (s_b^2) considers the squared deviations of the group means about the mean for the whole data set $(\bar{X}_t)$. These squares, however, are also weighted in terms of the number of observation (n_j) wihin each group, and the equation reads:

$$s_b^2 = \frac{\sum\limits_{j=1}^{k} n_j(\bar{X}_j - \bar{X}_t)^2}{k - 1} = \frac{\text{between-group sum of squares}}{\text{between-group degrees of freedom}} \tag{9.23}$$

The degrees of freedom are now one less than the number of groups, $(k - 1)$. The total sum of squares is, conveniently, the sum of the within- and between-groups squares, and hence requires no separate calculation. Note, however, that this comparability does not extend the variances, which *cannot* be added to give the total variance. The required F ratio is always found from:

$$F = \frac{s_b^2}{s_w^2} \tag{9.24}$$

The first step in the analysis must be to estimate the group and total means from which the various deviations can be calculated. Tables 9.11 and 9.12 demonstrate how these figures are treated to give the necessary variances. Because of the size of the samples not all individuals are shown, but the principles should be clear.

Table 9.11 Abbreviated data sheet of monthly potential evapotranspiration (in cm) for four British regions (Source: *British Rainfall* 1961–68. (HMSO))

	North-West Scotland		North-East England		South-East England		South-West England	
i	a	b	a	b	a	b	a	b
1	5.49	0.6400	4.79	0.0036	5.40	0.0064	5.49	0.0036
2	4.61	0.0064	5.41	0.3136	5.76	0.0784	5.82	0.0729
3	5.06	0.1369	5.07	0.0484	6.22	0.5476	5.77	0.0484
.	.	.	.	.	.	.	.	.
.	.	.	.	.	.	.	.	.
20	4.76	0.0049	4.41	0.1936	5.51	0.0009	5.49	0.0036
	93.81	4.523	97.03	3.204	109.54	3.068	110.97	2.366

$$\bar{X}_j = 4.69 \qquad \bar{X}_j = 4.85 \qquad \bar{X}_j = 5.48 \qquad \bar{X}_j = 5.55$$

$$\sum_{j=1}^{r} \sum_{i=1}^{k} (X_{ij} - X_j)^2 = 4.523 + 3.204 + 3.068 + 2.366 = 13.16$$

$a = X_{ij}$; $b = (X_{ij} - \bar{X}_j)^2$; $\bar{X}_j$ = group means; X_{ij} = individual observations.

Table 9.12 Inter-group variations used in the analysis of variance test of regional evaporation transporation rates

	North–West Scotland	North–East England	South–East England	South–West England
$\bar{X}_j$	4.69	4.85	5.48	5.55
n_j	20	20	20	20
$n_j(\bar{X}_j - \bar{X}_t)^2$	4.05	1.682	2.312	3.362
	$\bar{X}_t = 5.14$			

$$\sum_{j=1}^{k} (\bar{X}_j - \bar{X}_t)^2 = 4.05 + 1.682 + 2.312 + 3.362 = 11.406$$

$\bar{X}_t$ = mean of all observations; $\bar{X}_j$ = group means.

For convenience and by convention the results of such work are summarized in an ANOVA (analysis of variance table), which also provides the final F ratio. Table 9.13 shows how such a table appears using the example data.

The degrees of freedom are, for the greater variance estimate $v_1 = 3$, and for the lesser $v_2 = 76$. From the tables in Appendix VI for $\alpha = 0.01$ a critical F value of 4.07 is obtained. Some interpolation is needed for the latter, but as the test statistic falls so clearly within the rejection region no great accuracy is necessary and H_0 is emphatically rejected. Accuracy is, however, necessary in attaching quantities to v_1 and v_2. If, for example, they had been mistakenly allocated in the example to give $v_1 = 76$ and $v_2 = 3$, then the critical F value would have been 26.3 and H_0 consequently accepted.

Table 9.13 An ANOVA table summarizing the results of the one-way analysis of potential evapotranspiration

Source of variation	Sum of squares	Degrees of freedom	Variance	F ratio
Between-groups	11.406	$4-1 = 3$	3.802	
Within-groups	13.160	$80-4 = 76$	0.173	
	24.566	79		21.98

The decision to reject H_0, however, implies that the four samples have been drawn from different hydro-meteorological populations. Furthermore the group means show a strong north to south trend and conform closely to the expected pattern of evapotranspiration. Thus the partitioning of the complete data set becomes an explanatory exercise, and the between-groups variance becomes the 'explained' component in that regional climatic variation is supposed to bring about these group contrasts. The within-groups variance, by contrast, remains the 'unexplained' and random variation about each group mean. Hence, using qualitative expressions, we may write:

$$F = \frac{\text{between-groups variance}}{\text{within-groups variance}} = \frac{\text{explained variance}}{\text{unexplained variance}}$$

This treatment of data variation is a fundamental one and will reappear under a similar guise in many of the following chapters.

9.14 CONCLUSIONS

Earlier chapters have discussed the preliminary treatment of statistical data and, useful though such procedures are, it is largely by hypothesis testing that geographical problems can be examined. It must be stressed, however, that is is the task of the geographer and not the test to make a sensible interpretation of the results.

Of the tests available the non-parametric forms are particularly useful in being distribution-free and relatively simple in execution. The more efficient use of data implicit in the parametric tests is a quality brought at the expense of stringent data requirements and lengthier arithmetic procedures. Indeed, the use of computers becomes indispensible when even modest data sets are employed. But the specific requirements of each test should not be overstressed at the expense of the universal specification of random and independent observations.

As far as individual test are concerned this chapter has attempted to suggest the range and variety, from applications of the familiar normal distribution to the less familiar statistics of, for example, the Kruskal–Wallis test. Many other tests exist, each with their own needs and methods; Siegel (1956) discusses a large number of them. The difficulty of having so many tests is the problem of selecting the most appropriate one for any problem. The layout of this chapter following the scheme in Table 9.2 may help to reduce the risk of misapplication. On the other hand it is easy to move from one test to another as they all have the same strategy and sequence of execution:

(1) formulation of the geographical/research hypothesis;
(2) choice of statistical test;
(3) formulation of null and alternative hypotheses;
(4) selection of significance level and definition of rejection region;
(5) calculation of test statistic;
(6) acceptance or rejection of H_0.

REFERENCES

Batley Community Development Project (1974). *Batley at Work: the Rise and Fall of a Textile Town*, Discussion Paper 1.

Beresford, M., and Hurst, J. G. (1971). *Deserted Mediaeval Villages*, Lutterworth, London.

Central Statistical Office (1981). *Regional Trends*, HMSO, London.

Meteorological Office (1961–1968). *British Rainfall*, HMSO, London.

Pitty, A. F. (1969) 'A scheme for hillslope analysis: initial considerations and calculations'. Uni. of Hull Occasional Papers 9, Hull.

Siegel, S. (1956). *Nonparametric Statistics for the Behavioural Sciences*, McGraw-Hill, New York.

Vlassoff, M., and Vlassoff, C. (1980). 'Old age security and the utility of children in rural India', *Pop. Studies* **34**, 487–500.

RECOMMENDED READING

Armstrong, R. W. (1976). 'The geography of specific environments of patients and non-patients in cancer studies', *Econ. Geog.* **52**, 161–170. The χ^2 test has been used widely by geographers in recent years. This is but one of a very large number of good applications.

Baker, A. R. H. (1973). 'Adjustments to distance between farmstead and fields: some findings from the southwestern Paris Basin in the nineteenth century', *Canadian Geogr* **17**, 259–275. Baker uses both the χ^2 and the Kolmogorov–Smirnov tests in this historical/geographical study.

Baker, E. J., and Yuk, L. (1975). 'Alternative analyses of geographical contigency tables', *Prof. Geogr* **27**, 179–188. Though this paper does not deal with specific geographical data, it is an excellent general discussion of the problem.

Blaikie, P. M. (1971). 'Spatial organisation of agriculture in some north Indian villages', *Trans. Inst. Br. Geogrs* **52**, 1–40. A good example of the use of the Kolmogorov–Smirnov test.

Blalock, H. M. (1960). *Social Statistics*, McGraw-Hill, New York. The latter half of this book covers, in detail, a variety of parametric and non-parametric tests, which though dealt with from a sociological point of view retain their importance for geographers.

Gould, P. (1971). 'Is *statistix inferens* a geographical name for a wild goose?', *Econ. Geog.* **46**, 439–448. A thought-provoking analysis of the state and philosophy of inferential statistics (of which hypothesis testing is an integral part) in geography — indispensible reading for those interested in the broader and fundamental issues of this chapter.

Hayter, R. (1979). 'Labour supply and resource-based manufacturing in isolated communities', *Geoforum* **10**, 163–177. Here the χ^2 test is used to study industrial and labour activities.

Jones, K., and Kirby, A. (1980). 'The use of chi-square maps in the analysis of census data', *Geoforum* **11**, 409–417. An interesting application of χ^2 methods, if not in the strictest traditions of hypothesis testing.

Mackay, J. R. (1958). 'Chi-square as a tool for regional studies', *Ann. Ass. Am. Geogrs* **48**, 164. This brief paper contains a salutory warning for all of us who might not pay due attention to the requirements of specific tests.

Moroney, M. J. (1951). *Facts from Figures*, Penguin, Harmondsworth. Chapter 19 takes the reader through elementary to advanced forms of analysis of variance. Well worth consulting, though it goes well beyond the one-way analysis of variance dealt with here. Neither does it assume, or require, access to an electronic computer.

Norcliffe, G. B. (1974). 'Territorial influences in urban political space: a study in Kitchener–Waterloo', *Canadian Geogr* **18**, 311–329. A thorough exposition of the manner in which the *t*–test for differences between means can be explored to maximum benefit.

Siegel, S. (1956). *Nonparametric Statistics for the Behavioural Sciences*, McGraw-Hill, New York. There are few books which deal thoroughly and exclusively with the expanding area of nonparametric statistics. This is indisputably the best, providing not only a veritable catalogue of methods (using sociological and psychological examples), but also a clear account of the principles involved in their use.

Swan, B. (1979). 'Areal variations in textures of shore sands, Sri Lanka', *J. Trop. Geog.* **49**. 72–85. A geomorphological example of the use of the *t*–test for differences between the means of two samples. Well laid out and easily understood.

EXERCISES

1. The following table lists some statistical tests and the scales of measurement and number of samples appropriate to them. Fill in the gaps to complete the matrix.

Test	Scale of measurement	Samples (k)
χ^2 two-sample	?	2
Kruskall–Wallis	ordinal	?
?	ordinal	2
t–test of means	?	2
?	interval/ratio	3 or more

2. A study of the housing character of two areas provided the following frequencies in the random sample of properties:

	Detached house	Semidetached house	Terrace house	Purpose-built flat	Converted flat
Town A	16	35	45	4	11
Town B	21	28	32	12	3

Use the χ^2 two-sample test to determine the significance of the differences between the housing in the two towns. Select your own significance level.

3. The statistical distribution of mean annual temperature at a climatological station revealed the following frequencies when categorized according to the standardized values:

z range	Frequency	z range	Frequency
> 2.0	1	0 to −0.5	14
2.0 to 1.5	1	−0.5 to −1.0	9
1.5 to 1.0	7	−1.0 to −1.5	5
1.0 to 0.5	8	−1.5 to −2.0	3
0.5 to 0.0	10	< 2.0	2

Use the χ^2 one-sample test to determine if the observed distribution is significantly non-normal. In this instance the expected frequencies will not be identical for each cell and must be found from $n \times p$, where n is the total number of observations (60) and p is the probability of an event within that range. The latter is found from the z table in the manner indicated in Chapter 7.

4. The spatial location of stream junctions in the North Yorkshire Moors, England, is summarized in the following table of frequencies:

Junctions per km square	Frequency
0	60
1	17
2	12
3	7
4	2
5	2

Use the Kolmogorov–Smirnov test to determine if the scatter of junctions is spatially random. i.e. Follows a Poisson distribution. Consult a topographic map of the North Yorkshire Moors (the Ordnance Survey 1:63360 tourist edition for the area is best) and suggest reasons for your findings.

5. Diseases are often acknowledged to present distinct spatial patterns. In this example lung cancer mortality is examined for two contrasting regions, the traditionally industrial area of North-East England and the more rural region of the County of Devon. Does the Mann–Whitney test indicate a statistically different death rate between these areas? Select your own significance level. The data are measured as average death rate per 10,000 persons due to lung cancer (1954–58). Source: G. M. Howe, (1963) *National Atlas of Disease in the U.K.*

NE region		Devon region	
Newcastle	5.16	Exeter	3.76
Tynemouth	4.13	Plymouth	3.33
Other urban areas	3.70	Other urban areas	3.97

6. Prior to 1974 western Europe could be divided into three economic spheres: the European Economic Community (EEC), the European Free Trade Area (EFTA) and non-aligned nations. Do the following real annual percentage wage increases, i.e. taking increases in cost of living into account, permit any distinction to be drawn between economic growth of the groups? The Kruskal–Wallis test is most appropriate, but as the number of observations within the groups is more than five the H statistic is distributed as χ^2 with two degrees of freedom. Source: *Atlas of Europe*, Bartholomew & Warne, London, (1974)

EFTA		EEC		Non-aligned	
United Kingdom	2.3	West Germany	6.3	Eire	4.0
Norway	3.3	Netherlands	4.9	Finland	3.7
Sweden	4.4	Belgium	5.0	Spain	7.0
Denmark	4.4	Luxembourg	4.5	Iceland	2.2
Austria	4.2	France	4.2		
Portugal	2.9	Italy	3.8		

7. The following average slope measurements (in degrees) were made on different rock types but in the same area:

Limestone	Gritstones
32.1, 29.4, 33.0	17.8, 15.8, 12.5
27.3, 19.0, 14.4	15.5, 15.1, 12.2
21.1, 25.5, 9.1	13.1, 10.6, 9.3
10.5, 10.5, 11.0	5.5
14.2	

Use the t–test for the difference between means to assess the validity of the assertion that slopes on these two rock types differ.

8. In addition to the data in question 7, further surveys were conducted on chalk slopes. Using the analysis of variance procedure, do all three groups differ significantly in the measured average slope? There are sufficiently small volumes of data to permit the Kruskall–Wallis test to also be used. The interested reader might care to compare the results for the parametric and non-parametric tests. Source: data adapted from Pitty, (1969)

Average slopes over chalk transects
6.3, 8.3, 9.4, 8.8, 4.8
3.9, 3.5, 5.5

Chapter 10

Methods of Correlation Analysis

10.1 CORRELATION ANALYSIS

As we would expect a great deal of geographical analysis involves studying the relationships between two or more variables, either through time or in different places. This chapter focuses on those statistical techniques that enable us to measure and determine the strength of a relationship between two variables. To measure this we can use the concept of the coefficient of correlation, which as Table 10.1 shows can take a number of different forms. Once again the decision as to which one to employ will depend on the type of data we are dealing with, and in particular the scale of measurement in which each variable is expressed. Obviously, the strength and efficiency of each of these different types of correlation analysis varies, and it is generally recognized that the product-moment or Pearson's correlation coefficient (r) is the most powerful. For example, it has been calculated that Spearman's rank correlation is only 91 per cent as efficient as Pearson's r. This means that if, in a sample of 100 cases from a bivariate normal population, the product-moment coefficient is significant, it will require a sample of 110 cases of the same data to achieve the same level of significance for the rank order coefficient. But of course each correlation coefficient has its merits, depending on the type of data you wish to examine.

Table 10.1 Types of correlation coefficients and their relation to data characteristics

Type	Measurement scale	Data characteristics
Product-moment (r)	Interval/ratio	Use with both scales
Spearmans rank (r_s)	Ordinal	Both variables must be expressed
Kendall's tau (τ)	Ordinal	as ranked data for these two tests
Biserial (r_b)	Nominal	One dichotomous variable and one variable that has more than two values
Phi coefficient (ϕ or r_ϕ)	Nominal	Both variables must be dichotomous

Regardless of which type of correlation coefficient we select there are certain general terms that can be used to describe the relationship between two variables. First, values of correlation coefficients can only vary from -1.0 to $+1.0$, with

152

both these extreme values representing a perfect relationship between the two variables, while a value of 0.0 indicates the absence of any statistical relationship. In addition, if we are measuring linear relationships (see Chapters 11 and 12 for further details), as illustrated in Figure 10.1, further descriptions can be made. Thus, Figure 10.1(a) shows a perfect positive relationship ($r = +1.0$), which means that an increase in one variable is matched by an increase of equal degree in the second variable. Figure 10.1(b) shows the same situation for a perfect negative correlation ($r = -1.0$), while figure 10.1(c) indicates the scatter of points that arise when there is no correlation.

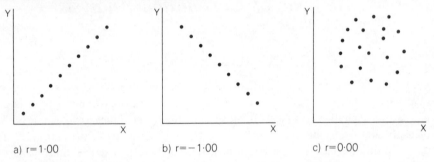

a) r=1·00   b) r=−1·00   c) r=0·00

Figure 10.1 Scattergraphs and correlation coefficients

10.2 PRODUCT-MOMENT CORRELATION

Like all statistical tests the product-moment correlation coefficient is based on certain assumptions concerning the data to which it can be applied. First, we need data measured on either the interval or ratio scale; and second, the two variables should fit a normal distribution. Being more specific, the two variables should have a bivariate normal distribution, in which both their marginal and conditional distributions are normal. This means that the individual frequency distributions or marginal distributions of the two variables, X and Y, should be normal. In turn the conditional distribution is normal when for any value of variable X, variable Y is normally distributed, and vice versa. However, a glance through the geographical literature will show that, in practice, many people relax these assumptions, especially when dealing with fairly large samples. Small samples bring their own problems and need to be treated more cautiously, as we shall see later in this chapter.

The product-moment correlation coefficient is based on the idea of covariance, a statistical term closely related to the variance we discussed in Chapter 5. As we saw earlier the variance can be used to describe the variation of one set of observations about their mean value. The covariance, however, goes a stage further and describes the correspondance or co-variation of two variables together. The covariance is calcuated by summing the product of the individual values of the two variables from their respective means:

$$\text{covariance} = \frac{\Sigma(X - \bar{X})(Y - \bar{Y})}{N} \qquad (10.1)$$

Consequently, like the variance and standard deviation, it is an absolute measure, and its value will vary according to the units X and Y are measured in. Such variations in magnitude can be avoided by making the covariance dimensionless through the use of an index or coefficient. In this instance we use the correlation coefficient, which is obtained by dividing the covariance of X and Y by the product of the standard deviations of the X and Y variables:

$$r = \frac{\Sigma xy}{\sqrt{(\Sigma x^2 \ \Sigma y^2)}} \tag{10.2}$$

where $x = X - \bar{X}$ and $y = Y - \bar{Y}$.

A practical method of calculating the correlation coefficient is to use equation 10.3:

$$r = \frac{N\Sigma XY - (\Sigma X)(\Sigma Y)}{\sqrt{[N\Sigma X^2 - (\Sigma X)^2] \times [N\Sigma Y^2 - (\Sigma Y)^2]}} \tag{10.3}$$

This enables use to divide the calculation into a number of simple steps as shown in Table 10.2. This example examines the suggestion that drainage basin area and mean river discharge are closely correlated. By taking a random sample from the publication *Surface Water : United Kingdom* (HMSO, 1978) it is possible to examine this proposition, based on a sample of 31 pairs of observations. The individual observations are not listed here, but Table 10.2 provides a section of the appropriate worksheet and the sub-totals necessary to execute equation 10.3.

Table 10.2 Product-moment correlation analysis of drainage area with river discharge

(X) Drainage area (km²)	(Y) Mean annual discharge (cumecs)	XY	X²	Y²
648.0	8.75	5670.00	419904.00	76.56
59.6	1.17	69.73	3552.16	1.37
21.4	0.53	11.54	457.96	0.28
2180.0	43.50	94830.00	4752400.00	1892.23
.	.	.	.	.
217.0	7.60	1649.20	47089.00	57.76
21257.5	244.58	341539.31	31183178.00	4590.72

When we substitute the sub-totals from Table 10.1 into equation 10.3 we have:

$$r = \frac{31 \times 341539.31 - (21257.5 \times 244.58)}{(31 \times 31183178 - 21257152^2) \times (31 \times 4590.72 - 244.58^2)}$$

$$= +0.827$$

If we recall that correlation coefficients may vary only within the range + 1.0 to −1.0, then this result of +0.827 does indeed suggest a close association between the two variables. But such a simple qualitative assessment is by no means the end of the exercise. Thus, we can calculate the square of the correlation coefficient; this is known as the coefficient of determination (r^2) and measures the proportion of variance explained. For example, in our study of drainage basins the coefficient of determinations would be 0.683, which would mean that 68.3 per cent of the total variance is explained in the correlation of the two variables. Conversely, this also implies that in our study some 31.7 per cent of the variance remains to be explained by the examination of other variables.

10.3 SIGNIFICANCE TESTING IN CORRELATION ANALYSIS

It is accepted practice to use correlation coefficients as purely descriptive devices which summarize the numerical association between pairs of variables. However, it is often important also to assess the statistical significance of the coefficient. When sampling from bivariate populations there is always the possibility that entirely spurious correlation coefficients may be derived. This is particularly the case for small samples. In this context, significance testing provides information on the probability of coefficients having arisen by chance alone. If that probability is remote then the derived coefficient may be held to reflect a genuine and non-random association. Should the probability be high, however, the correlation may be regarded as spurious.

Interest focuses on the H_0 of no difference between the observed correlation (r) and some hypothesized correlation (ρ_0). The greater the difference between r and ρ_0 (the latter is generally zero) the larger will be the value of the test statistic, though sample size is also critical and more confidence can be placed in a correlation of 0.4 derived rom a sample of 100 than in 0.8 from a sample of four.

The sampling distribution of Pearson's r is not normal under any circumstances, but may be made so by Fisher's z-transformation, given by:

$$Z_f = 1.1513 \log_{10} \frac{(1 + r)}{(1 - r)} \tag{10.4}$$

Now Z_f is normally distributed with a mean of μ_f and a standard deviation of σ_f, where:

$$\mu_f = 1.1513 \log_{10} \frac{(1 + \rho_0)}{(1 - \rho_0)} \tag{10.5}$$

$$\sigma_f = \sqrt{\left(\frac{1}{n - 3} \right)} \tag{10.6}$$

from which the standardized value (z_f) is found:

$$z_f = \frac{Z_f - \mu_f}{\sigma_f} \tag{10.7}$$

the significance of which can now be assessed in the usual fashion from z tables.

However, H_0 frequently specifies that $\rho_0 = 0.0$. In other words, and despite what preconceptions exist concerning the character of the bivariate population, it is an assumption of zero correlation that is under test. But when $\rho_0 = 0.0$, the distribution mean (μ_f) is also zero and the significance of r depends on its own magnitude and the sample size only (this should be clear from equations 10.5–10.7). And although the assumption of zero correlation is not obligotory it is so frequent that tables have been drawn up giving critical r values for specified sample sizes to avoid repetition of the calculations. One such table is given in Appendix VIII and its use requires only sample size n and significance level α. If the observed correlation r exceeds the critical value (r_{crit}), then H_0 of zero correlation can be rejected. On occasions when a non-zero correlation is specified under H_0 there is no alternative to mathematical estimation.

We may now assess the significance of the correlation derived in the earlier example. H_0 specifies no difference between the observed and a zero correlation. Additionally, we are anticipating that discharge and drainage area are positively correlated and one increases in correspondence to the other. Hence H_1 confirms that r must be greater than zero. The test is, then, one-tailed and at the 0.01 significance level the critical z value is $+2.33$ and not ± 2.58 as it would be for the two-tailed test. We may proceed thus:

$$Z_f = 1.1513 \log_{10} \frac{(1 + 0.827)}{(1 - 0.827)} = 1.1786$$

$$\mu_f = 0.0$$

$$\sigma_f = \sqrt{\frac{1}{31 - 3}} = 0.189$$

Therefore

$$z_f = \frac{1.1786 - 0.0}{0.189} = 6.24$$

As a result H_0 may be confidently rejected and the correlation confirmed to be real and not to be a chance event with no underlying reality.

We may also check the correlation's validity by consulting the published tables of critical values. This is done here simply to demonstrate their use and only one of these two significance tests is necessary. The tables in Appendix VIII are so arranged that the sample size n determines the row, and the significance level and 'tailedness' of the test direct us to the correct column. In this example attention is turned to row 31 and the column for a one-tailed test significance level of 0.01; the resulting critical value of 0.416 is easily exceeded by the observed value and, again, H_0 is rejected.

10.4 CORRELATION ANALYSIS OF ORDINAL DATA

In some circumstances we may wish to assess statistically the relationship between variables measured on the ordinal scale, and at this level of measurement two types of correlation coefficients are available, Kendall's tau and Spearman's rank (Table 10.1). Both measures are equally powerful, although Spearman's rank is more easy to compute and is therefore used most frequently. Given this slight advantage we shall therefore concentrate our attention on Spearman's r, which demands that observations be rank ordered and that at least five pairs of values be used. The test statistic requires that you obtain the differences in the ranks of the two variables and then sum the squared differences using equation 10.8:

$$r_s = 1 - \frac{6\Sigma D^2}{N(N^2 - 1)} \qquad (10.8)$$

Equation 10.8

$D =$ differences between ranks of corresponding values of X and Y

$N =$ number of pairs of X, Y values

Under certain circumstances you may encounter ranked data that contains three or more observations with the same rank value. When these conditions occur, and when such pairs of ties account for more than one-quarter of the observations, the Spearman's formula yields a spuriously high coefficient of correlation. In such cases as this, a correction factor should be applied by using equation 10.9:

$$r_s = \frac{\Sigma X^2 + \Sigma Y^2 - \Sigma D^2}{\sqrt{(\Sigma X^2 \Sigma Y^2)}} \qquad (10.9)$$

$$\Sigma X^2 = \frac{N^3 - N}{12} - \Sigma T_x$$

$$\Sigma Y^2 = \frac{N^3 - N}{12} - \Sigma T_y$$

$N =$ number of pairs of rankings

$TX, TY =$ number of tied observations at a particular rank for X and Y respectively

In geographical analysis Spearman's rank correlation is used fairly frequently and in a variety of situations, with both ordinal and interval data. For example, it may be used with interval data when the conditions of normality, required by product-moment correlation, are not met. Similarly, it is used when the detailed accuracy of interval type data is in some doubt, but the information is still suitable for providing ranked data. This may sometimes be the case when using data from suspect, untested data sources, or in certain less developed economies with poor data bases. Thus, Doornkamp and King (1970) used Spearman's rank correlation in a study of drainage basin characteristics in Uganda, where the scale and accuracy of the base maps did not allow precise measurements to be made. This test is also often applied to interval data because it is far easier to compute than Pearson's (r), especially when large numbers are involved and computers are not available.

However, it should be stressed that rank correlation becomes particularly important within the research carried out by geographers on cognitive-perception and behavioural studies in general. It is in such work that ordinal data is the common scale of measurement (Gould and Ola, 1970). This can be demonstrated by the example given in Table 10.3, which compares the environmental dislikes of a sample of people living in the London borough of Stockwell with those of people in the rest of Greater London. The various environmental dislikes are ranked by each sample, and Spearman's rank correlation can be used to measure whether there is any relationship between the perceptions of the two study areas. Applying equation 10.8 we have:

$$r_s = 1 - \frac{6 \times 88}{11(11^2 - 1)} = +0.6$$

In terms of these dislikes a value of -1.0 would have indicated that the two areas had completely different perceptions of urban environmental problems, whereas identical sets of rankings would have given a coefficient of $+1.0$.

Table 10.3 Comparison of environmental dislikes between Stockwell and Greater London (Source: Madge, C., and Willmott, P., *Inner City Poverty in Paris and London*, 1981)

Dislikes	Ranked Stockwell	Ranked Gt. London	Differences in Ranks (d)	d^2
Immigrants	1	2	−1	1
Dirt, litter	2	7	−5	25
Lack of facilities	5	4	1	1
Crime, vandalism	3	5	−2	4
Lack of open space	8	10	−2	4
Children and young people	4	6	−2	4
Council services	6	3	3	9
Traffic problems	7	1	6	36
Schools	10	8	2	4
Noise	9	9	0	0
Shopping facilities	11	11	0	0
				88

Since our example is based on a sample we can proceed to test the significance of our result — the null hypothesis H_0 being that there is no correlation between the environmental dislikes of people living in Stockwell and in the rest of London.

When we are testing the significance of Spearman's r there are two methods available, depending on the size of the sample. Thus, for smaller samples, it is best to use prepared tables (Appendix IX). For larger samples, however, r_s is converted into a t value, where:

$$t = r_s \sqrt{\left(\frac{N-2}{1 - r_s^2}\right)} \tag{10.10}$$

with the degrees of freedom equalling $(N-2)$. Our r_s value of $+0.6$ is significant at the 0.05 per cent level and we can therefore reject the null hypothesis and establish that there is a correlation between the perceptions of the two samples.

10.5 MEASURES OF CORRELATION FOR NOMINAL DATA

Very often we may want to examine the relationship between two variables measured on the nominal scale, in which case we need to consider other types of correlation statistics (Table 10.1). The two most useful measures are the 'point biserial coefficient' and the 'phi coefficient', each of which deals with slightly different data sets, as can be illustrated with some simple examples.

Suppose we want to examine the relationship between two variables, one measured on a continuous scale, while the other is a dichotomous variable which can only take the values of 1 or 0. In our example, variable Y represents population totals in a sample of 12 villages in South-West England, and variable X indicates the presence or absence of a village post office (Table 10.4). The continuous variable (Y) can therefore be divided into two sub-groups depending on the value of X, i.e. villages with a post office $(X = 1)$ and those without $(X = 0)$. The point biserial coefficient is then given by equation 10.11, in which the means of the two sub-groups need to be calculated $(Y_0$ and $Y_1)$, together with the standard devation of Y:

$$r_b = \frac{\bar{Y}_1 - \bar{Y}_0}{\hat{s}_Y} \sqrt{\left[\frac{N_1 N_0}{N(N-1)} \right]} \qquad (10.11)$$

Equation 10.11

$N_0 =$ number of observations with an X value of 0
$N_1 =$ number of observations with an X value of 1

$$\bar{Y}_0 = \frac{\Sigma Y_{0i}}{N_0} \; ; \; \bar{Y}_1 = \frac{\Sigma Y_{1i}}{N_1}$$

$$\hat{s}_Y = \sqrt{\left[\frac{N\Sigma Y^2 - (\Sigma Y)^2}{N(N-1)} \right]}$$

The computational procedure is shown in Table 10.5 in which ΣY^2 is provided by $\Sigma Y_0^2 + \Sigma Y_1^2$, and $\Sigma Y = Y_0 + Y_1$; from this we derive:

$$\hat{s}_Y = \sqrt{\frac{10(1208500) - (3130)^2}{10(10-1)}} = 159.4$$

On applying equation 10.11 we have:

$$r_b = \frac{263.3 - 387.5}{159.4} \sqrt{\frac{6 \times 4}{90}}$$

$$= -0.39$$

Table 10.4 Calculation of point biserial coefficient between village population and existence of post offices in Wiltshire (Source: W. Wiltshire Structure Plan, 1979)

Villages	(Y) Population	(X) Post Offices
Bowden Hill	200	1
Corsley	200	1
Bishops Cannings	250	—
Poulshot	250	—
Compton Bassett	250	1
Seend Cleeve	300	—
Hilmarton	300	1
Hormingsham	320	1
Worton	310	1
Broughton Gifford	750	—

Table 10.5 Calculation of $\bar{Y}_0$ and $\bar{Y}_1$ from Table 10.4

Y_0	Y_0^2	Y_1	Y_1^2
250	62500	200	40000
250	62500	200	40000
300	90000	250	62500
750	562500	300	90000
—	—	320	102400
—	—	310	96100
1550	777500	1580	431000
$\bar{Y}_0 = \dfrac{1550}{4} = 387.5$		$\bar{Y}_1 = \dfrac{1580}{6} = 263.3$	

As with the other measures of correlation we can assess the significance of our result by applying the t test. The null hypothesis in this case is that the presence of a post office is not correlated with village population. Indeed, the point biserial coefficient of -0.39 indicates a negative relationship, although the application of the t test (equation 10.12) shows that with a t statistic of -1.39 and $N-2$ degrees of freedom, the value is not significant. We can therefore accept the null hypothesis and look for other variables, such as distance to the nearest large centre, that may condition the survival of village post offices.

$$t = r_b \sqrt{\left(\frac{N-2}{1-r_b^2} \right)} \tag{10.12}$$

The nature of the point biserial statistic, relating as it does a continuous variable with nominal data, means that certain assumptions do underly its application. Thus, the values of the continuous variable shoud be normally distributed. Furthermore, the two sub-samples of the dichotomous variable should not be vastly different in terms of the number of observations in each sub-set. It has been shown that the more equal these sub-groups are, the more accurate the test becomes. This particular test is relatively well used in geographical analysis, and received early attention from those geographers interested in the empirical testing of central placy theory. For example, Berry and Garrison (1958), in their classic study of central places in Snohomish county in the USA, used the point biserial coefficient to examine the relationship between the size of central places and other functional characteristics.

Obviously, the above test is only applicable if one variable is measured on an interval or ratio scale, and when we want to asses the correlation between two dichotomous variables other statistics need to be considered. Under these conditions we can use the phi coefficient, which represents an extension of the χ^2 test, but only uses 2×2 contingency tables (see Chapter 9). However unlike χ^2 values, the phi coefficient can only vary between -1.0 and $+1.0$. The two tests also differ in another fundamental way, in that the phi coefficient, like other correlation statistics, tells us about the strength of a relationship between two variables, whereas the chi-square test shows whether there is any significant association between the variables.

There are two methods of calculating phi (ϕ), according to whether the χ^2 value has already been derived. If this is not the case then we need to apply equation 10.13:

$$\phi = \frac{AD - BC}{\sqrt{[(A + B)(C + D)(A + C)(B + D)]}} \tag{10.13}$$

(see Table 10.6 for definitions). In the example in Table 10.6 we want to determine whether there is any correlation between the retail centres used for food shopping by car and non-car owning households. The null hypothesis H_0 is that there is no such relationship. From equation 10.13 we have:

$$\phi = \frac{(295 \times 83) - (95 \times 27)}{\sqrt{(390 \times 110 \times 322 \times 178)}}$$

$$= +0.44$$

Table 10.6 Calculation of the ϕ coefficient between car ownership and shopping behaviour

	City shops		Local shops		Total
Households with a car	(A)	295	(B)	95	390
Households without a car	(C)	27	(D)	83	110
		322		178	500

This value of the coefficient indicates some degree of association. Unfortunately there is no method of finding confidence limits for ϕ, and in order to test its significance we need to convert the coefficient into a χ^2 value by using equation 10.14:

$$\chi^2 = \phi^2 N \tag{10.14}$$

where N is the total number of observations; or

$$\phi = \sqrt{\left(\frac{\chi^2}{N} \right)} \qquad (10.15)$$

If we do this we obtain a chi-square value for our example of 96.8, which well exceeds the critical value of 6.64, determined by checking in Appendix III. This indicates that our correlation is significant at the 0.01 per cent level and that car ownership does affect distance travelled to food stores.

The value of ϕ can be calculated from a chi-square value by using equation 10.15, but this is not generally recommended since this method always gives a positive value for ϕ. It Should also be pointed out that the actual layout of the contingency table can affect whether ϕ is positive or negative. In order to have the positive value the table should be constructed so that A and D (see equation 10.13 and Table 10.6) represent the frequencies of individuals who possess both traits or neither trait; while B and C represent the frequencies of individuals who possess one trait but not the other. Therefore, in order to make practical use of the ϕ coefficient it is essential that we inspect the data we are testing very closely to derive a sensible and meaningful correlation. In addition to these difficulties care also needs to be taken over the size of frequencies in the contingency table, since the marginal frequencies can restrict the ultimate value of ϕ. For example, a data set with individual, or marginal frequencies, of 100, 100, 164 and 40 restricts the possible range of ϕ to between -0.5 and $+0.5$. As you can see our example suffers partly from this problem since we have rather unequal marginal frequencies, that give only a moderate sized correlation but one which is highly significant in terms of its χ^2 value. However, despite these problems the ϕ coefficient is used within geography when assessments are being made of any possible correlation between two variables measured on the nominal scale (for an early example see Robinson, 1950).

10.6 PROBLEMS OF CORRELATION TECHNIQUES IN GEOGRAPHICAL ANALYSIS

Since the late 1950s the use of correlation techniques in geographical analysis has been fairly widespread, both as a descriptive tool and as an inferential statistic assessing the significance of a relationship. As well as the statistical assumptions that underly the use of the various correlation tests, their application to geographical problems is further conditioned by the size of the spatial units on which much of our data are based. In this respect we should recognize that correlation coefficients measure the relationship between variables relative to the scale of the spatial units from which the observations are drawn. The problem is that most spatial units are modifiable, and that in many areas of study the geographer may have a choice of such units. Thus, if you are using census data the choice of areal units ranges from small enumeration districts through to a county level of analysis. In an early study of this problem McCarty *et al.* (1956), in a study of industrial production in the USA, found that when data were measured at a county level correlation coefficients were lower than when the same data were correlated at the larger scale of the state. Indeed, we now know that the correlation coefficient between two variables varies in most cases when the data are aggregated into larger areal units. A detailed study by two statisticians, Yule and Kendal (1950), of the relationship between wheat and potato yields for English counties in 1936 found that, when the counties were grouped into larger regional units, the correlation coefficient increased.

Such variations are not, however, entirely related to the question of spatial scale; but as Duncan *et al.* (1961) have demonstrated they are also affected by the problem of spatial contiguity. This refers to the concept where adjacent areal units are likely to resemble each other more closely than units in other areas. In this section, however, we shall concentrate on the problems of modifiable areal units, while the difficulties caused by spatial contiguity or autocorrelation are examined in the following chapter.

One of the first geographers to attempt to solve the problems associated with modifiable spatial units was Robinson (1956), who suggested weighting the calculations for the correlation coefficient by the size of the areal units involved. However, later work by Thomas and Anderson (1965) showed that such weightings do not on the whole provide adequate solutions to the problem of different sized units, and are only relevant in two specific circumstances. The first case is where the density of X and Y values for the initial and the aggregrate spatial units have exactly the same total distribution; while the second case is a slight modification of the first, with the values being in exact proportion to one another. Such events are regarded as chance occurrences, and for this reason arcal weighting should be regarded as a solution only for these special cases. The use of weightings in correlation analysis has been further complicated by the work of Curry (1966), who in direct contrast to Robinson suggested that it is the smallest units that should be given the greatest weighting.

A further solution suggested by Robinson (1961) was to modify the areal units themselves and turn them into regular, equal sized hexagons, rather than use the irregular unit found in both human and physical geography. However, for most work this is totally impractical both in terms of the time required and also because such units do not provide a meaningful solution to the nature of the problem.

These past attempts have largely failed to solve the difficulties of variable areal units, and their effects on correlation techniques. Indeed, it was the failure of these somewhat rigid mathematical solutions that prompted geographers recently to take a more realistic view of the problem. This new approach has taken the form of inquiring into the role of spatial scale in correlation analysis, accepting its affects and attempting to incorporate them into the overall research design. Such a change in emphasis was to some extent stimulated by the work of social geographers and their debates over scale problems which became known as the 'ecological fallacy'. This concept describes the situation where patterns of individual behaviour are inferred from larger scale aggregate patterns, and is a problem long recognized by urban ecologists.

In this particular case the scale of analysis moves from that of variable sized areal units, usually census tracts or enumeration districts, down to the level of the individual household. For example, suppose in an analysis of crime patterns in central London we found a high statistical correlation between proportions of coloured immigrants and levels of crime within census enumeration districts. This correlation analysis does not mean, however, that all coloured immigrants are criminals, since all we have shown is that at one level of areal aggregation (the enumeration district) there is a relationship between these two variables, and that immigrants tends to live in those parts of London with high crime rates. Indeed, a further complication is that some of these crimes may well have been committed by persons from outside these areas. It is such problems of interpretation that have led many social geographers to take a more behavioural approach and study the individual as a decision-maker. However, while this may be a satisfactory solution in some circumstances it is not such a panacea for all geographers, since interest in the effect of distance, together with the characteristics of the spatial units, remains a central theme.

In physical geography, while such scale problems have in the past been less debated, they nevertheless still exist and have begun to receive some attention. Thus, Schuman and Lickty (1965) stressed the importance of both time scale and spatial scale on the statistical interrelationships between stream channel variables. Their paper has prompted further analysis of the scale problem, such as Slaymaker's (1972) study of scale changes and rates of soil creep. More important, however, in this context is the work of Penning-Rowsell and Townshend (1978) on the role of spatial scale in affecting stream channels. Their study demonstrates that the variables affecting the slope of stream channels vary markedly between the broad regional scale and the local scale, as measured by changes in the correlation coefficients.

Two important points emerge from our brief discussion of scale problems and correlation. First, spatial scale is an inherent property in most geographical analysis and we therefore need to understand its effect on statistical tests such as correlation; and owing to considerable work in the 1950s and 1960s, we now have a clear idea about how variations in the size of areal units can produce variations in correlation coefficients (Blalock, 1964). Leading on from this is the fact that statistical inferences between variables should be made within the context of the areal units used. Therefore, provided we interpret the results of our correlation analysis within the correct spatial framework, then scale problems can be accommodated in a positive fashion, rather than being negated as some of the early solutions sought to do.

REFERENCES

Berry, B. J. L., and Garrison, W. L. (1958). 'The functional bases of the central place hierarchy', *Econ. Geog.* **32**, 145–154.

Blalock, H. M. (1964). *Causal Inferences in Non-Experimental Research*, Chapel Hill, Univ. of Carolina Press.

Curry, L. (1966). 'A note on spatial association', *Prof Geog.* **18**, 97–99.

Department of the Environment (1978). *Surface Water: United Kingdom, 1971–73*, HMSO, London.

Doornkamp, J. C., and King, C. A. M. (1970). *Numerical Analysis in Geomorphology*, Arnold, London.

Duncan, O. D., Cuzzort, R. P., and Duncan, B. (1961). *Statistical Geography: Problems in Analyzing Areal Data*, Free Press, New York.

Gould, P. R., and Ola, D. (1970). 'The perception of residential desirability in the western region of Nigeria', *Environ. Plan.* **2**, 73–87.

McCarty, H. H., Hook, J. C., and Knos, D. S. (1956). *The Measurement of Association in Industrial Geography*, University of Iowa Press.

Penning-Rowsell. E. C., and Townshend, J. R. G. (1978). 'The influence of scale on the factors affecting stream channel slope'. *Trans. Inst. Brit. Geogrs* (N.S.) **3**, 395–415.

Robinson, A. H. (1956). 'The necessity of weighting values in correlation of areal data', *Anns Ass. Am. Geogrs* **47**, 379–391.

(1961) 'A correlation and regression analysis applied to rural farm densities in the Great Plains', *Ibid.* **51**, 211–221.

Robinson, W. S. (1950). 'Ecological correlations and the behaviour of individuals, *Am. Soc. Rev.* **15**, 351–357.

Schumm, S. A., and Lichty, R. W. (1965). 'Time space and causality in geomorphology'. *Am. J. Sci.* **263**, 110–119.

Slaymaker, M. O. (1972). 'Patterns of present sub-aerial erosion and landforms in mid-Wales', *Trans. Inst. Brit. Geogrs* **55**, 47–67.

Thomas, E. N., and Anderson D. L. (1965). 'Additional comments on weighting values in correlation analysis of areal data', *Anns. Ass. Am. Geogrs* **55**, 492–505.

Yule, G. U., and Kendal, M. G. (1965). *An Introduction to the Theory of Statistics*, 14th edn, Griffin, London.

RECOMMENDED READING

Alker, M. R. (1969). 'A typology of ecological fallacies' in M. Dogan and S. Rokkan (eds) *Quantitative Ecological Analysis in the Social Sciences* MIT Press, Cambridge, Mass., 69–86. Discusses the range of problems associated with individual behaviour and inferences drawn from an analysis of areal units.

Leigh, R. (1969). 'Analysis of the factors affecting the location of industry within cities', *Canadian Geogr.* **13**, 28–33. Shows a frequent application of correlation coefficients within a descriptive framework, in this case by comparing different measures of manufacturing.

Melton, M. A. (1958). 'Correlation structure of morphometric properties of drainage systems and their controlling agents', *J. Geol.* **66**, 442–461. Presents a good example of the use of correlation linkage diagrams as a first stage in understanding interrelationships.

EXERCISES

1. It is often suggested that rainfall and altitude are correlated. Use the Pearson product-moment correlation coefficient to examine the validity of this statement for the

following data which were obtained from rainguage stations in North-East England. (Data source: Northumbrian Water Authority records)

Mean annual rainfall (mm)	Altitude (m)
640	46
620	11
670	53
810	183
785	104
730	111
730	76
1014	373
1055	375
895	267
810	213
747	99
860	177
1050	266
810	198
1199	290
945	122
972	327
856	267
771	162
854	222

2. Assess the statistical significance of the correlation obtained in question 1 using: (i) tables of critical correlation coefficients; (ii) the equations for correlation coefficient significances. Select your own significance level.

3. For the following random sample of American States calculate the Pearson product-moment correlation between the two variables of physicians and personal income. Use either the necessary equations or tables of critical values to determine the statistical significance of the correlation coefficient at the 0.05 level. (Data source: *US Statistical Abstract* 1965–66, (US Dept. of Commerce)).

State	Physicians (per 100,000 people)	Personal income ($ per person)
Massachussets	196	3,023
Vermont	108	2,340
New York	207	3,242
Maryland	158	3,014
Minnisota	145	2,625
Missouri	149	2,628
West Virginia	103	2,007
North Carolina	100	2,028
Alabama	79	1,910
Utah	127	2,340
Iowa	117	2,595
Louisiana	114	2,061
Texas	111	2,346
South Dakota	73	2,055
Nevada	107	3,289
Ohio	136	2,816

State	Physicians (per 100,000 people)	Personal income ($ per person)
New Hampshire	143	2,570
Maine	125	2,245
California	178	3,196
Arizona	131	2,310
Michigan	141	3,009

4. Do the following data support any suggestion that the number of midwives per person in each of the English health authority regions correlates with mortality at childbirth? Because there are only fourteen health authority regions the Spearman rank correlation should be used. (Data source: *Regional Trends* 1983, HMSO)

Region	Perinatal mortality (per 1,000 births)	Midwives (per million of population)
Northern	13.2	71.0
Yorkshire	13.9	72.4
Trent	11.3	96.2
East Anglia	10.4	102.4
NW Thames	10.7	50.2
NE Thames	11.1	65.0
SW Thames	12.1	64.2
SE Thames	10.7	55.6
Wessex	9.9	60.0
Oxford	9.4	72.2
South Western	11.4	43.7
West Midlands	12.8	80.9
Mersey	12.4	88.5
North Western	12.4	92.4

5. A study of median particle size at fifteen randomly selected points along a debris surface profile yielded the data presented below. Use these observations to establish the Spearman rank correlation between the variables and its statistical significance at the 0.01 level.

Slope angle (degrees)	Median particle size (mm)
3.0	0.7
5.5	1.1
7.5	1.3
9.0	1.3
9.0	2.8
13.0	3.8
17.0	3.9
13.5	2.8
16.5	4.7
17.5	3.2
19.0	5.5
21.0	1.5
24.0	6.9
25.0	7.3
26.0	6.8

6. Two random samples (10 each) of Republican and Democratic States (according to leading party by votes cast for US representatives in 1978) were examined in terms of their unemployment rates. Use the point biserial coefficient to measure the correlation between the continuous (unemployment) and the dichotomous (leading party) variables. Denote the Republican states by one and the Democratic states by 0. (Data source: *Statistical Abstract of the United States* 1982–83, US Dept. of Commerce)

	Republican states		Democratic states	
State	Unemployment (% of total pop.)	State		Unemployment (% of total pop.)
Idaho	5.3	Alabama		7.1
Illinois	5.5	California		6.2
Maine	7.2	Florida		8.0
Nebraska	3.2	Indiana		6.4
New Hampshire	3.1	Lousiana		6.7
North Dakota	3.7	Massachusetts		5.5
South Dakota	3.5	Mississippi		4.5
Utah	4.3	Oregon		6.8
Virginia	4.7	Pennyslvania		6.9
Wyoming	2.8	Wisconsin		4.5

Use the 0.01 level to estimate the statistical significance of the coefficient found for these data.

7. Use the ϕ coefficient to establish the degree and statistical significance of the correlation between pass and fail rates in driving tests for men and women. The data consist of absolute frequencies for a sample taken in the United Kingdom in 1981. (Data source: *Social Trends* 1982, HMSO)

	Pass	Fail
Male	537	478
Female	427	588

8. Many plants have a preference for particular types of rock and their associated soils. Does the data set given below support the hypothesis that the incidence of *Nardus stricta* (Mat grass) correlates with the presence (or absence) of calcareous rocks? The data are based on a random sample of one hundred 10 km squares from the Vale of York, England. Each square is allocated to a cell in the contigency table according to the absence or presence of *N. stricta* and by the proportion of calcareous rocks. (Data source: Perring, F. H. and Walters, S. M., *Atlas of British Flora,* 1962)

	Chalk and limestone more than 5 per cent by area	Chalk and limestone less than 5 per cent by area
N. stricta present	21	32
N. stricta absent	´28	19

Chapter 11

Simple Linear Regression

11.1 THE LINEAR REGRESSION MODEL

In chapter 10 we examined methods of measuring the strength of a relationship between two variables, with the emphasis being placed on the extent to which the data sets varied relative to each other. Thus, if we had a positive association, then the two sets of observations would increase together. Simple regression techniques take this type of analysis a stage further by measuring the form of a relationship between two variables. One variable is considered to be dependent on the other and therefore controlled by it. For example, urban land values could be seen as a function of distance from city centre and vary accordingly. In statistical terms land values may be seen in this example as the dependent variable (labelled Y) and distance from the city centre to the independent or controlling variable (labelled X). Unfortunately, as we shall see, not all relationships are so clearly framed, since in some situations it is impossible to distinguish between dependent and independent variables. In addition, not all the problems that we analyse are based around linear relationships, and in these circumstances we need to consider nonlinear regression techniques (Chapter 12).

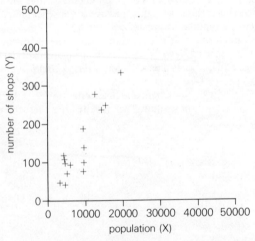

Figure 11.1 Plot of shops and population for Cornwall (1971)

168

Examine the data presented in Figure 11.1, which shows the variation between shop numbers and population for a group of English towns. We are interested in studying the relationship between these two variables, and in particular in examining how shop numbers (the dependent variable) are conditioned by population (the independent variable). By using the linear regression model, as stated in equation 11.1, we can fit a straight line to the scatter of points, and thus be in a position to understand the form of the relationship between these variables. The logic underlying linear regression is based on the criterion of 'least squares', the aim of which is to minimize the square of deviations about the fitted line. Taking Figure 11.2 as an example, we can say that the regression line is constructed in such a way that: the positive and negative deviations of individual points from the line must total to zero ($\Sigma v = 0$), and the sum of the squared deviations must be smaller than those from any other line (Σv^2 is a minimum).

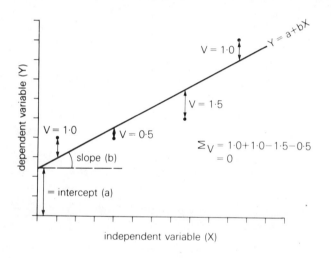

Figure 11.2 Characteristics of the linear regression line

If we take a closer look at the regression model (equation 11.1) we can see that as well as X and Y it contains three other components:

$$Y = a + bX + e_i \qquad (11.1)$$

Equation 11.1
$Y =$ dependent variable
$X =$ independent variable
$a =$ intercept on Y axis, the value of Y at the point where $X = 0$
$b =$ slope coefficient
$e =$ error term

The values of a and b are therefore required before we can proceed to fit a least squares regression line to our data, and the values may be computed from equations 11.2 and 11.3 respectively:

$$a = \frac{(\Sigma Y)(\Sigma X^2) - (\Sigma X)(\Sigma XY)}{N\Sigma X^2 - (\Sigma X)^2} \tag{11.2}$$

$$b = \frac{N\Sigma XY - (\Sigma X)(\Sigma Y)}{N\Sigma X^2 - (\Sigma X)^2} \tag{11.3}$$

(see Table 11.1 for definitions).

It is worth noting here that these two equations also use information derived from calculating the product-moment correlation coefficient (equation 10.3); a fact that may save time when you are interested in obtaining both statistics for the same data.

Table 11.1 Calculation of regression equation between shop numbers and town size (Cornwall, 1971)

	Population (X)	X^2	Shops (Y)	Y^2	XY
Bodmin	9204	84713616	95	9025	874380
Bude	5629	31685640	94	8836	529126
Falmouth	17883	319801664	271	73441	4846293
.	.	.	.	.	.
.	.	.	.	.	.
Truro	14830	219928896	233	54289	3455390
	125442	1453923072	2127	412911	22648580

Table 11.1 shows an example of this — shop numbers are tabulated against town size. The calculations are as follows. From equation 11.2 and 11.3 we have:

$$a = \frac{(2127)(1453923072) - (125442)(22648580)}{(15 \times 1453923072) - 15735695360}$$

$$= 41.4$$

$$b = \frac{(15)(22648580) - (125442)(2127)}{(15 \times 1453923072) - 15735695360}$$

$$= 0.0120$$

Therefore the regression equation is:

$$Y = 41.4 + 0.0120X$$

It is very apparent from these calculations that for large samples and large number sets the calculation of the regression coefficients by hand is a very time-consuming task. In addition, with such large numbers it is often easy to make simple arithmetical mistakes. For both these reasons it is obviously far easier to use computer facilities if they are available, either by writing your own program or using package programs.

As an alternative to these equations we can also calculate the a and b coefficients by using the following formulae:

$$b = \frac{\Sigma xy}{\Sigma x^2} \qquad (11.4)$$

$$a = \bar{Y} - b\bar{X} \qquad (11.5)$$

where $x = X - \bar{X}$ and $y = Y - \bar{Y}$ for each observation and a is calculated through using the coefficient b, with the means of X and Y.

By fitting the calculated regression to our data points from Figure 11.1, we can see that there is a fairly close relationship between shops and population, as we can have only a small amount of scatter about the line (Figure 11.3). Thus, it can be shown that if all the observations fell on the line then variable Y would be perfectly related to variable X. However, in our example this is not the case, and a certain amount of what is called 'unexplained variance' occurs. To understand this, examine Figure 11.4, which shows the explained and unexplained components of a regression analysis. The vertical distance between $\bar{Y}$ (the mean of Y) and the regression line represents the variation from $\bar{Y}$ explained by variable X.

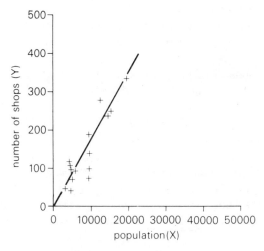

Figure 11.3 Regression of shops and population
in Cornwall (1971)

In contrast, the additional vertical distance to Y, represented by the difference between the regression line and point P in Figure 11.4, is that part of Y unexplained or unaccounted for by variable X. This unexplained component is termed the residual of Y, and is defined as the difference between the actual and the estimated values of Y or $(Y_i - \hat{Y}_i)$ as predicted by the regression model. In statistics, estimated values are usually indicated by placing a hat over them thus: $\hat{Y}$; and so the explained proportion of a regression function can be represented by:

$$\hat{Y} = a + bX \qquad (11.6)$$

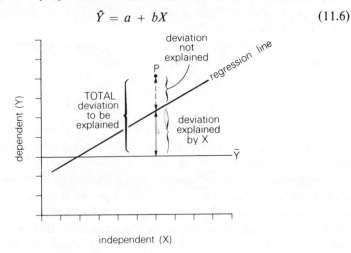

Figure 11.4 Explained and unexplained components of linear regression

The deviations of observations about the regression line as shown in Figure 11.3 and 11.4 are also known as residuals from regression and can take either positive or negative values. They are defined as the differences between the actual and predicted values of Y; that is the residual of Y_i is $Y_i - \hat{Y}_i$. This gives the absolute value of a residual of observation Y_i, but due to scale differences such absolute values are very rarely used. As a valid alternative geographers usually make use of a standardized residuals, which are given in terms of a standardized, normal distribution of residuals. The deviation of these, and their use within geographical analysis, are topics discussed in Sections 11.2 and 11.3.

It should also be stressed that, like all parametric statistics, the linear regression model has a number of underlying assumptions. First, we assume the data are measured on either the interval or ratio scales. Second, as is shown in Figure 11.5, the probability distributions of observed Y's have the same variance for all values of X. Third, the random variables Y_i are statistically independent, in the sense that a large value of Y_1 does not tend to make Y_2 large; this is discussed further in Section 11.4. Finally, the random error term (e) in equation 11.1 indicates that in geographical analysis we are mainly dealing with partial relationships and not precise mathematical ones. In our work, as we shall see, many of the problems are of a multivariate nature, and consequently the independent variable does not entirely account for all the variations in Y. Furthermore, in some cases, especially in human geography, we are dealing with chance or random elements, which introduce some degree of unpredictability into statistical relationships. Therefore the regression model describes only an average, while the various disturbances encountered in the relationship are incorporated as part of the error term. It therefore represents that which is not accounted for by the regression equation.

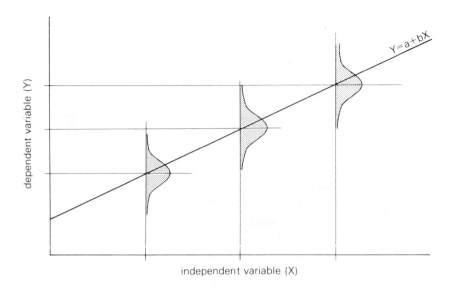

Figure 11.5 Representation of condition of constant variance of observations about regression line

11.2 SIGNIFICANCE TESTING IN SIMPLE REGRESSION

Regression equations are estimated from samples drawn from unknown populations. As such they are subject to sampling variations and it is important to identify the reliability of sample estimates a and b of the population parameters α and β, and of the estimates of Y (usually termed $\hat{Y}$) made from them. A number of alternatives are available to do this. All yield the same result for identical data, but the method here selected has the advantage of being adaptable to many circumstance (it will be used again in nonlinear and multiple regression), while its fundamental principles have already been demonstrated in the section dealing with one-way analysis of variance (Section 9.13). In the latter case the total variance was apportioned in two parts; the within- and between-groups variances. In regression analysis this sub-division changes only slightly to yield two aspects of the variation in the dependent variable; that 'accounted for' by the regression line (in effect the variance of $\hat{Y}$) and the variance of the residuals. These variances are obtained by dividing the relevant sums of squares by their degrees of freedom. Table 11.2 indicates how these quantities may be estimated.

From these quantities both regression ($s_{\hat{Y}}^2$) and residual (s_e^2) variances are found. The test statistic for the significance of the regression model is distributed as F (the variance ratio). In this case the ratio required is that of regression to residual (error) variance. Generally:

$$\text{Variance} = \frac{\text{sum of squares}}{\text{degrees of freedom}}$$

$$s_{\hat{Y}}^2 = \frac{\Sigma(\hat{Y} - \bar{Y})^2}{k} \tag{11.7}$$

$$s_e^2 = \frac{\Sigma(\hat{Y} - Y)^2}{n - k - 1} \tag{11.8}$$

Equations 11.7 and 11.8

$\hat{Y} =$ estimated Y values
$\bar{Y} =$ mean of observed Y values
$Y =$ individual Y values
$n =$ number of observations
$k =$ number of predictors
(always 1 in simple regression)

---

which combine to give:

$$F = \frac{s_{\hat{Y}}^2}{s_e^2} \tag{11.9}$$

Table 11.2 Explanation of regression equation analysis of variance

Source of deviation	Sum of squares	General description	Degrees of freedom
Total	$\Sigma(Y - \bar{Y})^2$	Sum of squared deviations of observations from sample mean	$n - 1$
Regression	$\Sigma\hat{Y} - \bar{Y})^2$	Sum of squared deviations of predictions from sample mean	k
Error	$\Sigma(\hat{Y} - Y)^2$	Sum of squared differences of observed and predicted values	$n - k - 1$

The null hypothesis under test is that of no explanation of the variability of Y (the dependent variable) in terms of X (the predictor). In this sense regression variance can be regarded as the 'explained' variance and the error or residual variance as 'unexplained'. Consequently, the greater F ratios are provided by higher proportions of explained variances. The associated degrees of freedom and predetermined significance level are used to determine critical F values from the tables in Appendix VI.

This method of significance testing tells us a great deal about how regression analysis actually works. Thus, for example, when two variables are zero correlated the best fit line through the scatter of points will be horizontal ($b = 0.0$) passing through the data centroid ($\bar{X}, \bar{Y}$). Under these conditions the sum of squares of observations about the regression line is certainly a minimum, but at the same time the residual or error (unexplained) sum of squares is clearly equal to the total sum of squares. The regression sum of squares is zero (see Figure 11.6) as $\hat{Y}$ is always equal to $\bar{Y}$. However, as soon as the regression coefficient has any degree of slope the equality between the total and error sum of squares vanishes and the difference is made up by the regression sum of squares (Figure 11.7). The problem is one of deciding at what point the proportion of regression variance is statistically significant. With a decrease in the scatter of points about the regression line we should find that the regression (explained) sum of squares grows at the expense of the residual (unexplained) sum. The consequent F ratios reflect this balance and we can use the F tables to determine if the proportion of 'explained' variance is significant.

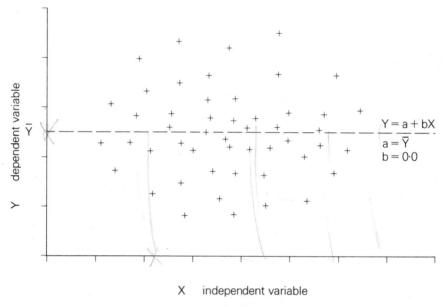

Figure 11.6 Regression line through scatter of points made by uncorrelated bivariate sample

The results of this form of test are best set out in an ANOVA table of the type used in the following example. A useful 'by-product' of this method is that both the product-moment correlation and the coefficient of explanation can be easily extracted from such a table. The coefficient of explanation is determined from the ratio of regression to total sum of squares. The correlation coefficient is the square root of the latter:

$$r = \sqrt{\left(\frac{\text{regression sum of squares}}{\text{total sum of squares}}\right)} = \sqrt{\left[\frac{\Sigma(\hat{Y} - \bar{Y})^2}{\Sigma(Y - \bar{Y})^2}\right]} \quad (11.10)$$

Let us now examine how these principles work in practice by testing the significance of the regression model of shops and population. Once again the use of even elementary computer programs will save time; nevertheless all of the following tasks are easily performed with the aid of pocket calculators. All operations are performed solely on the dependent variable and its attributes such as mean and variance. But in all analysis of variance problems only two of the three sums have to be calculated because, as is shown in Table 11.3, total = error + regression sums of squares. By reference to the observed Y values, the estimated values that correspond to them ($\hat{Y}$) and their deviations from the sample mean ($\bar{Y}$), equations 11.7 and 11.8 can be evaluated. In this way, and without going into the arithmetic processing of the numbers, we can construct the appropriate ANOVA table (Table 11.3).

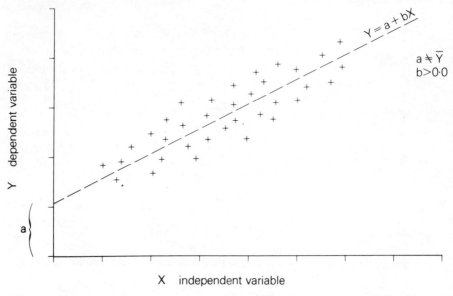

Figure 11.7 Regression line through scatter of points made by correlated bivariate sample

From Table 11.3 we can see that the null hypothesis of no explanation of the dependent variable by the regression model is rejected. We conclude, therefore, that the proportion of explained variance is significantly high. An important point to bear in mind is that, with this scheme, we are testing the whole regression model. It is possible, although it is not performed here, to test the two coefficients (a and b) separately; a number of computer programs treat the problem in that manner, providing t statistics for both estimates.

Table 11.3 ANOVA table for significance testing of the regression model of Cornwall shops/population

Source	Sum of squares	Degrees of Freedom	Variance
Regression	58359	1	58359
Error	52943	13	4073
	111302	14	

$F = 14.33$; $F_{\text{critical}} = 4.67$ (at the 0.05 significance level)

We may also calculate the coefficients of correlation and explanation from equation 11.10 and the data contained in Table 11.3. Hence:

$$r^2 = \frac{58359}{111302} = 0.524$$

$$r = +0.724$$

As a result we see that 52.4 per cent of the observed variation in the dependent variable is 'explained' by variation in the independent variable. Readers will have noted that the square root of r^2 could have been ±0.724 as the square of both give $+0.524$. The appropriate sign is determined by the sign of the regression coefficient b, which in this case was positive.

Lastly, but importantly, the term 'explained' is used here in a statistical sense and it is the task of the geographer to provide the theoretical and geographical arguments that substantively explain the behaviour of the dependent variable in terms of the independent variable.

11.3 CONFIDENCE LIMITS IN SIMPLE REGRESSION

The general idea of confidence limits has already been introduced in Chapters 7 and 8. where we looked at the reliability of sample means. Those principles can be extended, with modification, to regression analysis in order to examine the reliability of estimates of Y made using regression equations.

The first and most important aspect that governs the reliability of any regression estimate is the variance of the residuals about the regression line itself, i.e. the quantity given by $s_e{}^2$ in equation 11.8, the square root of which is the standard error of the residuals (s_e). From the information contained in Table 11.3 this can be estimated as:

$$s_e = \sqrt{4073} = 63.82$$

An alternative method would have been to use the formula:

$$s_e = s_Y{}^2 (1 - r^2) \tag{11.11}$$

in which $s_Y{}^2$ is the variance of the dependent variable Y and r^2 the coefficient of explanation. Such rapid methods act as valuable checks on our results and can signal the presence of any mistakes, although due allowance should be made for rounding errors.

If we are using small samples (less than 30 observations) it is a good idea to apply a correction factor to counteract the bias due to underestimation of variances. The corrected 'best estimate' of the population standard error, usually denoted by $\hat{\sigma}_e$, is easily found from:

$$\hat{\sigma}_e = s_e \sqrt{\left(\frac{n}{n-k-1}\right)} \tag{11.12}$$

> ### Equation 11.12
>
> s_e = standard error of residuals
> n = number of observations
> k = number of predictors

For our example:

$$\hat{\sigma}_e = 63.82 \times \sqrt{(15/13)} = 68.55$$

This, however, is a measure of the spread of the observed points about the regression line. Because that line is itself not known with certainty and will vary from sample to sample, s_e alone cannot be used to establish the reliability of predicted values of Y.

More specifically it is the estimates of the coefficients a and b that are in doubt, as both are based on samples drawn from an unknown population. Both are subject to sampling error and will differ from the corresponding population parameters α and β. The results of the sampling errors are illustrated in Figures 11.8 – 11.10. Variations in estimates of the intercept term cause the regression line to move in a vertical yet parallel fashion represented by the grey area on the diagram. (Figure 11.8) Sampling variations in the regression coefficient (the slope of the line) cause a rotational effect about the data centroid through which all lines will pass, but with differing slopes. The combined effect of these two sources of error is to create curved — to be precise, hyperbolic — confidence limits. One obvious result is that error margins become greater as we move away from the data means. All values of $\hat{Y}$ are points on the regression line, and in establishing their confidence limits we also establish the confidence limits of the regression line at that point.

Because the confidence limits are curved different values occur at each point along the line. The standard error of the estimate of Y ($SE_{\hat{Y}}$) at any point X_k along the independent scale can be calculated using equation 11.13:

$$SE_{\hat{Y}} = \hat{\sigma}_e \sqrt{\left[\frac{1}{n} + \frac{(X_k - \bar{X})^2}{\Sigma(X - \bar{X})} \right]} \qquad (11.13)$$

> ### Equation 11.13
>
> $\hat{\sigma}_e$ = standard error of residuals
> n = number of observations
> X_k = selected X value
> $\bar{X}$ = mean of variable X
> X = individual observations of X

Depending on the availability and type of calculator the quantities needed in this equation can be obtained with differing degrees of efficiency; we already have $\hat{\sigma}_e$ and the only notable arithmetic is involved in deriving $\Sigma(X - \bar{X})^2$, the sum of squared deviations of X about their mean. This leaves only X_k values to be selected; we will consider only two cases, one for $X_k = \bar{X}$ and another at the extreme end of the observed range where $X_k = 18,000$.

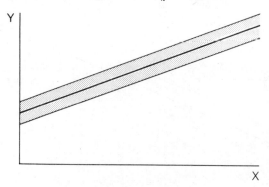

error region for intercept terms

Figure 11.8 Error range for sampling variations in the intercept term a; shaded area around regression line indicates range of variation

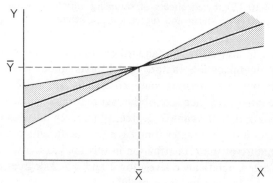

error region for regression slope

Figure 11.9 Error range for sampling variations in regression coefficient b

In the first case $X_k = \bar{X} = 8362.8$, and so:

$$SE_{\hat{Y}} = 68.55 \sqrt{\left[\frac{1}{15} + \frac{(8362.8 - 8362.8)^2}{4.049 \times 10^8} \right]}$$

$$= 68.55 \sqrt{(0.0667 + 0.0)} = 17.71$$

For the second case $X_k = 18,000$, and so:

$$SE_{\hat{Y}} = 68.55 \sqrt{\left[\frac{1}{15} + \frac{(18000 - 8362.8)^2}{4.049 \times 10^8} \right]}$$

$$= 68.55 \sqrt{(0.067 + 0.2294)} = 37.30$$

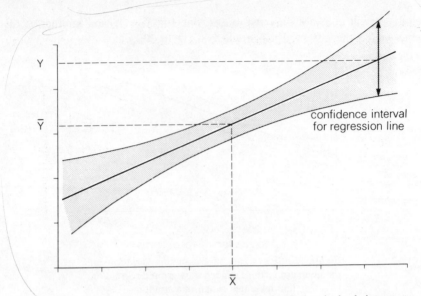

Figure 11.10 Combined effects of sampling variation in both intercept term and regression coefficient

The next step is to convert these standard errors into confidence limits by using the appropriate multiplication factor. Many texts quote the t-distribution to accomplish this but, for reasons outlined below, we prefer to use the F-distribution. The latter takes account of the fact that the confidence interval needs to accommodate not only the sample degrees of freedom but also the fact that the line itself is based on two estimates (one each for coefficients a and b). However, because of the nature of the F-distribution in this application we take the critical value (at the selected significance level) for 2 and $n - 2$ degrees of freedom and adjust it by $(2F_{crit})^{\frac{1}{2}}$. In the current example $n - 2 = 13$, and we shall use the 0.05 critical value in order to obtain the 0.95 (95 per cent) confidence limits. From Appendix VI, F_{crit} is 3.80, which after multiplication and square rooting becomes 2.76 (the corresponding t statistic would have been 2.16 and would have given narrower confidence intervals). In general our confidence limits can be written as:

$$\hat{Y}_k \pm SE_{\hat{Y}} \times (2F)^{\frac{1}{2}} \tag{11.14}$$

For the two X_k values of 8623.8 and 18,000, and using the regression equation established earlier, we estimate $\hat{Y}$ values of 141.75 and 257.4 respectively.

The 95 per cent confidence limits for these estimates can now be evaluated:

$$141.75 \pm 17.71 \times 2.76$$
$$= 141.75 \pm 48.88 = 92.87 \text{ to } 190.63 \text{ shops}$$

and

$$257.4 \pm 37.30 \times 2.76$$
$$= 257.4 \pm 102.95 = 154.45 \text{ to } 360.35 \text{ shops}$$

These figures tell us that we may be 95 per cent confident that the population, i.e. 'true', regression line lies between those limits at those points along the X scale (see Figure 11.11).

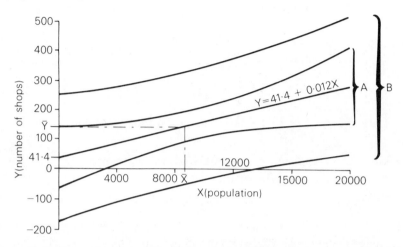

Figure 11.11 Regression line, with 95 per cent confidence limits for the line (A) and individual predictors (B)

The most striking feature of these results is the very great difference, more than a factor of 2, between the widths of the confidence intervals. This arises despite the close correlation found between the two variables and is a timely reminder of the dangers of using small samples. These curved confidence limits can be regarded as an 'envelope' within which the true (population) regression line lies with selected probabilities (in this case 0.95). From the form of equation 11.13 we can see that as the sample size n increases the envelope must become narrower, until it vanishes at the point where the whole population is included in the analysis and there can be no doubt regarding the parameters. At this stage, however, individual observations will still vary about the regression line as X does not completely determine the behaviour of Y. Because of the almost inevitable scatter of observed points about even the population regression line, the confidence limits for *individual* estimates of Y cannot shrink to zero and must remain. These confidence limits may be determined by a simple development from equation 11.13, namely:

$$SE_{\hat{Y}_i} = \hat{\sigma}_e \left[1 + \frac{1}{n} + \frac{(X_k - \bar{X})^2}{\Sigma(X_i - \bar{X})^2} \right]^{\frac{1}{2}} \tag{11.15}$$

where $SE_{\hat{Y}_i}$ is used to denote the standard error for an individual prediction of Y.

Clearly, curved limits will again result, though these will be necessarily wider than before by a factor proportional to $\hat{\sigma}_e$ owing to the inclusion of the 1.0 within the brackets. Using once again values for X_k of 8362.8 and 18,000, we can examine the consequences of this change. For $X_k = 8362.8$:

$$SE_{\hat{Y}_i} = 68.55 \sqrt{\left[1 + \frac{1}{15} + \frac{(8362.8 - 8362.8)^2}{4.049 \times 10^8} \right]}$$

$$= 68.55 \sqrt{(1 + 0.0667 + 0.0)} = 70.80$$

and for $X_k = 18,000$:

$$SE_{\hat{Y}_i} = 68.55 \sqrt{\left[1 + \frac{1}{15} + \frac{(18000 - 8362.8)^2}{4.049 \times 10^8} \right]}$$

$$= 68.55 \sqrt{(1 + 0.0667 + 0.2294)} = 78.04$$

Expression 11.14 can now be rewritten to provide confidence limits for the individual predictions, and becomes:

$$\hat{Y}_k \pm SF_{\hat{Y}_i} \times (2F)^{\frac{1}{2}}$$

in which the F value is again that for 2 and 13 degrees of freedom at the 0.05 level. The required figure remains 2.76 and the 95 per cent limits become now, for $\hat{Y}_k = 141.75$:

$$141.75 \pm 70.81 \times 2.76$$
$$= 141.75 \pm 195.44 = -53.69 \text{ to } 337.19 \text{ shops}$$

and for $\hat{Y}_k = 257.4$:

$$257.4 \pm 78.04 \times 2.76$$
$$= 257.4 \pm 215.39 = \quad 42.01 \text{ to } 472.79 \text{ shops}$$

We should note that the great width of these intervals brings about the 'impossible' range of predictions of -53.69 to 337.19 shops but, as a purely statistical device, the conclusion drawn is that we are 95 per cent confident that individual observations of the dependent variable will lie between the specified limits at those X_k points.

These new prediction limits are far wider than their counterparts found above (Figure 11.11) and reflect the wide scatter of points about the fitted regression line. In some respects such widths are impractical and we have already drawn attention to the straying of the confidence limits into negative shop regions. Confidence limits for individual predictions of Y tend to be only shallowly curved and, as the sample size increases, tend to become straight and parallel to the regression line at distances almost wholly determined by the standard error of the residuals ($\hat{\sigma}_e$). Indeed, in many cases where sample sizes are large and correlations high they can be assumed, for all practical purposes, to behave in this fashion, and many texts overlook the slight degrees of curvature that may be present and draw the standard error lines as straight and parallel to the regression line.

The differences between individual observed Y's and their predicted counterparts on the regression line are known as residuals. The spread of residuals about the line is measured directly by their standard error ($\hat{\sigma}_e$) which, as we have seen, influences both forms of confidence interval in equation 11.13 and 11.15. These intervals, consequently, combine uncertainty concerning the regression parameters with residual variation about the best-fit line. This former uncertainty is best reduced by use of large samples. Little, however, can be done about residual variance because any one variable cannot determine wholly the behaviour of its dependent counterpart. Hence it remains, no matter how large the sample, and is reduced only by the use of better or additional predictor variables (multiple regression). The next section shows how the residuals of a regression analysis can be helpful both in establishing the validity of the model and in guiding subsequent studies with multiple predictors.

11.4 ANALYSIS OF RESIDUALS

In the previous sections reference was made to residuals, these being the differences between observed values of Y and those predicted by the regression model. Such measures can provide a valuable extention to regression analysis by helping us to account for that part of the unexplained variance between X and Y. As previously stated many geographical problems involve a consideration of more than simple bivariate relationships. Thus, with our example relating numbers of shops to population we found a significant, but only a partial positive, relationship, suggesting that perhaps we need to consider other variables in addition to population.

An analysis of the residuals from our original regression line may help in the search for such new variables. However, before we can use residuals in this way it is usual practice to standardize their values. One of the assumptions of regression analysis is that such residual terms are normally distributed about the regression line with constant variance irrespective of the point along the line (see Figure 11.5). If these assumptions are presumed correct then standardizing the residuals is a simple task performed by reference to the best estimate of the standard error of the residuals ($\hat{\sigma}_e$). Draper and Smith (1966) refer to the transformed residuals as 'unit normal deviates' which are given by:

$$\text{Standardized residual} = \frac{Y_i - \hat{Y}}{\hat{\sigma}_e} \tag{11.16}$$

This is also the form of residual standardization used in the popular SPSS computer package. The reader is warned, however, that different conventions are employed by other package programs, the results of which may not always agree.

Geographers can take the analysis of residuals a stage further by mapping out standardized residuals and examining their spatial distributions. The use of residuals in this form has been discussed in a number of papers, but their application is especially well illustrated by Clark (1967) in his work on farming patterns in New Zealand, and McCarty's (1959) study of electoral geography. In addition, a paper by Thomas (1968) reviews the full extent to which residuals can be used in map form. His work highlights three important areas of use: first, the formation and modification of hypotheses concerning the existence of spatial associations between variables, and the search for new variables; second, the establishment or modification of regional boundaries, so that by using economic variables we may define economic regions based on the groupings of residuals; third, the identification and selection of specific study areas for intensive field work, so that we may select those areas which have high residual values to conduct field work to establish why the regression model fails to predict accurately variations in the dependent variable.

11.5 AUTOCORRELATION AND RESIDUAL ANALYSIS

In many statistical procedures the arithmetic aspects of the process are complete after the establishment of the test's significance. In regression analysis this is not so and attention has to be paid to the residuals. More precisely, there are preconditions concerning residual behaviour that need to be fulfilled before the regression equation can be finally regarded as both unbiased and the best possible estimate of the population parameters. All of these points are discussed by Draper and Smith (1966) and Poole and O'Farrell (1971), but they may be usefully summarized here.

(1) The probability distribution of Y should, for any value of X, be normally distributed about the regression line, as explained already in Section 11.1.
(2) In addition the observed Y values should have zero mean (itself a direct consequence of their normality) and a constant variance about the line (see Figure 11.5). The latter is the requirement of homoscedasticity. If the scatter of the residuals is not constant about the line the data are said to be heteroscedastic.
(3) The residuals must not be serially correlated (autocorrelated).

When all of these requirements are met the values of a and b can be regarded as the best linear unbiased estimates of the unknown population parameters α and β. For both autocorrelation and homoscedasticity simple graphical plots of the scatter of points about the fitted line will often indicate severe infringements of (2) and (3). Alternatively residuals (e) can be plotted against $\hat{Y}$, as has been done in Figures 11.2(a) and (b).

There are also two statistical tests that can be used to examine any visual impressions gained from residual graphs. Both were originally devised for use with time series data, but may be usefully employed for residuals sequenced in terms either of $\hat{Y}$ or X. The tests identify degrees of autocorrelation ranging from the perfect negative to the perfect positive cases. Positive autocorrelation is seen in long residual sequences of similar magnitude and same sign, as in Figure 11.13(a). Negative autocorrelation, less common in geographical data, is indicated by rapid and regular sequential alternation from one side of the regression line to the other (Figure 11.13(b)). Zero autocorrelation lies midway between these extremes and is shown by a random sequence. We shall examine both a parametric and a non-parametric method of assessing autocorrelation; namely the 'runs' and Durbin–Watson tests.

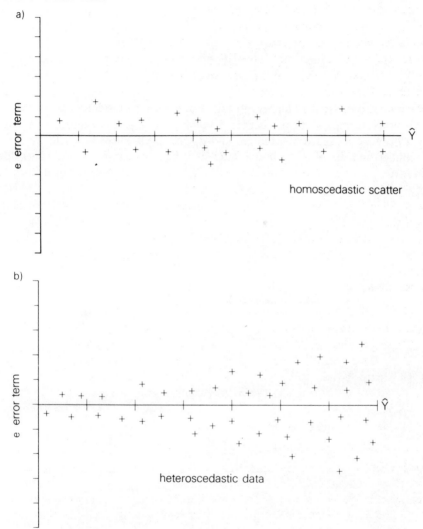

Figure 11.12 Schematic representations of (a) homoscedastic (constant variance) and (b) heteroscedastic (changing variance) conditions

11.6 THE 'RUNS' TEST

This test treats only the signs of the residuals. The number of 'runs' or sequences of the two signs, irrespective of length, are counted. For example, the sequence + + + + + − − − − has two runs, while + − + − + − has six. An overabundance of runs indicates negative autocorrelation and a dearth betrays the presence of positive autocorrelation. The sampling distribution of r (the number of runs and not to be confused with the correlation coefficient) is normally distributed with a mean (μ_r) and a standard deviation (σ_r) defined by equation 11.17 and 11.18:

$$\mu_r = \frac{2n_1 n_2}{n_1 + n_2} + 1 \tag{11.17}$$

$$\sigma_r = \sqrt{\left[\frac{2n_1 n_2 (2n_1 n_2 - n_1 - n_2)}{(n_1 + n_2)^2 (n_1 + n_2 - 1)} \right]} \tag{11.18}$$

where n_2 and n_1 are the number of positive and negative residuals respectively. In real terms μ_r represents the expected number of runs if a sequence of length $n_1 + n_2$ were arranged at random, and σ_r^2 is its variance. But with these two quantities and the observed r the latter may be re-expressed as a z value thus:

$$z_r = \frac{r - \mu_r}{\sigma_r} \tag{11.19}$$

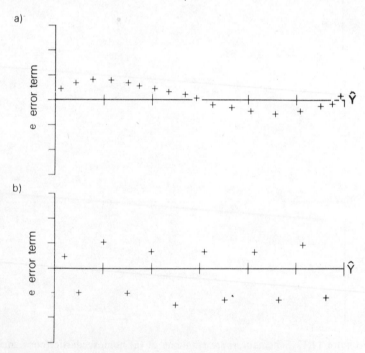

Figure 11.13 Representation of (a) positive and (b) negative autocorrelation of residuals

This standardized score may be treated exactly as any other z value, and if it falls within the predetermined rejection region then the null hypothesis of no autocorrelation may be rejected. The test may be one- or two-tailed but, rather confusingly, a negative z value denotes positive autocorrelation and vice versa. For the current example the following results are achieved.

Table 11.4 gives the residuals of shop numbers ranked in ascending order of $\hat{Y}$. From these results the mean and standard deviation of r under H_0 of no autocorrelation are as follows:

$$\mu_r = \frac{2(7 \times 8)}{7 + 8} + 1 = \frac{112}{15} + 1 = 8.47$$

$$\sigma_r = \sqrt{\frac{2 \times 7 \times 8(2 \times 7 \times 8 - 7 - 8)}{(7 + 8)^2 \times (7 + 8 - 1)}} = 1.857$$

$$z_r = \frac{7 - 8.47}{1.857} = -0.792$$

The two-tailed critical values of z (we have made no *a priori* assumptions concerning the presence of either form of autocorrelation) are, at the 0.05 level, ± 1.96. The test statistic therefore falls within the acceptance region and our null hypothesis of no autocorrelation must be accepted. We conclude that the sequence is random.

Table 11.4 Data for runs test of regression residuals

$\hat{Y}$	e	Run
59.196	173.804	1
84.276	−39.276	2
90.012	19.988	3
98.100	4.900	
98.784	−58.784	
102.384	−32.384	
104.460	−8.460	4
108.948	−14.948	
151.848	−56.848	
157.920	31.08	5
159.324	−27.324	6
160.476	−84.476	
220.956	22.044	
255.996	15.004	7
273.624	56.376	

$r = 7; n_1 = 7; n_2 = 8.$

11.7 THE DURBIN–WATSON TEST

This test uses the numerical value of the residuals and no information is lost by data reduction. The derivation of the test statistic d is relatively simple:

$$d = \frac{\text{sum of successive squared differences}}{\text{sum of squared residuals}}$$

$$= \frac{\Sigma(e_i - e_{i-1})^2}{\Sigma e_i^2} \tag{11.20}$$

In the case of positive autocorrelation the adjacent errors do not differ widely and d tends towards zero. If, on the other hand, negative autocorrelation is present then the differences between adjacent values are large and d tends towards its maximum value of 4.0. Zero autocorrelation is indicated by $d = 2.0$.

The decision to accept or reject H_0 of zero autocorrelation is not easily arrived at. The tables in Appendix X provide bounds for the critical regions in the d-distribution, but their application is not straightforward. The bounds are obtained by reference to sample size n, the number of independent variables k and the significance level a. Upper (d_u) and lower (d_l) values are given and are used to define two rejection regions for H_0. One is the rejection region due to positive autocorrelation, the other for significant negative autocorrelation. In addition, a zone of acceptance of H_0 is defined about $d = 2$, leaving two areas where the test is indeterminate. Figure 11.14 perhaps clarifies the matter. Data used earlier may be reapplied here. Normally, of course, such repetition is unnecessary. Both the absolute and the standardized residuals can be used; the latter are chosen here as being smaller and they are more manageable from a purely arithmetic point of view. From the figures in Table 11.5 the following statistic is obtained:

$$d = \frac{17.775}{11.262} = 1.578$$

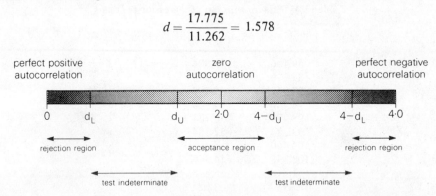

Figure 11.14 Rejection and acceptance regions for the Durbin–Watson statistic

On consulting Appendix X and using $n = 15$ and $k = 1$ (number of predictors), we find the following values at the 0.05 level: $d_u = 1.361$ and $d_l = 1.077$. Using the method of defining acceptance and rejection regions outlined in Figure 11.14, we obtain the scheme given in Table 11.6, from which we see that the test statistic falls within the 'acceptance' region. Hence H_0 of no autocorrelation is again accepted. The size of the indeterminate region, it should be noted, decreases with increasing sample size, and indeterminacy is less common with larger data sets.

11.8 SPATIAL AUTOCORRELATION

Finally we must attend to the specifically geographical question of spatial autocorrelation. In the case of non-spatially organized data it is assumed that autocorrelation of the residuals is absent. When the data are drawn from spatially continuous units, such as the American States or English counties, then a new problem arises: are the residuals spatially autocorrelated? Expressed otherwise, do the residuals tend to group together, the negatives in one area and the positives in another? If they are ordered in this manner then the assumption of residual independence has been infringed irrespective of the conclusions of the runs test or Durbin–Watson test, both of which are looking at different aspects of residual behaviour.

Table 11.5 Standardized residuals used to estimate the Durbin–Watson d statistic for the shops/population regression model

Standardized e	$(e_i - e_{i-1})^2$	e^2
2.535		6.426
	9.659	
−0.573		0.328
	0.748	
0.292		0.085
	0.049	
0.071		0.005
	0.863	
−0.858		0.736
	0.149	
−0.472		0.222
	0.121	
−0.123		0.015
	0.009	
−0.218		0.048
	0.323	
−0.829		0.687
	1.644	
0.453		0.205
	0.726	
−0.399		0.159
	0.694	
−1.232		1.518
	2.415	
0.322		0.104
	0.011	
0.219		0.048
	0.364	
0.822		0.676
	17.775	11.262

While we are certainly entitled to a purely visual judgement of the degree of spatial autocorrelation, an objective assessment is always preferable. The example offered here pays attention only to the signs of the residuals. The data are hence a nominal conversion from interval/ratio scale measurements with the consequent loss of detail. But, as Geary (1954) and Dacey (1968) have suggested, this method is less complicated than its parametric counterparts.

Table 11.6 Rejection/acceptance regions for the Durbin–Watson d statistic for $n = 15$ and $k = 1$

Range of d	Definition
0 – 1.077	Reject H_0 (positive autocorrelation)
1.077 – 1.361	Test indeterminate
1.361 – 2.639	Accept H_0 (non-significant autocorrelation)
2.639 – 2.923	test indeterminate
2.923 – 4.00	Reject H_0 (negative autocorrelation)

Maps and the manual manipulation of at least some of the data in this exercise is unavoidable, though the final numerical processing may be left to a computer. The first step is to map out the residuals using, say, black for the positive residual spatial units and white for the negative. Each of the units will be contiguous with a number of others; California, for example, is contiguous with Oregon, Nevada and Arizona. Each contact will be one of three types, either black to black (BB — positive to positive), black to white (BW — positive to negative) or white to white (WW — negative to negative). The units must be taken in turn and the number of each type of contact gradually accumulated until all units have been considered. On completion it is clear that each contact will have been counted twice and the total for the three contact types and the number of contacts (ΣL_k) all need to be divided by 2 to give:

$$X = (\Sigma BB)/2 \qquad (11.21)$$
$$Y = (\Sigma BW)/2 \qquad (11.22)$$
$$Z = (\Sigma WW)/2 \qquad (11.23)$$
$$L = (\Sigma L_k)/2 \qquad (11.24)$$

where L_k is the number of contacts made by each unit k.

It is an easy matter to estimate the expected values for X, Y and Z under an H_0 of no spatial autocorrelation (or spatial randomness) by using the binomial theorem and distribution (see Chapter 7). If the residuals are indeed random then the probability of a positive and negative residual (q_1 and q_2 respectively) must be equal and, as only these two outcomes are possible, we have:

$$q_1 = q_2 = 0.5$$

It follows that the probabilities (p) for the three twofold connections are:

$$p(BB) = q_1{}^2 = 0.25$$
$$p(BW) = 2q_1 q_2 = 0.5$$
$$p(WW) = q_2{}^2 = 0.25$$

as BB and WW can occur in only one form while BW can appear as either BW or WB, i.e. either of the two units can be positive or negative.

To find the expected frequencies (E) under H_0 we need only multiply each of these probabilities by L. Thus:

$$E(X) = q_1{}^2L \qquad (11.25)$$
$$E(Y) = 2q_1q_2L \qquad (11.26)$$
$$E(Z) = q_2{}^2L \qquad (11.27)$$

These may now be compared with the observed values, and a chi-square table constructed from which the one-way chi-square statistic can be calculated.

A simple example is provided by the regression model that describes the rainfall in the 48 contiguous American States by reference to the States' mean altitude. The equation, significant at the 0.01 level, is:

$$Y = 45.33 + 0.062X$$

where Y is annual rainfall in inches and X is the mean altitude in feet. If we wish to examine the question of spatial autocorrelation of the residuals they must first be mapped (Figure 11.15). From this map the necessary data must be abstracted manually. The counting procedures can be tedious but must be carried out with care if representative results are to be obtained. In particular, not all states have the same number of contacts (Florida has only 2 but Missouri has 8) and short boundaries are easily overlooked. From Figure 11.15 the number of BB, BW and WW contacts recorded were 60, 74 and 76 respectively, from which:

$$X = 60/2 \quad = 30$$
$$Y = 74/2 \quad = 37$$
$$Z = 76/2 \quad = 38$$
$$L = 210/2 = 105$$

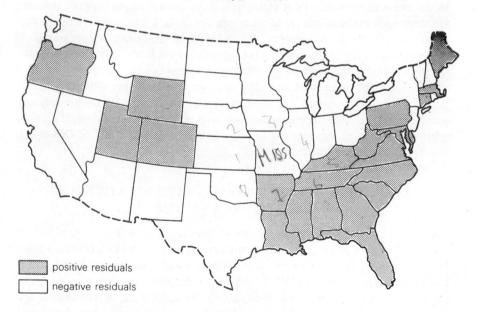

positive residuals
negative residuals

Figure 11.15 Map of residuals from the rainfall–altitude regression model for US states

Using the principles outlined above the expected values for X, Y and Z are:

$$E(X) = 0.25 \times 105 = 26.25$$
$$E(Y) = 0.50 \times 105 = 52.50$$
$$E(Z) = 0.25 \times 105 = 26.25$$

With this information the one-way chi-square test may be performed, the results of which appear in Table 11.7 and from which the H_0 of no spatial autocorrelation can be rejected.

Table 11.7 Chi-square test for spatial randomness of US States

	X	Y	Z
Observed frequencies (O)	30	37	38
Expected frequencies (E)	26.25	52.50	26.25
$(O-E)^2/E$	0.54	4.58	5.26

$\chi^2 = 10.48$; critical $\chi^2 = 5.991$ (at the 0.05 significance level with 2 degree of freedom).

We can conclude that the State residuals are arranged in a non-random, grouped pattern and are not independent of each other. This supports the purely visual impression gained from Figure 11.15, but the position will not always be so clear.

Finally, it must be stressed that the presence of non-random residuals should not be viewed as a wholly negative conclusion. The inevitable regional groupings that arise will often suggest ways in which the predictive model may be improved. In this example the wetter than anticipated south-eastern states suggests that they are more open to the influences of the warm and moist Caribbean air streams than are the interior states to the north. In this sense the residuals express variation of the dependent variable after removal of the primary predictor, in this case altitude,. Correspondingly, the presence of spatial randomness does not mean that no other variables are operating, only that they are non-spatial at the State level.

The work in this section has been treated in a relatively simple yet effective manner. Students with an unusually high level of mathematical competance are referred, however, to the extensive writings on this subject by Cliff and Ord (1981) which they may find interesting.

11.9 CLOSED NUMBER SYSTEMS IN CORRELATION AND REGRESSION ANALYSIS

Geographers make frequent use of closed number sets, such as proportions or percentages, often in an attempt to bring varying sized areas to a common scale. The term 'closed numer set' refers to the situation where all the data sum to either 1 (if they are proportions) or 100 (in the case of percentages). However, the use of percentages can raise potential problems in the application of correlation and regression techniques. Thus, Krumbein (1962) has demonstrated how correlation coefficients based on percentages can vary quite markedly from those based on open number sets for the same data. Indeed the use of percentages in regression raised two significant problems.

The first of these, which relates to the fact that the range of possible values is fixed between 0 and 100, has been explored by Wrigley (1973). He illustrates the full extent of the problem by quoting some work on the Irish economy by O'Sullivan (1968). In his study a linear regression model was developed relating the percentage immigrant population of each county (the dependent variable, Y) against accessibility of the arterial road network (the independent variable, X). Wrigley demonstrated how, using the derived regression equation, and within the specified limits of the independent variable, it was possible that for the very high values of X, counties could have a Y value of less than 0 per cent. Clearly, however, to find counties with a negative percentage of immigrants is nonsensical, but such are the problems when percentages are used in regression analysis.

Fortunately there are solutions to such problems. For example, we could state that our regression equation refers only to the range of values examined in the original data. This works in many cases, but is clearly not suitable for the example we have just mentioned. In such circumstances we can turn to a second solution involving the transformation of the percentage values into an infinite ratio scale, using a logistic or logit transformation:

$$L_j = In \frac{P_{cj}}{(100 - P_{cj})} \tag{11.28}$$

Equation 11.28

P_{cj} = percentage value at observation j
L_j = logit value for observation j

For Wrigley's example, the ratio $P_{cj}/(100 - P_{cj})$ defines the odds of being an immigrant in county j, and as the percentages increase from 0 to 100, so too do these odds increase from zero to infinity. In addition, the natural log (In) of these odds increases from minus infinity to plus infinity. This transformation can therefore modify the dependent variable to conform with the assumptions of linear regression, such that the errors are normally distributed, and Y can take any value between plus and minus infinity.

The procedure for transforming percentages using a logit equation can be illustrated by taking a simple example, relating percentages of car ownership to percentages of families with incomes over £5000 for five hypothetical counties. In this example both the dependent and independent variables are expressed in percentages (Figure 11.16(a)). It is also possible from the regression equation, $Y = -10.5 + 0.924X$, to obtain negative percentages of car ownership for low values of the independent variable measuring income. However, if we apply the logit transformation then this problem can be resolved. For example, if the percentage of households in county j is 5, then using equation 11.28 we have $5/(100 - 5) = 0.05$, and its natural log is -2.9957. By transforming all the values in this fashion we can replot the data and use this logit transformation, so that the predicted values of Y will never be less than zero (Figure 11.16(b)). Thus, our new regression equation is logit $Y = -0.743 + 1.30$ logit X, and for counties with low proportions of householders with incomes over £5000 predicted levels of car ownership are above zero.

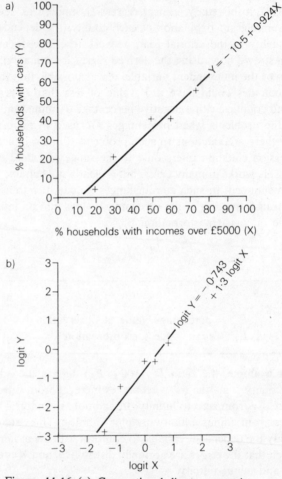

Figure 11.16 (a) Conventional linear regression model relating percentage car ownership with percentage incomes; (b) logit transformation of data in (a)

The second problem concerns the use of percentages within closed number systems, as illustrated in Table 11.8. In this example the relationships between the three variables are fixed, such that if the values of two are known then these determine the value of the third. Thus, the correlation between any two variables can be predicted without measuring the covariance (see Chapter 10). In a three variable set as in Table 11.8 the problem is an extreme one, and essentially we are analysing the same thing twice. However, the larger the number of classes the less is the extent to which correlations between any two are fixed. As Johnston (1978) has shown, it is often difficult to identify such fixed correlations, and he demonstrates the case of social area analysis. For example, once the percentages of households in social classes 1 – 4 are known, those in social class 5 are fixed for each enumeration district. Furthermore, the inclusion of all those variables introduces bias into the correlation analysis, since we may be measuring the same variable more than once. Unfortunately there are no satisfactory mathematical solutions to the problem of closed number systems in correlation and regression analysis. If possible they should be avoided; but if they have to be used then obviously considerable attention should be given to the structure of the variables.

Table 11.8 Three-variable closed number system (employment in Northern Ireland)

Employment areas	employed in (per cent)			
	Agriculture	Manufacturing	Service	Total
Craigavon	2.4	51.8	45.8	100
Downpatrick	7.3	36.7	56.0	100
Dungannon	5.5	45.0	49.5	100
Enniskillen	8.2	34.2	57.6	100
Londonderry	2.2	44.5	53.3	100
Mean percentages	5.1	42.4	52.5	100

11.10 USE OF LINEAR REGRESSION MODELS IN GEOGRAPHY

Regression analysis is one of the most widely used statistical techniques in geography. It is used in three main ways: first to establish a predictive model based on the regression equation, second to test a model or hypothesis, and third to describe relationships between variables.

A review of the literature shows that all three uses are fairly widespread, but probably most emphasis has been given to using regression in a descriptive and explanatory fashion. For example, Haggett (1964) used regression to describe the relationship between physical and economic variables and forest cover in Brazil. Similarly, Doornkamp and King (1971) used the technique to explore aspects of drainage basin morphometry. In a number of such studies geographers have compared variations in the intercept and slope coefficients, to contrast changing relationships over time and space. Thus, Johnston (1981) has used the regression coefficients to examine trends in voting patterns over time, relating the slope coefficients to the values of r^2 (the coefficient of determination), to test an ecological model of voting behaviour. Obviously, it is possible to assess statistically the differences in such coefficients, as was demonstrated by Garner (1966) who assessed the differences between regression slopes, in a study of shopping centre charactistics, using the F-test.

Such uses of regression are to be found throughout geography, a popularity that is largely based on the flexible nature of the technique. More recently, however, some geographers have taken a more critical look at regression analysis and have suggested a certain amount of misuse of the technique (Mark and Peucker, 1978). This work cast doubt on the use, in certain circumstances, of regression for describing the relationship between variables.

The problems seem to arise when the independent variable is subject to measurement error. In most uses of regression we relax this condition and assume that all the unexplained or 'residual' error is ascribed to the dependent variable (Y), and that the independent variable (X) is error-free. Mark and Peucker argue that this is not the case in many geographical studies and that, in these circumstances, the alternative technique of 'functional' or 'structural' analysis should be used. This technique, unlike regression, divides the 'unexplained' error between both the dependent and independent variables. As yet such a technique has been little used in geography and as such is outside the scope of this book. However, it is something that researchers will perhaps be giving more attention to in the future. Fortunately, these problems with regression are not important when the coefficients of determination (r^2) are high, i.e. close to 1, because the error terms of both X and Y must be relatively small.

A final and most important problem concerns the identification of dependent and independent variables in regression analysis. So far we have assumed that little difficulty exists in identifying which variable is dependent on the other. However, in some situations this so-called process–response relationship is not entirely clear, especially if we are merely using regression to explore the relationships between variables. Under such circumstances we may end up plotting two regression lines of Y on X and X on Y. However, more crucial is the fact that statistical dependence does not necessarily imply that there is a geographically valid, physical relationship between two variabales. For example, a high correlation coefficient, or a close fitting regression line, can either indicate that there is a link between two variables or that the two variables are responding and dependent on a third factor. If this is the case our analysis needs to take into account this other variable; usually this can be done through the use of multivariate techniques. Indeed, as we shall see in the following chapters, many geographical problems are of a multivariate nature and can only be fully analysed by using more complex techniques.

REFERENCES

Clark, W. A. V. (1967). 'The use of residuals from regression in geographical research', *New Zealand Geog.* **23**, 64–67.
Cliff, A. D., and Ord, J. K. (1981). *Spatial Processes: Models and Applications,* Pion, London.
Dacey, M. F. (1968). 'A review on measures of contiguity for two and k-colour maps', in B. J. L. Berry and D. F. Marble (eds) *Spatial Analysis: a Reader in Statistical Geography,* Prentice-Hall, Englewood Cliffs.

Doornkamp, J. C., and King, C. A. M. (1971). *Numerical Analysis in Geomorphology,* Arnold, London.

Draper, N. R., and Smith, H. (1966). *Applied Regression Analysis,* Wiley, New York.

Garner, B. J. (1966). *The Internal Structure of Retail Nucleations,* North Western University Studies in Geog. No. 12.

Geary, R. C. (1954). 'The contiguity ratio and statistical mapping', reprinted in B. J. L. Berry and D. F. Marble (eds) *Spatial Analysis: a Reader in Statistical Geography,* Prentice-Hall, Englewood Cliffs.

Haggett, P. (1964). 'Regional and local components in the distribution of forested areas in south east Brazil: a multivariate approach', *Geog. J.* **130,** 365–377.

Johnston, R. J. (1978). *Multivariate Statistical Analysis in Geography,* Longman, London.
(1981) 'Regional variations in British voting trends, 1966–1979: tests of an ecological model', *Regional Studies* **15,** 23–32.

Krumbein, W. C. (1962). 'Open and closed number systems in stratographic mapping', *Bull. Ann. As. Petrol Geol.* **46,** 2229–2245.

Mark, D. M., and Peucker, T. K. (1978). 'Regression analysis and geographic models, *Geogr,* Canadian **22,** 51–64.

McCarty, H. H. (1959). *'Toward a more general economic geography',* Econ. Geog. **35,** 283–289.

O'Sullivan, P. M. (1968). 'Accessibility and the spatial structure of the Irish economy', 'Reg. Studies **2,** 195–206.

Poole, M. A., and O'Farrell P. N. (1971). 'The assumptions of the linear regression model', *Trans. Inst. Brit Geogrs* **52,** 145–158.

Thomas, E. N. (1968). 'Maps of residuals from regression: their characteristics and uses in geographic research', in B. J. L. Berry and D. F. Marble (eds) *Spatial Analysis: A Reader in Statistical Geography,* Prentice-Hall, Englewood Cliffs.

Wrigley, N. (1973). 'The use of percentages in geographical research', *Area* **5,** 183–186.

RECOMMENDED READING

Blalock, H. M. (1960). *Social Statistics,* McGraw-Hill, New York. Although this book was written with sociologists primarily in mind, Chapters 17 and 18 are equally relevant to geographical problems and present a very clear picture of the philosophy and methodology of correlation analysis.

Clements, D. W. (1978). 'Utility of linear models in retail geography', *Econ. Geog.* **54,** 17–25. This paper uses correlation analysis to examine how retail sales in specific urban and rural areas of the USA are related to socioeconomic factors. The exercise is performed on both 1960 and 1970 data and the contrasting results are used to examine temporal trends in economic activity.

Ezekial, M. and Fox, K.A. (1941) *Methods of Correlation and Regression Analysis,* 2nd edn, Wiley, New York. A valuable and instructive text whose usefulness stems not only from the author's skill but also from its origins in a pre-computer age.

Harman, J. R., and Elton, W. M. (1971). 'The Laporte, Indiana, precipitation anomoly', *Anns Assc. Am. Geogrs* **61,** 468–480. In this paper the authors set out to use correlation analysis to establish the environmental factors that determine regional vegetation patterns.

Norcliffe, G. B. (1977). *Inferential Statistics for Geographers,* London. Chapters 7 and 11 contain an interesting discussion and examples of parametric and non-parametric correlation.

Penning-Rowsell, E. C., and Townshend, J. R. G. (1978). 'The influence of scale on the factors affecting stream channel slope', *Trans Inst. Br. Geogrs,* (N.S.) **3,** 395–415. The principal concern here is with the role of scale in geomorphological processes.

Nevertheless, the paper contains some very good examples of the manner in which correlation analysis can be used to unravel seemingly complex interrelationships and to identify those that may be most important.

Pinch, S. P. (1978). Patterns of local authority housing allocation in Greater London between 1966 and 1973', *Trans. Inst. Br. Geogrs* (N.S.) **3**, 35–54. The acclaimed purpose of this paper is to '. . . use correlations to gauge the extent to which rates of housing provision are allocated on a level commensurate with needs . . .' The paper also uses some multivariate methods that are not covered until a later point in this book, but the correlation studies are clearly presented and thoroughly discussed.

EXERCISES

Refer to question 1 in Chapter 10 and use the rainfall/altitude data to carry out the following tasks.

1. Plot the scatter of points on a graph, being careful to place the dependent variable on the vertical axis and the independent variable on the horizontal axis.

2. Construct the least-squares regression line and draw it through the completed graph from question 1.

3. Determine the statistical significance (at the 0.05 level) of the regression equation.

4. At the data centroid $(\bar{X}, \bar{Y})$ determine the 95 per cent confidence limits of (i) the regression line, and (ii) individual predictions of Y.

5. Repeat question 4, but for the point on the regression line where X (altitude) is 300 m.

6. Estimate the Durbin–Watson statistic for serial correlation of this regression line and establish its significance at the 0.05 level.

Chapter 12

Nonlinear Regression

12.1 INTRODUCTION

Thus far we have assumed in both correlation and regression that the relationship between two variables is linear and a scatter diagram of the data would produce an approximation to a straight line. But such linearity cannot always be found and it is important to know how to deal with nonlinearly related variables. In general the techniques used under these circumstances are more sophisticated than for the linear case, and geographers have been hesitant in employing them. Indeed Gould (1970) has gone so far as to suggest that geographers 'seem to be stuck in a linear rut'. This over-reliance on the linear model imposes an unnecessary constraint and may cause us to overlook powerful nonlinear associations. To take the extreme case, Figure 12.1 demonstrates how a linear correlation of zero disguises a notable nonlinear trend.

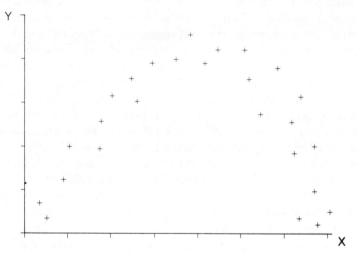

Figure 12.1 Scatter diagram of a nonlinear relationship

Nevertheless nonlinear methods are not necessarily difficult and range from simple adaptations of the procedures already discussed to highly demanding mathematical techniques. This chapter will emphasize the former and opens by

199

examining methods for fitting simple curves through scatters of points. The range of forms that can be treated in this way is surprisingly great and should encompass a sufficient variety to satisfy the needs of most geographers. The chapter concludes by examining some of the more advanced methods used when the choice of nonlinear association needs to be more extensive.

12.2 FIRST STEPS IN CURVE-FITTING

The most efficient method of detecting nonlinearity is by the simple expedient of plotting the scatter of data points for the two variables. In this way any nonlinear tendencies will become apparent. However, to assume linearity when trends are unclear is to overlook a potentially fruitful area of analysis, and it must be remembered that the character of some relationships only emerges over a wide range of values; consider, for example, the impression that would be gained if sample observations were taken only over one of the extreme ends of the scatter in Figure 12.1. If a nonlinear pattern is expected but does not appear, a more extensive sample, if possible, should be used to clarify the data.

The identification of nonlinearity marks only the start of an analysis that demands much thought on the part of the geographer. In linear regression the choice of model was restricted to the straight line, described by $Y = a + bX$, in which only the slope and the intercept term of the line could vary. In contrast there are a startling variety of curves from which to choose. This choice has important implications, and a major problem is posed by the need to select the most appropriate model for each case. This may be accomplished through either of two strategies. On the one hand it might be expected that two variables will cobehave in a specified manner. Such deductive reasoning, as it is termed, may create expectations concerning the most appropriate form of curve to describe the data. The linear model, to take the fundamental case, assumes that equal increments in the independent variable produce equal responses in the dependent variable at all points along the scale. This is not always the case. We know that population density does not decline linearly with distance from city centre. An assumption is often made that with equal increments of distance the density declines by a fixed percentage rather than a fixed absolute amount; for example, over the first kilometre density may decline by 10 per cent of its initial value, then over the next kilometre by 10 per cent of that new value, and so on with each kilometre. Thus the response of the dependent to the independent variable changes by an ever decreasing increment rather than remaining constant. One of the exponential curves would describe this well-known form of behaviour, and these are reviewed later in the chapter.

This deductive approach has the advantage of directing the researcher immediately to a specific curve or family of curves. Its disadvantage is that its efficacy depends on the soundness of the deductive reasoning underlying the choice. If the latter is flawed then the deduced curve may well be inappropriate. Hence such an approach is ill-advised unless the geographical background to the problem is well understood. The alternative is the empirical approach. This is popular in many branches of geography because it makes no such presuppositions concerning the most suitable curve and leaves us free to search over the whole range of available forms for the most suitable. This unrestricted aspect of the search for the most suitable curve presents the method's most serious problem, as it is impossible, without the enormous effort of fitting many curves to the same data, to be genuinely confident of having found it.

In practice a compromise procedure has developed in geography with attention confined, with a few exceptions, to a limited selection of curves. This is not necessarily a disadvantage as the flexibility of these curves and their ability to summarize different trends is considerable. Three popular curves can be identified:

Simple power curve	$Y = aX^b$	(12.1)
Simple exponential curve	$Y = ae^{bX}$	(12.2)
Simple logarithmic curve	$Y = a + b \log X$	(12.3)

It can be seen that all three are similar to the linear expression in possessing one predictor and one dependent variable whose behaviour is described by two coefficients (here termed a and b). Only the exponential equation differs by its inclusion of the base of natural logs (e). It is important not to confuse this 'e', which always represents the constant $2.7183 \ldots$, with the symbol e for the error term in regression models. These three curves also share the important characteristic of being 'linearizable' through the transformation of either or both of the variables. More importantly, the required transformation having been made, traditional linear least-squares regression can be used to find the values for the coefficients a and b. This process produces a least-squares straight line, but by 'detransforming' the variables the curve corresponding to the straight line can be studied.

We can take the case of the logarithmic curve to demonstrate how this linearizing principle works. The curve for $Y = 1.0 + 0.5 \log X$ is plotted in Figure 12.2(a). If, however, Y is plotted against the logs of X instead of untransformed X the curve is straightened. This has been done in Figure 12.2(b), where the original X values are included in brackets beneath their log equivalents to demonstrate how the axis is rescaled and 'distorted'. Finally we can take the term $\log X$ and regard it as a 'new' variable; let us call it V. Equation 12.3 becomes:

$$Y = a + bV \qquad (12.4)$$

which is a direct re-expression of the equation of a straight line. In this fashion, with the common log of X acting as a 'new' independent variable, the least-squares estimates for coefficients a and b can be established using the conventional methods outlined in Chapter 11.

For the power and exponential curves the linearizing procedures differ only slightly from this scheme. Power curves are straightened by log-transforming both X and Y variables, while exponential curves plot as straight lines when Y alone is logged. It must be remembered, however, that the geographer's raw material consists of scatters of points and not perfect curves. The curves fitted through these points merely generalize those scatters and trends. Thus the successful use of the curve depends on the two variables cobehaving in the specified manner (log, power or exponential). Checks on these specifications are easily made by plotting the variables, transformed as necessary on a graph. The presence of a power relationship, for instance, can be confirmed by examining the scatter diagram if logs of both variables are used (see Figure 12.4). If the resulting scatter is linear the analysis may proceed using the power model. Persistent nonlinearity suggests that another, more appropriate, model should be found.

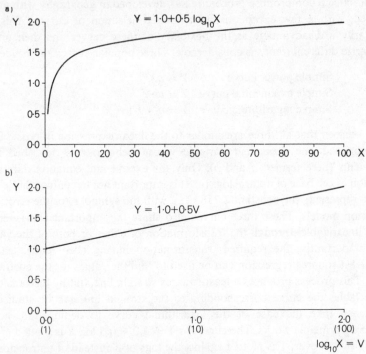

Figure 12.2 Linearization of the logarithmic curve; in the lower graph (b) logs of the independent variable are plotted (their arithmetic equivalent are given in brackets)

A final word of warning at this stage; as there is no logarithm for either zero or any negative number such observations cannot, in their 'raw' state, be treated by the above methods. One way round this problem, though it is not ideal, is to add a constant term to the offending variable to take it above zero. Thus if X ranges between 0 and -14.5 we might employ the equation:

$$Y = a + b \log (X + 15.0)$$

In this way X is incremented prior to being log-transformed and the results may be plotted and treated, provided always that the increment is not forgotten in any later analysis. It must also be added that the use of nonlinear regression does not reduce the importance of attaching the terms dependent (Y) and independent (X) to the correct variables. Thus, in the next example where river discharge and stream velocity are the variables the former was adjudged to determine the latter, and they are annotated in the equation accordingly; X being discharge and Y velocity.

12.3 THE SIMPLE POWER CURVE $Y = aX^b$

In this first worked example we shall consider the simple power expression and the parts played by the two coefficients. Figures 12.3 (a) and (b) show that the curvature of the line is determined by coefficient b, while a is the value of Y when X is 1.0 and determines the position of the curve on the graph. Both functions are slightly different from their counterparts in the linear model. We can see also how variations in the values attached to these constants give rise to a variety of curved forms. If b is a greater than 1.0 and positive, the curve is concave-upwards; if less than 1.0 it is convex-upwards. If, however, b is negative then a curve (a hyperbola in this case) is produced that is asymptotic to both axes, i.e. while it continues to approach the axes contact is never made.

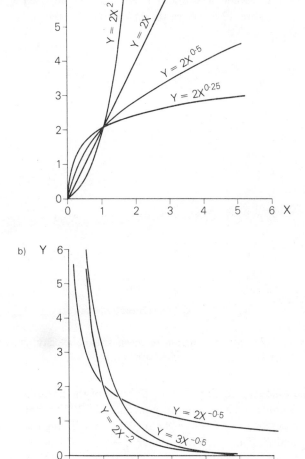

Figure 12.3 Forms of different power curves

This curve can be linearized by log-transforming both variables. Consequently a power relationship between variables may be identified by plotting the logs of the dependent variable Y against those of the independent variable X. If the resulting scatter of points is linear the power relationship is worth investigating further.

Power expressions are frequently used to summarize associations in geomorphology as, for example, in the hydraulic geometry examined by Leopold *et al.* (1964), and it is from this area that the example is drawn. The data are concerned with stream discharge and average water velocity for a river gauging site on the Shebandowan River, Ontario. It has been proposed that discharge–velocity relationships may be generalized by a simple power curve in which velocity is the dependent variable Y, controlled by discharge X. This suggestion is strongly supported by the plot of the variables in Figure 12.4, where the log–log plot also reveals a distinctly linear trend as opposed to the nonlinear scatter shown in Figure 12.5, where the raw data are used.

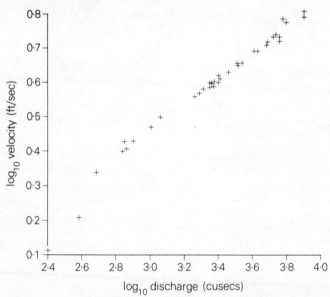

Figure 12.4 Scatter diagram of mean stream velocity against discharge

In all of the following sections the coefficients of the linearized curves will be estimated by simple adaptations of conventional least-squares methods. In Chapter 11 we demonstrated one of the methods of deriving these values for the coefficients a and b. It is, however, a characteristic of statistical methods that there may be several different ways of solving the same problem, and regression analysis is one such case. The examples that follow adopt an alternative scheme of coefficient estimation to that used in Chapter 11. The volume of pure arithmetic is a little greater, but this is offset by much simpler equations at the final stages.

Another difference is that the principal equations (equations 12.5 and 12.6 below) operate not on the raw data but on their individual differences from the means. These differences will, throughout, be denoted by lower case variable characters; thus x_i is to be interpreted as $(X_i - \bar{X})$ and y_i as $(Y_i - \bar{Y})$, although we will dispense, as usual, with the use of subscripts in the accounts.

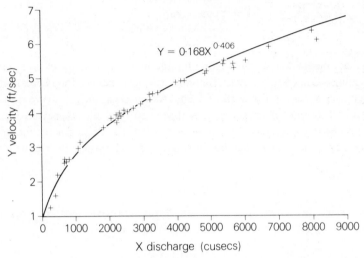

Figure 12.5 Fitted curve describing the mean stream velocity discharge-relationship

The following examples are confined to applications in nonlinear, but linearizable, cases; the method is, however, equally valid to the genuinely linear form and could have been used on the data in Chapter 11.

The sequence of mathematical operations is easy to follow and involves:

(1) estimation of variable means ($\bar{X}$ and $\bar{Y}$);
(2) subtraction of these means from the individual observations to provide the x and y values;
(3) use of these values in the two equations for least-squares values of a (the intercept term) and b (the regression coefficient).

The fundamental equations used for estimating least-squares values for coefficients a and b are those introduced in Section 11.1, whereby:

$$b = \frac{\Sigma xy}{\Sigma x^2} \tag{12.5}$$

Equation 12.5
b = regression coefficient
Σxy = sum of all xy products
Σx^2 = sum of all x^2
$x = X - \bar{X}; y = Y - \bar{Y}$

$$a = \bar{Y} - b\bar{X} \qquad (12.6)$$

Equation 12.6
$a =$ intercept term
$\bar{Y} =$ mean of variable Y
$\bar{X} =$ mean of variable X
$b =$ regression coefficient

These equations are fine as they stand for the linear model; but if both variables have been logged in order to attain that linearity, as has been done here, then there are major consequences for both of them. All observations of Y are now set to $\log_{10}Y$ and all those of X to $\log_{10}X$. Because coefficient a is expressed in units of Y, that too becomes logarithmic— let us term it simply $\log_{10}a$ for the moment. By the laws of operations with logs outlined in Section 2.3, the power equation 12.1 becomes linearized to:

$$\log_{10}Y = \log_{10}a + b \log_{10}X \qquad (12.7)$$

whose linearity becomes clearer if the terms are redesignated with $\log_{10}Y$ becoming U, $\log_{10}X$ becoming V and $\log_{10}a$ now a_1, so that:

$$U = a_1 + bV \qquad (12.8)$$

Having reached this point we can now see what the implications for equations 12.5 and 12.6 are as X's and Y's, x's and y's disappear to be replaced by V's, U's, v's and u's respectively; hence they now read as:

$$b = \frac{\Sigma uv}{\Sigma v^2} \qquad (12.9)$$

Equation 12.9
$b =$ power coefficient
$\Sigma uv =$ sum of all uv products
$\Sigma v^2 =$ sum of all v^2
$u = \bar{U} - \bar{U}; v = V - \bar{V}$

$$a_1 = \bar{U} - b\bar{V} \qquad (12.10)$$

Equation 12.10
$a =$ log of coefficient a
$U =$ mean of logs of Y
$V =$ mean of logs of X
$b =$ power coefficient

Table 12.1 is set out with part of the data set and shows how the various quantities needed to evaluate these equations can be derived. Reading across each row we have: the raw data, the logarithmic transformations of the data, the differences of each transformed observation from their means and, finally the squares and products of the latter. The final column totals can then be used in equation 12.9 and 12.10 to give the coefficients of the transformed linear model. These are:

$$b = 2.1328/5.253 = 0.406$$

$$a_1 = 0.5925 - 0.406 \times 3.369 = -0.7754$$

These estimates can now be substituted into equation 12.7 to yield:

$$\log_{10} Y = -0.7754 + 0.406 \times 3.369$$

In this form a straight line can be drawn through the points in Figure 12.4. In order to move back to the original power curve the expression must be antilogged where necessary. Hence a_1 is antilogged directly, while the estimate of b, though undergoing no numerical change, becomes the index for X and not a multiplying constant. The power expression would thus become:

$$Y = \text{antilog}_{10}(-0.7754) \ X^{0.406} = 0.168 X^{0.406}$$

Log Y and log X simply revert now to Y and X; the reverse of the procedure carried out to linearize the model (again consult Section 2.3 for the mathematical operations governing such changes).

Table 12.1 Calculations and some of the data used in evaluating the power curve relating water velocity to river discharge

Y Velocity (ft/s)	X Discharge (ft³/s)	$\log_{10}Y$ (U)	$\log_{10}X$ (V)	u ($U-\bar{U}$)	v ($V-\bar{V}$)	v^2	vu
1.64	377	0.2148	2.576	−0.3777	−0.793	0.629	0.2995
2.51	695	0.3997	2.842	−0.1928	−0.527	0.277	0.1016
1.25	249	0.0969	2.396	−0.4956	−0.973	0.947	0.4822
4.41	3,190	0.6444	3.504	0.0519	0.135	0.018	0.0070
.	.	.	.	.	.	.	.
.	.	.	.	.	.	.	.
		21.9225	124.653			5.253	2.1328

$\bar{U} = 0.5925$; $\bar{V} = 3.369$; $n = 37$.

If selected values of X are now substituted into the power equation, corresponding estimates of Y can be calculated from which the curve can be plotted and superimposed on the scatter of points of the raw data (Figure 12.5). For example, the point on the curve corresponding to a discharge of 4000 cubic feet per second (cusecs) is estimated as follows:

$$Y = 0.168 \times 4000^{0.406}$$

$$= 0.168 \times 29.003 = 4.87 \text{ ft/s}$$

As the power coefficient b is less than 1.0, the curve must be convex-upwards in accordance with the principles demonstrated in Figure 12.3(a). Similarly, when the discharge is 1.0 cusecs the estimated velocity will be 0.168 ft/s (the value of the coefficient a).

12.4 THE SIMPLE EXPONENTIAL CURVE $Y = ae^{bX}$

Despite the inclusion of a new and unfamiliar constant (e, the base of natural logarithms) the simple exponential curve is readily linearizable. Figure 12.6 illustrates the principal characteristics of this curve, in which coefficient b governs the degree of either positive or negative slope, with a now resuming its role as an intercept term and having the value of Y when X is zero. The simple negative exponential curve is asymptotic to the X axis. Thus in Figure 12.6 the downward sweep of the negative curves will never bring it into contact with the zero base line.

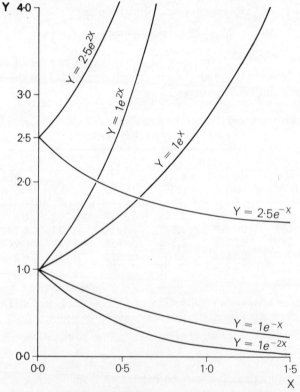

Figure 12.6 Forms of exponential curve with positive and negative exponents

To identify an exponential relationship, logs of Y should be plotted against the untransformed values of X. If a linear scatter of points results then the exponential equation will summarize the association between the variables.

Krumbein (1938) used exponential curves to describe pebble size changes along beaches, but the most popular applications have appeared in human geography. The work of Clark (1951) in seeking to describe population density decline away from urban centres offers a good example and provides the basis for the following study. The data consist of population densities for electoral wards in and around Newcastle-upon-Tyne, England. The independent variable X is distance of the ward from the centre of Newcastle.

In the process of linearizing this expression the independent variable X undergoes no numerical transformation, but the whole e^{bX} part of the equation needs to be rearranged. At the same time both Y and the intercept term (which is expressed in units of Y) need to be log-transformed to give:

$$\log_{10} Y = \log_{10} a + (b \log_{10} e)X \qquad (12.11)$$

Equation 12.11

$\log_{10} Y = $ common log of the variable Y
$\log_{10} a = $ common log of the intercept term
$b = $ exponential coefficient
$\log_{10} e = $ common log of the constant e
$X = $ untransformed variable X

This intimidating expression can be simplified by allocating new symbols to its components, thus:

$$U = a_1 + b_1 X \qquad (12.12)$$

Equation 12.12

$U = \log_{10} Y$
$a_1 = \log_{10} a$
$b_1 = b \log_{10} e$
$X = $ variable X

Once more, in this form, the equation is that of a straight line; it is mathematically identical to those reviewed earlier except that the symbols are new. We may use the conventional linear regression methods to evaluate least-squares values for a_1 and b_1. These estimates are made using adaptations of equations 12.5 and 12.6:

$$b_1 = \frac{\Sigma xu}{\Sigma x^2} \qquad (12.13)$$

Equation 12.13

$\Sigma xu = $ sum of all xu products
$\Sigma x^2 = $ sum of all x^2
$b_1 = b \log_{10} e$
$x = X - \bar{X}; u = Y - \bar{Y}$

$$a_1 = \bar{U} - b_1\bar{X} \tag{12.14}$$

> ### Equation 12.14
> $a_1 = $ log of intercept term
> $\bar{U} = $ mean of logs of Y
> $\bar{X} = $ mean of X
> $b_1 = b\log_{10}e$

Table 12.2 shows how the required quantities may be derived. Only variable Y is log-transformed to give the 'new' variable U and, as in earlier examples, the lower-case characters refer to differences between the individual observations and the mean. Using the appropriate column totals from Table 12.2 and substituting them first into equation 12.13 and then into equation 12.14, we obtain:

$$b_1 = -215.98/1501.2 = -0.1439$$
$$a_1 = 0.9318 - (-0.1439 \times 8.497) = 2.1544$$

which, by further substitution into equation 12.11, becomes:

$$\log_{10}Y = 2.1544 - 0.1439X$$

Table 12.2 Calculations and some of the data used in evaluating the exponential curve relating population density to distance from city centre (Source: *Census* 1971, County of Northumberland, HMSO)

Y Population density (people/hectare)	X distance (km)	$\log Y$ (U)	u $(U-\bar{U})$	x $(X-\bar{X})$	x^2	ux
70.26	1.90	1.847	0.915	−6.60	43.56	−6.039
71.97	3.48	1.857	0.925	−5.02	25.17	−4.810
56.98	4.71	1.756	0.824	−3.79	14.36	−3.123
39.76	2.22	1.599	0.667	−6.28	39.44	−4.189
.	.	.	.	.	.	.
.	.	.	.	.	.	.
.	.	.	.	.	.	.
	297.41	32.614			1,501.2	−215.98

$\bar{U} = 0.9318; \bar{X} = 8.497; n = 35.$

If, however, we wish to revert from the linear to the full exponential expression, both a_1 and b_1 have to be adjusted. The former may be achieved by antilogging, so that:

$$a = \text{antilog}_{10}a_1 \tag{12.15}$$

The adjustments for b_1 are more involved. Remembering that $b_1 = b\log_{10}e$, we find that $b = b_1/\log_{10}e$; so that:

$$b = b_1/0.4343 \tag{12.16}$$

For our example we therefore have:

$$a = \text{antilog}_{10}2.1544 = 142.69$$
$$b = -0.1439/0.4343 = -0.3313$$

with the full exponential equation now:

$$Y = 142.69e^{-0.3313X}$$

This expression can be used to estimate population densities at given distances from the city centre and to plot the position of the least-squares curve through the original data (Figure 12.7). We may notice that the coefficient b is negative, and therefore the population density Y must decrease with increasing distance. But because such curves are asymptotic to the X axis Y can never be negative. This property complies well with the behaviour of the variables as negative population densities are clearly impossible.

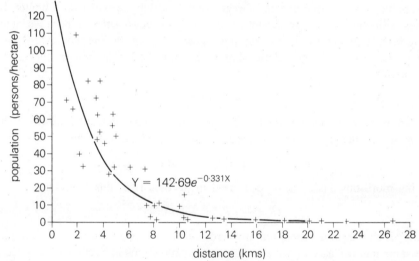

Figure 12.7 Exponential curve of population density decay around Newcastle-upon-Tyne, England

In evaluating Y from any X values it must be remembered that 'e' is always 2.7183 ... and the e^{bX} term is executed first. A sample of calculations is given in Table 12.3. Furthermore it should be recalled from Chapter 2 that any number raised to the power zero becomes 1.0. Hence the maximum population densities for this equation are encountered at the city centre where $X = 0.0$, and Y then assumes the value of a.

12.5 SIGNIFICANCE TESTING IN NONLINEAR REGRESSION

The curves obtained above are, in their linear forms, least-squares lines and as such may be tested for their statistical significance. This is as important as in the genuinely linear case, since it allows the reliability of the regression model to be assessed. Where curves are not estimated on the basis of the least-squares criterion any significance testing is inappropriate.

Table 12.3 Sequence of operations in estimating values on the exponential curve

X (km)	bX	e^{bX}	$Y(ae^{bX})$ (people/hectare)
0	0	1.0	142.69
2	−0.662	0.516	73.63
5	−1.655	0.191	27.25
10	−3.310	0.037	5.28
25	−8.275	0.0003	0.043

All significance tests proceed as if the data were linearly related, so the transformed and not the raw data must be used throughout such work. All the assumptions and requirements of regression modelling specified in Chapter 11 apply with equal force and cannot be overlooked because either or both variables have been transformed. As an example we can consider the line estimated for the water velocity–discharge relationship in Section 12.3. The power curve was provided by:

$$Y = 0.168X^{0.406}$$

and from this expression estimates of velocity ($\hat{Y}$) can be made. It is, however, the linearized model that forms the focus of our immediate attention:

$$\log_{10}\hat{Y} = -0.7754 + 0.406 \log_{10}X$$

In the more obviously linear form used in Section 12.3, this reads as:

$$\hat{U} = -0.7754 + 0.406V$$

from which the total, regression and error sums of squares can be derived in just the same manner as they were for the linear expressions in Section 11.2. Thus if we replace all Y terms with U the required equations become:

Total sum of squares (SST)	$= \Sigma(U - \bar{U})^2$	(12.17)
Regression sum of squares (SSR)	$= \Sigma(\hat{U} - \bar{U})^2$	(12.18)
Error sum of squares (SSE)	$= \Sigma(\hat{U} - U)^2$	(12.19)

The degrees of freedom associated with each of these quantities in order to estimate the corresponding variances are identical to the linear case; $n - 1$ for the total variance, 1 for the regression variance and $n - 2$ for the error or residual variance. Thus, and without repeating the procedures discussed in Section 11.2, Table 12.4 is a summary of the significance testing results. The only quantities necessary to prepare such a table are the observed values of U, the mean $\bar{U}$, the corresponding predicted values $\hat{U}$ based on equation 12.7, and the sample size n. When many observations are involved some form of computational assistance is helpful. The null hypothesis under test remains that of no explanation of the dependent variable in terms of the independent. In this example a very high F ratio (of explained to unexplained variance) of 1347.0 is obtained, and with a critical F value of only 7.41 the null hypothesis is clearly rejected and the model concluded to be valid.

Table 12.4 ANOVA table for testing the significance of the power expression describing the water velocity–discharge relationship

Source of variation	Sum of squares	Degrees of freedom	Variance
(SSR) Regression	0.8659	1	0.8659
(SSE) Error (residual)	0.0226	$37-2=35$	0.00063
(SST) Total	0.8885	$37-1=36$	

$F = 1374.0$; critical $F = 7.41$ (at the 0.01 significance level)

We may also proceed to estimate confidence bands about the line both for the line itself and for individual predictions ($\hat{U}$). Once more the methods outlined in Section 11.3 are used and applied to the linearized data. In this example both X and Y were log-transformed to give V and U respectively, and we must alter the symbols in equations 11.13 and 11.15. Thus the standard error of $\hat{U}_k$ (i.e. the standard error of the line at that point becomes:

$$SE_U = \hat{\sigma}_U \sqrt{\left[\frac{1}{n} + \frac{(V_k - \bar{V})^2}{\Sigma(V - \bar{V})^2} \right]} \tag{12.20}$$

> Equation 12.20
> SE_U = standard error of the line at point k
> $\hat{\sigma}_u$ = best estimate of the standard deviation of the residuals (SSE in Table 12.4)
> n = sample size
> V_k = selected value of V
> $\bar{V}$ = mean of transformed variable V
> $\Sigma(V - \bar{V})^2$ = sum of squared deviations of V (SST in Table 12.4)

As this equation is explained fully in Chapter 11 the detailed calculations are not shown here, but if we select an arbitrary V_k predictor value of 3.875 (in 'raw' terms equivalent to a discharge of 7500 cusecs) we find that the standard error (SE_U) of the line at that point is 0.00691. The value of the line at the same point is found from the least-squares linear equations

$$\hat{U}_k = -0.7754 + 0.406V_k = 0.7979$$

The 95 per cent confidence limits about this point are estimated by multiplying the standard error by the appropriately converted critical F value, i.e. $(2F)^{\frac{1}{2}}$, as explained in Section 11.3. With 2 and $n - 2 = 35$ degrees of freedom the critical F

value from Appendix VI is 3.28, which transforms to $(2 \times 3.28)^{\frac{1}{2}} = 2.56$. The 95 per cent confidence limits are given by:

$$0.7979 \pm 2.56 \times 0.00691$$

Thus we are 95 per cent confident that the log of the water velocity corresponding to a log discharge of 3.875 lies between 0.7802 and 0.8156. When antilogged these values become 6.028 and 6.540 ft/s respectively.

These figures represent confidence limits for estimates of the line. We may also, as we did in Section 11.3, estimate the corresponding limits for individual predictions of the log-transformed dependent variable $\hat{U}$. Equation 11.15 is used for this purpose; but because in the power equation both variables are logged, this is again acknowledged by a change of symbol with all X's replaced by V's:

$$SE_{Ui} = \hat{\sigma}_U \sqrt{\left[1 + \frac{1}{n} + \frac{(V_k - \bar{V})^2}{(V_i - \bar{V})^2} \right]} \qquad (12.21)$$

In all arithmetic respects the working of the equation is unaltered. If we again use $V_k = 3.875$ we obtain a standard error of 0.027. The same critical F value applies since we have not changed the confidence level, and we obtain 95 per cent confidence limits for individual observations of the U set at:

$$\hat{U}_k \pm 2.56 \times 0.027 = 0.7979 \pm 0.0691$$

We can, as a result, be 95 per cent confident that individual observations corresponding to a log discharge of 3.875 will vary between 0.7288 and 0.867.

These, again, are log values whose real velocity equivalents are 5.355 and 7.362 ft/s. As in the example in Chapter 11 these limits are wider than those for the line itself since they account for the variation of points about the line as well as errors in estimating the line. These limits are relatively narrow compared with those obtained in the earlier regression example based on shops in Cornwall. This difference is due to a higher correlation and reduced scatter of points about the regression line (see Figure 12.4) and to the larger sample size. It is useful to observe also that a correlation coefficient can be abstracted despite the nonlinear character of the correlationships using the ratio of regression to total sums of squares. In effect this is a ratio of explained to total variation and gives the coefficient of explanation (see Section 11.2), the square root of which is the correlation coefficient. Hence:

$$r = \sqrt{\frac{SSR}{SST}} = \sqrt{\left(\frac{\Sigma(\hat{Y} - \bar{Y})^2}{\Sigma(Y - \bar{Y})^2} \right)} \qquad (12.22)$$

In this form the calculation can be performed using the results included in any ANOVA table, irrespective of the linearity of the relationship. Remembering, however, that the dependent variable Y has here been log-transformed to create variable U, the expression is more properly written as:

$$r = \sqrt{\left(\frac{\Sigma(\hat{U} - \bar{U})^2}{\Sigma(U - \bar{U})^2} \right)} \qquad (12.23)$$

Using the results in Table 12.4 we now have:

$$r = \sqrt{\frac{0.8659}{0.8885}} = 0.9872$$

Because square roots can be either negative or positive, the sign of the correlation coefficient is taken from the regression model which, in this case, is positive.

If the raw arithmetic values of the dependent variable had been used to estimate the correlation coefficient in the standard manner outlined in Chapter 11, a lower figure would have been obtained. This amplifies a point made in the opening paragraph of this chapter where it was suggested that direct applications of linear analyses to nonlinear cases can underestimate or even totally obscure strongly correlated behaviour.

12.6 CURVE FITTING TO PHYSICAL FEATURES

A popular application of curve fitting has been in the realm of landform description. Hillslopes, river profiles and valley cross-sections have all been examined in this manner. The literature on this topic is particularly extensive for river long profiles. Because of this popularity these applications, while not falling exactly into the category of hypothesis testing, deserve brief examination. The problems of equation selection and procedural arithmetic are exactly as described above. The most important difference is that the continuum that constitutes the physical features, the hillslopes and river profiles for example, are reduced to a finite set of X and Y coordinate values; the precise number depending on the feature and the required degree of accuracy. The data may be obtained either from field surveys or accurate maps. The derived coordinates are then treated as conventional X and Y observations. Arbitrary origins must also be assigned, for elevation (Y) sea level often provides a ready zero datum, but horizontal distances (X) require some other, more arbitrary, zero point. River profile coordinates, for example, are usually defined by elevations above sea level and by distance downstream from the source. Because the curve is here used as a purely descriptive and not an inferential device, the usual interpretation of the X and Y 'variables' no longer holds, and no dependence or independence between them can necessarily be inferred.

With landforms reduced to a set of coordinate values different curves can be evaluated; but, as with all regression procedures, the coefficient values are dependent on units of measurement. More importantly, the problem of statistical significance is fraught with hazards owing to the questionable nature of the raw data which are unlikely to fulfil the major requirement of independence. Because of this problem derived curves are best used in a purely descriptive sense, the curves acting as mathematical approximations to real forms and not as regression curves. Statistics such as the correlation coefficient are consequently measures of goodness of description and no more. Significance testing is ill-advised under these circumstances and should not be attempted. A straightforward example will amplify these points and also allow us to use the logarithmic curve.

The long profile of the River Medway is a fine example of the type of feature to which curves have been fitted (Figure 12.8). Many workers have suggested that logarithmic curves may be used to approximate river long profile forms (Green, 1936; Brown, 1952). With the methods already outlined such a curve can be fitted. In this example the data consist of 21 observations of altitude above sea level (in hundreds of feet) and distance from the river source (in miles). The logarithmic curve is linearized by log-transforming the independent variable (X distance). If the new variable is termed V then the equations for the two coefficients are:

$$b = \frac{\Sigma vy}{\Sigma v^2} \qquad (12.24)$$

> Equation 12.24
>
> b = logarithmic coefficient
> Σvy = sum of all vy produces
> Σv^2 = sum of all v^2
> $v = V - \bar{V}; y = Y - \bar{Y}$

$$a = b\bar{V} - \bar{Y} \qquad (12.25)$$

> Equation 12.25
>
> a = intercept term
> b = logarithmic coefficient
> $\bar{V}$ = mean of transformed variable V
> $\bar{Y}$ = mean of variable Y

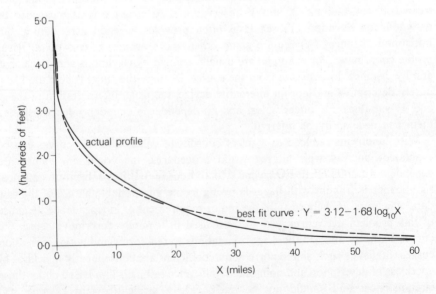

Figure 12.8 Logarithmic curve fitted to long profile of River Medway

No transformations are required of either estimate before it's subsition into the logarithmic expression:

$$Y = a + b \log_{10}X$$

Some of the raw data and their treatment are included in Table 12.5. From the appropriate column totals the estimates for a and b become:

$$b = -25.763/15.301 = -1.684$$
$$a = 2.293 - 0.4887 \times (-1.684) = 3.116$$

giving the logarithmic curve

$$Y = 3.116 - 1.684 \log_{10}X$$

The approximation of this curve to the real profile is shown in Figure 12.8.

Table 12.5 Calculations and some of the data used in fitting a logarithmic curve to the long profile of the River Medway

Y Elevation (00's ft OD)	X Distance (miles)	$\log_{10}X$ (V)	y ($Y-\bar{Y}$)	v ($V-\bar{V}$)	vy	v^2
4.75	0.04	−1.398	2.475	−1.887	−4.670	3.561
4.50	0.13	−0.886	2.207	−1.375	−3.034	1.891
4.25	0.28	−0.553	1.975	−1.042	−2.039	1.086
4.00	0.36	−0.444	1.707	−0.933	−1.592	0.870
.	.	.	.	.	.	.
.	.	.	.	.	.	.
48.15		10.263			−25.763	15.301

$\bar{Y} = 2.293$; $\bar{V} = 0.4887$; $n = 21$.

The residuals along the fitted curve highlight one of the greatest difficulties with this form of curve-fitting, and that is the very high degree of autocorrelation with long sequences of positive and negative residuals. The Durbin–Watson statistic is as low as 0.535, yet the coefficient of explanation is deceptively high ($r^2 = 0.87$). Such an excessive degree of residual dependence reflects the sequential character of the data which consist of observations taken along a physical continuum.

12.7 OTHER FORMS OF LINEARIZABLE CURVES

The logarithmic, power and exponential curves cover a wide range of forms but are by no means the only simple linearizable expressions available. For example, the curve generated by:

$$Y = a + \frac{b}{X} \tag{12.26}$$

is straightened by replacing X by its reciprocal. Linear regression methods for estimating a and b are then employed using reciprocals rather than raw values of

X. By the same token the curve of:

$$Y = a + bX^2 \tag{12.27}$$

is linearized by replacing X values by their squares. The list of such curves could be greatly extended; all are relatively simple and can be adapted for use by the methods outlined above. By this stage such forms should be sufficiently familiar to the reader to render a long list of possibilities unnecessary. The basic points to look out for are:

(1) two coefficients only
(2) one dependent variable (Y)
(3) one independent variable (X) which is either logged or raised to some power.

If these conditions are met then linearization and adaptation of linear regression methods may proceed. Both Hoerl (1954) and Daniel and Wood (1968) discuss these points and illustrate them further.

Nevertheless there are occasions, far from uncommon in geography, when we can use curves described by equations in which the predictor variable appears more than just once, as has been the case thus far. When this is the case the methods discussed above cannot be used. As an example we can take the 'quadratic' equation:

$$Y = a + bX + cX^2 \tag{12.28}$$

This equation generates the curve shown in Figure 12.9 which is, in fact, a parabola. The general term for this form of expression is a *polynomial*. We should note that X appears twice, under different indices, and each time with its own coefficient; b for X and c for X^2. It is an easy matter to continue adding terms to polynomials. The next in the sequence would be:

$$Y = a + bX + cX^2 + dX^3 \tag{12.29}$$

This is a cubic curve. We should note that powers of X are not repeated but continue to increase. As the power terms increase they impart to the resulting curve a greater flexibility (see Figure 12.9). Based on the order of the highest power such polynomials are known by their 'degree'. Equation 12.28 is a second-degree curve and 12.29 a third-degree curve. The higher the degree the greater the number of turning points or bends that the curve possesses. Obviously higher order curves will be able to follow slight changes in the trends of the data more closely. Only rarely, however, would we resort to very high degree curves, and despite the wide availability of computer programs to help us with this it is always a good idea to aim for an economy of terms; this certainly aids us at the important stage of analysis. Added to this is the very practical consideration that the improvements in approximation beyond fourth- or fifth-order polynomials are rarely sufficient reward for the effort involved in obtaining them.

What is most interesting about all polynomial curves of the type under discussion is that they are linearizable in each of the X terms. If the 'raw' X values is replaced by the same observations but now raised to the appropriate power, all the terms become linear. While there remains only one predictor variable the position has now become, from an arithmetic viewpoint, one of multiple regression with several predictors, and the various forms of multiple regression programs can be used to estimate least-squares values for the coefficients. In the case of equation 12.29, for example,the three 'predictor' variables fed into the data store would consist of the unaltered values of X, for which the coefficient b would be estimated, the values for X^2 for which coefficient c would be found, and the X^3 values would constitute the third predictor and coefficient d would describe its behaviour with regard to Y. If the X^2 terms are called Z and the X^3 terms W, the linearized model would now read:

$$Y = a + bX + cZ + dW$$

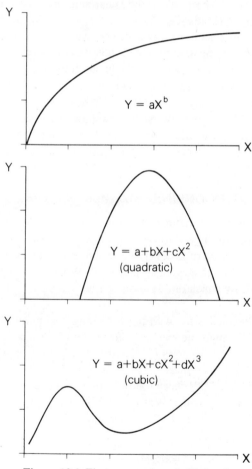

Figure 12.9 First-, second- and third-order polynominal curves

Furthermore, after estimation by this method none of the coefficients requires any further adjustment and can be inserted directly into the original form in equation 12.29. The companion volume referred to in the Preface covers the computational issues in more detail.

We are, however, prematurely moving into the area of multiple regression which belongs properly in the next chapter. We do not intend to talk at length on the arithmetic of least-squares estimates which can be very demanding indeed when several variables are involved. Most geographers having to deal with this type of problem will in any case be able to call on the large number of package programs available for either polynomial or multiple regressions.

All of the linearizing schemes outlined above are valuable because they allow the methods of linear regression, both simple and multiple, to be applied to the transformed data without the need to use complicated nonlinear methods. The most important task facing the researcher is to select the most suitable curve and identify the necessary transformations for linearizing the curve. Some programs, such as those in SPSS (Statistical Packages for the Social Sciences), have the additional advantage of permitting any specified transformation of the variable or variables to occur within the analysis. Such possibilities save a good deal of time in preprocessing data for otherwise purely linear programs.

Such methods are very good up to a point. Clearly, however, not all curves are linearizable, many certainly are not and there will be occasions when we may wish to test their utility on our data. We shall now look, necessarily briefly, at such nonlinearizable curves, how they can be evaluated and why we might need recourse to them.

12.8 NONLINEARIZABLE CURVES

The range and variety of linearizable curves is, as we have seen, very wide indeed, and if nonlinearizable forms are to be considered the choice of curves becomes bewildering. We must, then, of necessity be selective and the examples that follow have all appeared in the geographical literature. What is of critical importance here are the methods by which these curves can be examined and their parameters estimated.

The examples all explain why, with the wide range of curves already reviewed, more complex models might be sought. Clark's model of exponential population density decline around city centres implies that the highest populations must be found at the centres (see Section 12.4). For many western cities we know that this is not the case and that density decay follows a more complex pattern of change. In an attempt to model these changes Warnes (1975) used developments of the simple exponential curve in equation 12.2. In ascending order of development and flexibility these equations are:

$$Y = ae^{-cX^2} \tag{12.30}$$

$$Y = ae^{bX - cX^2} \tag{12.31}$$

$$Y = ae^{-bX + cX^2} \tag{12.32}$$

$$Y = ae^{-bX + cX^2 - dX^2} \tag{12.33}$$

The different characters of the curves resulting from these equations are shown in Figure 12.10, in which we see how different patterns of population density decline can be accommodated.

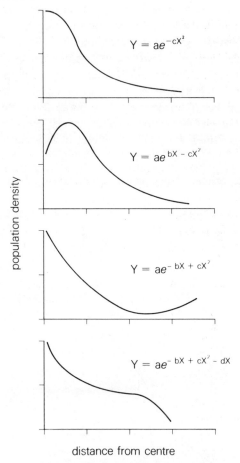

population density

distance from centre

Figure 12.10 Forms of high-order exponential curve (Warnes, 1975)

For similar reasons we find that Eyles (1969) was not content to use a simple logarithmic curve to describe the profiles of his Malayan study area. He preferred more flexible curves generated by expressions such as:

$$\log_{10} Y = e^{b + cX + dX^2} \qquad (12.34)$$

Another group of curves that generally cannot be linearized are the 'growth' curves. Two principal forms can be identified:

Gompertz curve $\quad \log_{10} Y = a - bc^X \qquad (12.35)$

Logistic curve $\qquad Y = \dfrac{1}{a + bc^X} \qquad (12.36)$

In both cases the coefficients *a, b* and *c* are positive, but *c* is always less than 1.0. The curves share the characteristic of being in the form of an attenuated *S* when plotted (see Figure 12.11), but the lower and upper limits of the curve are constrained between parallel bounds. This upper limit is not a feature encountered in the curves so far examined and makes both forms suitable for summarizing the behaviour of a variable that can operate only within fixed limits. Such behaviour is frequently seen in diffusion and growth studies. In the case of the diffusion of an idea, a philosophy or even an invention such as the radio or television the lower limit must be zero from which growth occurs to an upper limit provided by the population level or number of households. As soon as all people within the study area have adopted the philosophy or artefact growth must clearly finish. But between these two extremes it is often possible to distinguish three phases of adoption:

(1) a slow 'take-up' or innovatory phase as the philosophy or artefact becomes accepted;
(2) a phase of rapid adoption;
(3) as the upper limit is approached, a falling away of the adoption rate.

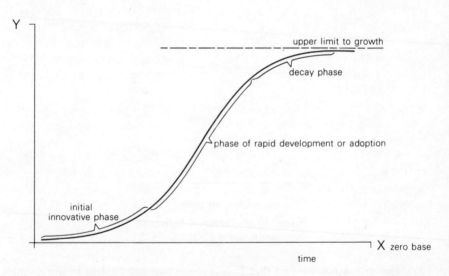

Figure 12.11 The logistic curve

All three phases can be distinguished in the changing slope of the growth curves in which *Y* is the adoption level and *X* is time (in months, days or even years). The precise form of the curve is determined by the values of coefficients *a, b* and *c*. These potentially useful curves have not been widely used in geography, perhaps because of their intractable nature. Berry (1971) mentions them in connection with adoption trends for television in the USA, but the best discussion of growth curves is found in Hagerstrand (1952). In this study the author examines the diffusion of cars and radios following their introduction into southern Sweden.

12.9 ESTIMATING NONLINEARIZABLE CURVES

The foregoing sections have indicated the range of curves that may be used to approximate scatters of points, but whether we are using simple or sophisticated equations the principal task is always to estimate their least-squares parameters. In all but the simplest of linearizable forms estimation is greatly assisted by computers, and most present-day literature is written with such assistance in mind. Nevertheless some texts, in particular that by Ezekial and Fox (1959), written before the widespread adoption of computers are very useful because of their clear application of purely manual methods of calculation. The strictly computational aspects of regression are dealt with in the companion volume of this series and the mathematical techniques for fitting nonlinearizable curves go far beyond the scope of this text. As a result much of what follows is expressed in rather general terms; but it will yield enough information to allow advanced programs to be considered for their possible advantages and limitations.

Least-squares estimates for non-linearizable curves can be gained by several means. Depending on both the type of curve and the data (how much, how many variables etc.) these methods meet with differing degrees of success. High-order equations with large multivariate data sets provide the worst problems, and the researcher should be alive to the fact that, even allowing for the notable achievements of modern computing, there will be some curves for which solutions cannot be found or situations where different programs give different solutions to the same problem. The relevant documentation for each program must always be studied with care before such work is carried out. Just one example should be sufficient to act as a warning.

A popular method of dealing with high-order nonlinearizable curves is to use one of the many programs that employ an 'iterative' approach to minimum sums of squares (see figure 12.12). However, these programs, for obvious reasons, will adjust coefficient estimates only so that they move 'downslope' towards regions of lower residual sums of squares. If the initial estimates are poor then the program can be started in a local least-squares hollow from which it cannot escape, leading to incorrect solutions which are perhaps widely different from those obtained at the point where the 'global' minimum occurs.

The accuracy of first estimates is clearly paramount, and one method of obtaining good values is to use one of the techniques of curve approximation discussed in texts such as Lipka (1939) or Hoerl (1954). Coefficient estimates derived by these 'quick' methods are not least-squares values, but they may be close enough to them to use as initial estimates. Unfortunately guidance is offered for only a limited range of curves and the researcher is frequently in the position of having to derive his own first estimates without assistance.

Another problem that can arise here is the inappropriateness of significance testing methods to many of the nonlinearizable curves. These curves frequently infringe the requirements of significance tests as set out in Chapter 11. Thus, while F ratios of regression to residual variance may be calculated they possess little other than descriptive force. But they remain valuable when comparing the usefulness of different curves applied to the same data when they can be used as measures of goodness of fit. Neither is it always very easy to determine the contribution towards explained sums of squares provided by each term in the equation, although some programs will list t statistics associated with each coefficient.

Some workers might justifiably feel, under these strictures, inclined to abandon the least-squares criterion altogether; indeed its usage results from the fact that it permits significance testing, an advantage enjoyed by no other criterion. If significance testing is abandoned then the technique becomes wholly descriptive in application and can be replaced by arithmetically less demanding forms. In this respect the curve approximations used by Lipka (1934) and Hoerl (1954), which are simpler to use and far quicker, might be confidently adopted. Further useful methods of curve approximation can be found in Mosteller and Tukey (1977) and Parsons (1974). The latter has a particularly useful method for approximating the Gompertz curve which was discussed in Section 12.8.

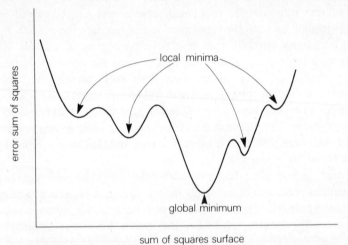

Figure 12.12 Global and local least-squares estimates on the sum of squares 'surface'

12.10 CONCLUSIONS

The linear model has many virtues: it is widely applicable and possesses statistical properties that permit the accuracy and reliability of correlation and regression analyses to be objectively assessed. Unfortunately not all pairs of variables co-behave in a linear manner, so we need nonlinear regression. But no sooner do we enter the realm of nonlinearity than all manner of hitherto nonexistent difficulties emerge. The first of these arises because, although a straight line always has the same 'shape', there are very many types of curve. Which do we use? And how do we arrive at the choice? Deductive and empirical approaches are both possible, but in their rigid application they offer serious constraints that are most effectively overcome only by compromise. Clearly it is impractical to examine a large number of expressions but, on the other hand, we may not wish to reject a deductively unacceptable curve if it offers an accurate description of events within either the observed or possible range of values.

Even if these difficulties are overcome the researcher is still confronted with the problem of estimating the least-squares coefficients. Much depends on the nature of the curve. Evaluation proceeds with relative ease if the expression is both simple and can be linearized. Otherwise the position is less straightforward and fraught with hazards. Nevertheless, a good nonlinear approximation to a scatter of points, or indeed to a landform, is an invaluable aid to geographers, helping to isolate the strengths and tendencies of relationships and forms. To describe such methods as wholly objective may be to overstate the case since different workers may, unless the relationships are very clear, prefer to use different curves for the same data. This difficulty is not helped by the invalidity of the significance testing techniques, though direct comparisons of residual sums of squares will indicate the best approximation for any choice of curves.

Nonlinear relationships are far from uncommon in geography and it is important not to sidestep the obvious difficulties of work in this area. Indeed, these very obstacles may work to our advantage for, as Mather and Openshaw (1974) have observed, nonlinear analysis means that 'one has to think before computing (which is often the reverse of normal practice!)'

REFERENCES

Berry, B. J. L. (1971). 'The geography of the United States in the year 2000', *Trans. Inst. Brit. Geogrs* **51**, 21–54.

Brown, E. H. (1952). 'The River Ystwyth, Cardiganshire: a geomorphological analysis', *Proc. Geol. Assoc.* **63**, 244–269.

Clark, C. (1951). 'Urban population densities', *J. R. Statist. Soc.* **114, A** 490–496.

Daniel, C., and Wood, F. S. (1968). *Fitting Equations to Data*, Wiley, New York.

Draper, N. R., and Smith H. (1966). *Applied Regression Analysis*, Wiley, New York.

Eyles, R. J. (1969). 'Depth of dissection of the West Malaysian landscape', *J. Trop. Geog.* **28**, 23–31.

Ezekial, M., and Fox, K. A. (1959). *Methods of Correlation and Regression Analysis*, Wiley, New York.

Gould, P. (1970). 'Is *statistix inferens* the geographical name for a wild goose?', *Econ. Geog.* **46**, 439–448.

Green, J. F. N. (1936). 'The terraces of southernmost England', *Proc. Geol. Assoc.* **92**, lviii–lxxxviii.

Hagerstrand, T. (1952). 'The propogation of innovation waves', *Lund Studies in Geography Series B* **IV**, Royal Univ. of Lund, Sweden.

Hoerl, A. E. (1954). 'Fitting curves to data', in *Chemical Business Handbook* J. H. Parry (ed.), McGraw-Hill, New York.

Krumbein, W. C. (1938). 'Sediments and exponential curves', *J. Geol.* **48**, 577–601.

Leopold, L. B., Wolman, M. G., and Miller, J. P. (1964). *Fluvial Processes in Geomorphology*, Freeman, New York.

Lipka, J. (1934). *Graphical and Mechanical Computation*, Wiley, New York.

Mather, P. M., and Openshaw, S. (1974). 'Multivariate methods and geographical data'. *The Statistician* **23**, 283–308.

Mosteller, F., and Tukey, J. W. (1977). *Data Analysis and Regression*, Addison-Wesley, Reading, Mass.

226

Parsons, R. (1974). *Statistical Analysis: a Decision-Making Approach,* Harper and Row, New York.

Warnes, A. (1975). 'Commuting towards city centres: a study of population and employmnent density gradients in Liverpool and Manchester, *Trans. Inst. Brit. Geogrs* **64**, 77–96.

RECOMMENDED READING

Blalock, H. M. (1960). *Social Statistics,* McGraw-Hill, New York. Chapter 18 of Blalock's lucid book includes a discussion of nonlinear methods and the logarithmic curve.

Daniel, C., and Wood, F. S. (1968). *Fitting Equations to Data,* Wiley, New York. Although this text is an advanced one there is a great deal that can be abstracted from it by less numerate geographers. There is also an iterative program available based on the material in the book.

Draper, N. R., and Smith, H. (1966). *Applied Regression Analysis,* Wiley, New York. Another text aimed principally at the mathematics student but containing also some instructive passages on the problems of nonlinear regression.

Ezekial, M., and Fox, K. A. (1959). *Methods of Correlation and Regression Analysis,* Wiley, New York. An excellent text covering both linear and nonlinear regression but written at a time before the advent of the modern computer. The approach adopted by the authors is refreshingly simple and informative.

Lipka, J. (1934). *Graphical and Mechanical Computation,* Wiley, New York. Chapter 6 of this book looks at curves and their equations in general and is informative in that respect. It also shows how rapid, if non-least-squares, coefficient estimates can be made.

Mosteller, F., and Tukey, J. W. (1977). *Data Analysis and Regression,* Addison-Wesley, Reading, Mass. As the title suggests a very specialized text but one without too much burdensome mathematical detail. Chapters 4, 5 and 6 discuss what the authors describe as 'straightening curves', i.e. linearizing curved trends.

Parsons, R. (1974). *'Statistical Analysis: a Decision-Making Approach',* Harper and Row, New York. Although aimed at the commercial student this is a useful text from many points of view. Chapter 25 gives a notably lucid discussion of logarithmic, polynomial and growth curves with good worked examples. Unfortunately the discussion is largely confined to fitting trends through time and not with respect to another geographical variable.

EXERCISES

1. The data set below consists of mean annual discharges and stream widths for rivers in the Yellowstone River Basin, Wyoming and Montana. Plot these observations on both arithmetic and log–log graphs. Calculate the least-squares simple power curve and estimate sufficient points on that curve to plot it through the arithmetic scatter of observations.

Mean annual discharge (ft^3/s)	Mean stream width (ft)
2,969	242
177	68
828	134
1,315	155
1,106	159
683	123
1,908	218

Mean annual discharge (ft^3/s)	Mean stream width (ft)
16	17
45	26
2,383	175
3,676	264

2. Using the curve generated in question 1, assess its statistical significance. For the discharge value of 1,143.07 cusecs estimate (i) the 95 per cent confidence limits of the regression line, and (ii) the 95 per cent confidence limits for individual predictions of width.

3. Use the following data set to estimate the least-squares simple power curve describing their association. Plot the scatter of points on both arithmetic and log–log graphs. The data consist of discharge observations (independent variable) and suspended sediment measurements (dependent variable) for the Powder River, Wyoming. (Source: Leopold and Maddock, 1953)

Discharge (ft^3/s)	Suspended sediment (tons/day)
5.9	3.8
11.2	87.0
13.7	126.0
104.0	1,750.0
160.0	1,650.0
520.0	11,400.0
125.0	1,800.0
430.0	10,600.0
575.0	28,000.0
910.0	51,000.0
320.0	9,300.0
195.0	4,500.0
375.0	6,400.0
560.0	18,300.0
665.0	28,200.0
780.0	34,300.0
795.0	45,500.0

4. Plot the following data on an arithmetic graph and then estimate the least-squares simple exponential curve that describes the pattern of points. Use selected values of the independent variable (X) to calculate enough points on the curve to enable you to plot it through the scatter of observations on the graph. The data consist of population densities in randomly selected electoral wards in and around the City of Kingston-upon-Hull, England, and the distances of those wards from the city centre. (Data source: *Census* 1961, HMSO)

Population density (people/hectare)	Distance from centre (kilometres)
49.51	2.5
32.29	2.9
51.77	2.3
93.82	1.6

Population density (people/hectare)	Distance from centre (kilometres)
25.55	5.4
67.47	1.8
59.99	1.7
44.42	1.1
57.17	4.3
19.31	4.8
33.17	4.0
15.73	4.3
13.76	7.3
10.84	6.3
2.81	7.9
0.54	13.4
0.14	11.7
0.32	15.4
1.58	14.8
0.33	19.3

5. Assess the statistical significance of the fitted curve in question 3 (but remember that this can only be performed on the linearized model). From the ANOVA table used to complete the first part of this question estimate also the 'nonlinear' correlation between the variables and its sign.

6. From the following data fit a simple logarithmic curve to the long profile of the River Avon, Dorset, and plot the profile and the fitted curve (both as continuous lines).

Height (m AOD)	Distance from source (km)
159	0.0
151	0.22
143	0.46
136	1.53
127	2.32
121	3.82
114	5.16
106	7.67
98	9.87
91	15.01
83	22.33
76	26.45
68	32.61
61	40.02
53	46.09
45	52.55
38	61.57
30	71.07
23	77.81
15	85.26
8	95.07
0	105.84

Chapter 13

Multiple Regression and Correlation

13.1 MEASURING MULTIVARIATE RELATIONSHIPS

Many of the problems studied by geographers are of a complex nature, often involving a consideration of a number of interacting variables. In such circumstances bivariate statistics are rather inadequate tools of analysis and multivariate techniques need to be used. As with other statistics the selection of a particular method depends on the type of data, the nature of the problem and the objectives of the research. Thus, some multivariate techniques are concerned with structural simplification, and summarizing a large number of variables or observations by a few parameters; others can often prove to be the key to exploring relationships, leading to the subsequent generation and testing of hypotheses.

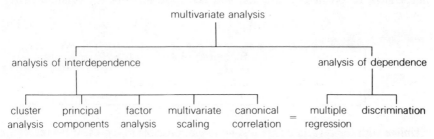

Figure 13.1 A typology of multivariate statistical techniques

As Figure 13.1 shows there is a wide range of multivariate techniques available, and in geography these are used in a variety of ways. First, we can distinguish those statistics concerned with the analysis of dependence, in which one or more variables are singled out and examined in terms of their dependence on others. For example, in multiple regression an attempt is made to explain the variation in one dependent variable via variations in the explanatory or independent variables. Similarly, canonical correlation techniques extend such analysis to examine interdependence between two main groups of variables, predictor and criterion variables. As such it provides something of a link to a second group of techniques which are concerned with the analysis of interdependence that exists between variables; this may range from being completely independent, to collinearity, with one variable being either a linear or nonlinear function of the others. Thus, we may use principal component analysis or factor analysis to transform our original group of variables to a new set which is uncorrelated, or to define groups of variables which are highly correlated together. Finally, we may use multivariate techniques as classification tools, where concern is with the grouping together of similar objects or variables. Techniques for classifying groups or individuals are termed cluster analysis, while discriminant analysis provides a mean of allocating individuals to the correct population of origin.

A great number of the multivariate techniques are based on the assumption that the data approximate to a multinormal distribution, and unlike with univariate statistics, other types of frequency distributions (e.g. Poisson) are not applicable. It is fortunate, however, that most multivariate statistics are fairly robust under conditions of departure from normality (Chatfield and Collins, 1980). Furthermore, it should be pointed out that very few of these statistics can deal satisfactorily with non-parametric types of data. The ability to use any of these multivariate methods is also conditional to having access to a main computer or mini- or microcomputer; and in many cases the necessary package programs.

13.2 MULTIPLE LINEAR REGRESSION

The multiple regression model is an extension of simple linear regression that was examined in Chapter 11. Multiple regression, however, attempts to predict and explain the variation of a single dependent variable Y, from a number of independent or predictor variables; and takes the form of the equation 13.1:

$$Y = a + b_1X_1 + b_2X_2 + \ldots b_iX_i \pm e \qquad (13.1)$$

Equation 13.1
$a =$ intercept value
b_1 to $b_i =$ partial regression coefficients
$e =$ error term

Unlike with the bivariate case it is no longer possible to represent the regression model in a graphical form, except where we have only two independent variables. Nevertheless, the concept of least-squares is still mathematically valid and used to obtain a 'best fit' surface in multidimensional space. This can be illustrated for the two-variable case in Figure 13.2, which represents a plane in three-dimensional space. The differences between the observed and expected values of Y are the error terms e, as was the case with bivariate regression. Least-squares estimates can also be derived, as shown in Table 13.1. This simple example also serves to demonstrate a further point, and that is the tedious nature of the required calculations. Indeed, much of the work using multiple regression techniques is only viable with the aid of computers. Therefore, in the remainder of this chapter reference will be made to commonly available package programs, namely SPSS and Minitab.

In multiple regression the slope coefficients b are known as partial regression coefficients, since they measure values that are obtained by controlling for each of the independent variables. This can be most clearly seen in the case of just two independent variables, where the multiple regression equation takes the correct form of:

$$Y = a + b_{01.2}X_1 + b_{02.1}X_2 \pm e$$

In this instance the subscript 0 represents the dependent variable, and $b_{01.2}$ is the partial regression coefficient representing the slope of the regression line of Y on X_1, holding X_2 at a constant value. Similarly, the coefficient $b_{02.1}$ is the slope value of the regression Y on X_2, controlling this time for X_1. Therefore each coefficient represents the amount of change in the dependent variable that can be associated with a variation in one of the independent variables.

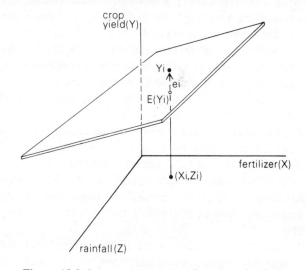

Figure 13.2 Least-squares regression plane in three-dimensional space

Table 13.1 Least-squares and multiple regression (calculation of a, b_1 and b_2 in equation 13.1).

$$b_1 = \frac{\Sigma(Y - \bar{Y})(X_1 - \bar{X}_1)\,\Sigma(X_2 - \bar{X}_2)^2}{\Sigma(X_1 - \bar{X}_1)^2(X_2 - \bar{X}_2)^2}$$

$$b_2 = \frac{\Sigma(Y - \bar{Y})(X_2 - \bar{X}_2)\,\Sigma(X_1 - \bar{X}_1)^2}{\Sigma(X_1 - \bar{X}_1)^2(X_2 - \bar{X}_2)^2}$$

$$\frac{-\,\Sigma(Y - \bar{Y})(X_1 - \bar{X}_1)\Sigma(X_1 - \bar{X}_1)(X_2 - \bar{X}_2)}{-\,\Sigma[(X_1 - \bar{X}_1)(X_2 - \bar{X}_2)]^2}$$

$$a = \bar{Y} - b_1\bar{X}_1 - b_2\bar{X}_2$$

It should be noted, however, that these partial regression coefficients are expressed in the particular units used to measure each of the independent variables (thus in our example, Figure 13.2, centimetres and lbs/acre); and as such cannot be directly compared with one another. To facilitate such comparisons the b coefficients can be transformed into beta weights (B). These are derived by applying equation 13.2, where s_{xi} is the standard deviation of the independent variable under consideration, s_y is the standard deviation of the dependent variable, and b_i is the unstandardized b coefficient of variable X_i:

$$B_i = b_i(s_{xi}/s_y) \qquad (13.2)$$

Beta weights therefore provide a means of measuring the relative changes in variables on a standard scale and are sometimes referred to as standardized partial regression coefficients. They indicate how much change in the dependent variable is produced by a now standardized change in one of the independent variables, when the others are controlled.

The use and derivations of these measures can be further illustrated by a worked example. This is an attempt to explain the variation in violent crimes in twenty US cities, in terms of population size and the proportion of coloured people. The results of the regression analysis give the equation $Y = 357 + 0.266X_1 + 40.2X_2$; where X_1 is the total population, X_2 is the percentage of coloured people, and Y is the number of detected crimes. Using this equation we can therefore estimate levels of violent crimes in different North American cities. The coefficient 0.266 shows the increase in crime rates for each change in population with the proportion of coloured people remaining unchanged. Similarly, if city population (X_i) is held constant, the coefficient 40.2 shows the increase in crime rates resulting from a change in the percentage of coloured people.

These partial regression coefficients are unstandardized and, therefore, direct comparisons are rather difficult. As we have seen, to make such comparisons we can convert the values into beta weights (equation 13.2). Thus, the beta value of b_1 is 0.388 and that of b_2 0.573. From these standardized coefficients it is now possible to see which of the two independent variables produces the most rapid change in the dependent variable. In our example, variable Y increases at a greater rate with an increase in variable X_2 (percentage coloured), than with the same rate of increase in X_1 (total population). Beta weights therefore provide one way in which the separate effects of individual variables in the regression equation can be assessed. As such they are frequently used by geographers to understand the structures of regression analysis, to compare the importance of different variables in different areas or in different time periods. An example of the latter is shown in Table 13.2, taken from a study by Henderson (1980) on the factors influencing industrial employment in Scotland between 1945 and 1970. From this it can be seen that site factors were the most important in explaining variations in incoming-employment into Scotland during the period 1945–51; but that in the late 1960s levels of female unemployment had become the most significant variable. A detailed examination of these beta weights could therefore go some way towards understanding the factors affecting the creation of new industrial employment in the Scottish economy.

Table 13.2 Variations in beta coefficients and the measurement of employment change, 1945–70

Explanatory variables	1945–51 All industry	1952–59 All industry	1960–65 All industry	1966–70 All industry
Female unemploy.			0.30	0.86
Total unemploy.	0.35	0.71		
% female unemploy.			0.21	
% total unemploy.	0.18			
Sites/premises	0.59	0.37	0.48	0.24
New advance factories			0.30	0.36
Image				0.32

The dependent variable is the actual level of employment provided by immigrant industry. All signs are positive unless otherwise shown. Only significant variables are included.

13.3 MULTIPLE AND PARTIAL CORRELATION

In the previous section attention was focused on the structure of the multiple regression equation and the importance of each of the independent variables. However, in many instances our initial interest may be in the explanatory power of the model, as measured by the multiple correlation coefficient. This is derived in the same way as the simple product-moment correlation, that is by the ratio of the sum of squares of the estimated Y values to the sum of squares of the actual values of Y. The only difference is that for the multiple case the estimates are based on several variables instead of on one, as shown in equation 13.3:

$$R = \frac{\Sigma(\hat{Y} - \bar{Y})^2}{\Sigma(Y - \bar{Y})^2} = \frac{\text{variation explained by all independent variables}}{\text{total variation of actual } Y \text{ values}} \quad (13.3)$$

The multiple correlation coefficient is usually denoted as $R_{0.1\ldots n}$, the subscript 0 indicating the dependent variable and the others the independent variables. The square of R is termed the coefficient of multiple determination and, as with the simple correlation, indicates the proportion of the variance in the dependent variable accounted for by the regression model. Thus, in the three-variable case $R^2_{0.12}$ is composed of r^2_{01} (the variance explained by X_1) plus $r^2_{01.2}$ (the additional variance explained by X_2). In our earlier example of crime rates in US cities the R^2 value was 58.9 per cent, suggesting a reasonable level of explanation. The significance of this can be tested using the F-test (see Section 13.5), although in the case of multiple regression the degrees of freedom are not $N - 1$ but N minus the number of independent variables used in the model.

In the same way that it was possible to examine the effects of individual variables in the regression equation using beta weights, we can measure the contribution of each variable to the overall level of explanation. This can be achieved through the use of partial correlation coefficients. These measure the correlation between the dependent variable and each of the independent variables, while holding the others constant. The partial correlation coefficient between the dependent variable and one independent variable X_1, holding X_2 and X_3 constant, would therefore be given as $r_{01.23}$. This notation can be extended simply by adding further subscripts to the right of the dot, each of these indicating the number of variables being controlled. The number of controlled variables are referred to as the order of the correlation. For example, a first-order partial correlation will have one control. ($r_{01.2}$), a second-order two ($r_{01.23}$) and so on. Thus, in this terminology, a simple bivariate correlation is termed a zero-order correlation, as there are no controlled variables. The first-order partial correlation can be calculated using equation 13.4:

$$r_{01.2} = \frac{r_{01} - (r_{02})(r_{12})}{\sqrt{(1 - r_{02}^2)}\sqrt{(1 - r_{12}^2)}} \tag{13.4}$$

Equation 13.4

r_{01} = correlation between Y and X_1
r_{02} = correlation between Y and X_2
r_{12} = correlation between X_1 and X_2

Using equation 13.4 we can therefore derive the partial correlation coefficients for our earlier example, concerning variations in violent crimes. The basic information we require is given in Table 13.3. Substituting this in the equation we obtain partial correlation coefficients for $r_{01.2}$ of 0.442 and $r_{02.1}$ of 0.662. It should be noted that higher-order partial correlations can be derived in a similar fashion by extensions to equation 13.4, although the calculations for third-, fourth- and higher-order partials are very time-consuming. Fortunately some package programs, such as SPSS, present partial correlations within their multiple regression printout, although as will be shown some caution is needed over the use of such results.

Table 13.3 Zero-order correlations

	Y	X_1	X_2
Crime levels (Y)	—		
Total Population (X_1)	0.552	—	
% coloured (X_2)	0.700	0.371	—

Partial correlation coefficients therefore provide a second method of measuring the separate influences of independent variables. Partial correlations are not, however, identical with beta coefficients, as they represent different measures of associations. Thus, the beta coefficient indicates how much change in the dependent variable is produced by a standarized change in an independent variable with the others held constant. In contrast, partial correlations measure the proportion of variation in the dependent variable accounted for by each independent variable in the regression model.

The relationships between zero-order correlations and partial correlations are not always obvious, and in many cases they may differ quite markedly. In some circumstances the difference may even involve a change in sign, between the two correlation coefficients. To demonstrate this point King (1969) used Olsson's study of migration, in which one of the hypotheses put forward was that the distance of migrant movement would be negatively related to the age of the migrant. However, the zero-order correlation coefficient between these two variables was highly positive (+0.814), thus refuting the hypothesis. When the analysis was extended to consider other variables, using multiple regression and partial correlation, the picture began to change somewhat. For example, when levels of income were taken into account, the partial correlation between distance and age was negative. In this study it was only when other variables were controlled that the full relationships could be explored. As we shall see, explaining partial correlations is no easy task and considerable care needs to be taken over their use (see Section 13.9).

By squaring the partial correlation coefficient we can obtain the amount of variance explained by individual variables within the regression model, when the effects of all other variables are controlled. For our study of violent crime the contribution of these variables is shown in Table 13.4.

Table 13.4 Relative importance of individual factors affecting levels of violent crime

Variables	Partial correlations	Explained variance
Total population	0.442	0.195
% coloured	0.662	0.438

One point that should be clarified is that the multiple correlation coefficient does not equal either the sum of zero-order or partial correlation coefficients. First, in the case of zero-order correlations, this is because of double counting in the sense that the separate correlations between Y and other variabales are in some instances accounting for the same variations in Y. Second, in terms of partial correlation, this difference is due to the omission of joint variance between the independent variables which is not considered within the partial coefficients. For example, in the crime study the two independent variables were slightly correlated (Table 13.3) and some of the variation in crime levels was associated with this joint variance between X_1 and X_2. The partial correlation coefficients relate, however, only to that part of the variance not jointly associated with X_1 and X_2. Indeed, intercorrelated independent variables are one of the major problems with the use of multiple regression and partial correlation in geographical analysis, as is discussed in the following section.

13.4 PROBLEMS OF MULTICOLLINEARITY

The multiple regression model makes the same demands and is based on the same underlying assumptions as discussed for simple linear regression in Chapter 11. However, meeting such requirements becomes far more difficult in the case of multiple regression, since a greater number of variables need to be considered. Take, for example, the assumption that the marginal frequency distributions are approximately normal; for multiple regression all of the independent variables need to be examined. In some instances data transformations are required and the whole regression equation takes on a rather different form, possibly with log values being introduced. While such transformations allow the linear model to be fitted, they can also in some circumstances complicate the process of interpretation (King, 1961).

In general, most studies do consider the problems of data normality and whether the relationships are linear in form, but other conditions rarely receive attention; such as that the conditional distributions of the residuals of Y (the dependent variale) should have zero means or that the variables contain no sampling errors.

The multiple regression model, however, also suffers from a further problem known as multicollinearity. This term refers to the situation where high correlation exists between the independent variables. Thus, if we have two highly intercorrelated independent variables, the second may essentially be explaining the same amount of variation in the dependent variable as the first, and the precision of estimation falls. This problem can be identified either by an examination of the simple zero-order correlation coefficients between the independent variables, or by comparing the simple and multiple correlation coefficients. This is illustrated by the following example, which attempts to explain the growth of superstores and hypermarkets in terms of total population and population density for a sample of twenty-five English counties. The zero-order correlation between the number of superstores and total population was +0.561, while the multiple correlation coefficient ($R_{0.12}$) was 0.588. Therefore, the addition of a further independent independent variable (population density) had very little impact on the value of R. This was partly due to multicollinearity, since the two independent variables (total population and population density) were highly related, with an r value of 0.856. Obviously, in this example if we want to explain the variation in the distribution of superstores at a county level then we should select other meaningful variables that are relatively uncorrelated with each other.

Multicollinearity also causes a considerable amount of ambiguity in the interpretation of the effects of individual variables. Usually under these circumstances very small differences in the total correlation R results in large variations in partial correlation coefficients; and in many cases the importance of each independent variable becomes difficult to estimate. Moreover, the greater the intercorrelation of the independent variables, the less is the reliability of their relative importance as indicated by partial regression coefficients. The point is that owing to multicollinearity, an increase in one independent variable is accompanied by changes in the others, and we cannot therefore regard them as being held constant (Kendall, 1975).

The problem of multicollinearity has not been widely discussed in the geographical literature, although King (1969) drew early attention to it, and recently Johnston (1978) has provided a more detailed discussion of its statistical consequences. In most cases difficulties arising from the existence of collinearity cannot be fully eliminated without considerable effort. However, by careful inspection and selection of the independent variables the problem may be kept to minimal proportions.

Depending on the severity of collinearity, three possible statistical techniques are available as aids in overcoming the problem. One possible solution is to use an objective method, such as stepwise regression, in the selection of independent variables. This is fully discussed in Section 13.7. Alternatively, if all variables are included then factor analysis or principal component techniques may be used to group together the independent variables in uncorrelated or orthogonal sets (see Chapter 15); after which the multiple regression model is applied to these new groupings of independent variables. Some package programs, such as BMDP, allow such a regression analysis on a set of principal components computed from the original X variables.

A third solution, is to use an alternative technique and to replace the least-squares regression model. Such an alternative is provided by a technique known as ridge regression (Hoerl and Kerrnard, 1968). This model, unlike least-squares regression, accepts that the standard regression coefficients are biased estimates. Ridge regression controls for the high interrelationships between the independent variables by calculating a new set of standardized regression coefficients. The new coefficients that are included in the final estimating equation are those which occur at the point where any further increase in the amount of bias does not substantially reduce the mean-square error of estimation in the coefficients (Moriarty, 1973). This technique is not, however, widely used by geographers and is not available in any of the standard package programs. Consequently, it is usually better to deal with multicollinearity by adopting one of the other two strategies.

13.5 SIGNIFICANCE TESTING IN MULTIPLE REGRESSION

In both simple and multiple regression it is important to assess the statistical significance of the model, and there are a number of ways in which this can be done. What follows is an extension of the analysis of variance methods used in simple regression (see Chapter 11). In any simple or multiple regression model the total sum of squares of the dependent variable about its mean can be apportioned between that accounted for by the regression model, the 'explained' sum of squares, and the error or residual sum of squares that remains 'unexplained'. These quantities are found by the same equations used in Chapter 11, and the corresponding variances by dividing by the appropriate degrees of freedom (see Table 13.5). The estimation of the regression sum of squares is, however, more protracted because of the number of predictors. The total sum of squares remains equal to the error plus regression terms.

Table 13.5 Sums of squares expressions and degrees of freedom for regression analysis

Source of variation	Sum of squares	Degrees of freedom
Regression	$\Sigma(\hat{Y}_i - \bar{Y})^2$	k
Error	$\Sigma(\hat{Y}_i - Y_j)^2$	$n - k - 1$
Total	$\Sigma(Y_i - \bar{Y})^2$	$n - 1$

k = number of independent variables; n = number of observations.

The following example uses meteorological data and is particularly interesting because the intercorrelation of the independent variables highlights the problems and benefits of multiple regression. The dependent variable is daily losses from an evaporation tank located in northern England. The variables hypothesized to control evaporation are atmospheric pressure, maximum daily temperature, daily sunshine hours, average daily relative humidity and mean daily wind speed. In their different ways each variable might be expected to influence evaporation. Simultaneously, they are correlated between themselves, as Table 13.6 demonstrates. The data are a random sample drawn from the period April to October 1980.

Table 13.6 Zero-order correlations of meteorological variables

	Evaporation	Air pressure	Max. Temp.	Humidity	Sunshine
Air pressure	−0.04				
Max. Temp.	0.55	0.18			
Rel. humidity	−0.55	−0.41	−0.44		
Sunshine	0.57	0.14	0.35	−0.54	
Wind speed	0.22	−0.36	−0.19	−0.12	0.29

The least-squares multiple regression equation expressing evaporation in millimetres per day is:

$$Y = 57.32 - 0.054X_1 + 0.275X_2 - 0.059X_3 + 0.145X_4 + 0.017X_5 \quad (13.5)$$

where X_1 is atmospheric pressure (millibars), X_2 is maximum daily temperature (°C), X_3 is daily mean relative humidity (per cent), X_4 is daily sunshine (hours) and X_5 is mean daily wind speed (knots).

Our initial interest might well be with establishing the significance of the whole regression model. Following the methods indicated in Chapter 11, Table 13.7 has been prepared, from which it is concluded that the model is a valid expression of the dependent variable's behaviour and the null hypothesis of no explanation is rejected. This conclusion is undeniably useful, but there are other questions that can be asked. In particular, there is no understanding of the relative importance of the five selected variables. It has been supposed, on *a priori* grounds, that they are all important; but by looking at the regression model in total we are unable to confirm that this is the case. The simple, or zero-order, correlations provide only the slightest insight into this problem, and the true association between the dependent and independent variables is disguised by the intercorrelation of the latter.

Table 13.7 ANOVA table of regression model for daily evaporation rates (five predictors)

Source of variation	Sum of squares	Degrees of freedom	Variance	F ratio
Regression	41.513	5	8.303	
Error	31.943	24	1.331	6.24
Total	73.456	29		

$R^2 = 0.565$; critical F (with 5 and 24 degrees of freedom at the 0.05 significance level) $= 2.62$.

As we have already seen the partial regression coefficients are dependent on the units of measurement and, consequently, they cannot be used to examine the relative importance of each predictor. But the problem can be approached in either of two ways. First, most computer packages include within their results t or F statistics for each of the predictor variables, and the popular SPSS system is a good example of this useage. These statistics can be used to sort out the significant and non-significant variables. Second, a useful approximation of relative importance can be determined from the beta weights of the variables (see Section 13.2). In this form the partial regression coefficients are free from the scale problem and are, in effect, standardized. The beta weights for the current example are given in Table 13.8. It is profitable to compare these figures with the partial regression coefficients and zero-order correlations.

Table 13.8 Beta coefficients for the evaporation regression model predictors

Predictor	Partial regression coefficient (b_i)	Standard deviation of variable X_i (s_{xi})	Beta weight (B_i)
Air pressure	−0.054	7.48	−0.256
Max. Temp.	0.275	2.13	0.368
Rel. Hum.	−0.059	8.96	−0.336
Sunshine	0.145	2.95	0.270
Wind speed	0.017	7.02	0.077

$s_y = 1.59$.

13.6 SIGNIFICANCE TESTING OF INDIVIDUAL VARIABLES BY FORWARD INCLUSION

Beta weights are a quick, but approximate, method of assessing the relative importance of predictor variables, and researchers will often need to know the precise statistical significance of each. Some, though not all, computer packages provide this information. The elimination of non-significant variables is important in order to ease the task of data collection and manipulation. The complete regression model for evaporation losses we know is significant, but non-significant variables, if they exist, can be identified by a simple extension of the analysis of variance method.

A worked example best conveys the principles of the process. Employing what is termed the forward inclusion method (there are others) the test starts by regressing the variable with the highest zero-order correlation against the dependent variable. From Table 13.6 this is found to be daily hours of sunshine ($r = +0.57$). The significance of the simple regression equation (equation 13.6) is tested in the usual manner, and from the results in Table 13.9 the null hypothesis of no explanation can be rejected. On this basis it may be asserted that hours of daily sunshine have an individual significant role, and:

$$Y = 1.583 + 0.305X_4 \qquad (13.6)$$

Table 13.9 ANOVA table of regression model for daily evaporation rates (best single predictor)

Source of variation	Sum of squares	Degrees of freedom	Variance	F ratio
Regression	23.521	1	23.521	
Error	49.936	28	1.783	13.19
Total	73.456	29		

$r^2 = 0.320$; critical F (with 1 and 28 degrees of freedom at the 0.05 significance level) $= 4.20$.

Now, there is little to be gained by regressing all five predictors individually against evaporation. Instead, and subject to conditions that will be explained, they are added sequentially to this first simple expression. The difficulty of selecting the best variable at each inclusion step is solved by reference to the partial correlations of the non-included variables against the dependent, but holding constant those predictors already within the model. Table 13.10 indicates that maximum daily temperature must be the next inclusion, providing the new least-squares equation:

$$Y = -3.2768 + 0.299X_2 + 0.229X_4 \qquad (13.7)$$

Because equation 13.7 includes a variable already established as significant (sunshine) and which is itself correlated with maximum temperature ($r = +0.35$), there is no advantage in testing the new expression; instead attention focuses on the increase in the regression sum of squares. Whenever a new variable is added to a regression equation, unless it has no correlation with the dependent, there is an increase in the regression (explained) sum of squares. The problem is to assess the significance of this increase under the null hypothesis of no improvement in explanation.

Table 13.10 First-order partial correlations of evaporation predictors

Variable	First-order partial correlation (sunshine constant)
Relative humidity	−0.3029
Maximum temperature	0.4554
Mean wind speed	0.0696
Atmospheric pressure	−0.1810

The required F ratio is provided by the variance of the increase in the regression sum of squares which, with one new variable, has only one degree of freedom, and the error variance *after* inclusion of the new variable. The quantities used in the calculation may be set out in a modified ANOVA table (Table 13.11) from which one observes that the null hypothesis of no improvement in explanation may be rejected at the 0.1 level, and the additional explanation due to the inclusion of maximum daily temperature is indeed significant. These quantities can all be estimated from the equations given in Table 13.5.

Table 13.11 ANOVA model of regression model improvement (one- and two-variable models)

Source of variation	Sum of Squares	Degrees of Freedom	Variance	F ratio
Regression ($k = 1$)	23.521	1		
Regression ($k = 2$)	33.828	2		
Change in above	10.307	1	10.307	
Error ($k = 2$)	39.628	27	1.468	7.02

$R^2 = 0.461$; critical F value (with 1 and 27 degrees of freedom at the 0.1 significance level) $= 2.90$.

A point to be noted at this stage is the manner in which the coefficient of explanation (R^2) changes with the inclusion. While the zero-order correlations of sunshine and maximum temperature with evaporation are +0.57 and +0.55 respectively, the R^2 value rises by only 0.141, from 0.320 to 0.461. This small change, apparently incommensurate with the addition of so well-correlated a variable, reflects the multicollinearity of the independent variables and their combined effect on evaporation is partly duplicative and not separately additive (see Section 13.4).

The next step is to examine the second-order partial correlations, now holding both sunshine and maximum temperature constant, to determine which variable to include next. The new set of partial correlations in Table 13.12 indicate that atmospheric pressure should be the third predictor (the sign of the partial correlation does not, in this sense, matter). The regression model, now with three predictors, is provided by:

$$Y = 35.297 + 0.238Y_4 + 0.320X_2 - 0.039X_1 \tag{13.8}$$

Table 13.12 Second-order partial correlations of evaporation predictors

Variable	Second-order partial correlation (sunshine and max. temp. constant)
Relative humidity	0.1876
Mean wind speed	0.2586
Atmospheric pressure	−0.2783

But, as we see from the results in Table 13.13, this improvement in description is not significant as the observed F ratio is smaller than the critical F value, and atmospheric pressure together with the remaining variables yet to be examined are all excluded from further consideration. As a result equation 13.7 is accepted as the most efficient means of predicting evaporation losses.

Table 13.13 ANOVA model of regression model improvement (two- and three-variable model

Source of variation	Sum of squares	Degrees of freedom	Variance	F ratio
Regression ($k = 2$)	33.828	2		
Regression ($k = 3$)	36.134	3		
Change in RSS	2.306	1	2.306	
Error ($k = 3$)	37.322	26	1.436	1.606

$R^2 = 0.492$; critical F value (with 1 and 26 degrees of freedom at the 0.1 significance level) = 2.91.

An important final point concerns the choice of significance level (α). Previously, and on admittedly arbitrary grounds, either the 0.05 or 0.01 levels were used. This needs to be reconsidered in the current context. Chapter 9 has already drawn attention to the nature of type I and type II errors. In the former case the null hypothesis is correct but rejected, and in the latter it is incorrect but accepted. The probability of a type I error is given by the value of α itself. In the present example the inclusion of a redundant variable (a type I error by this definition) is far preferable to the erroneous exclusion of a significant one (a type II error). Increasing the value of α to, say, 0.10 helps to accommodate this stipulation.

13.7 VARIABLE SELECTION IN MULTIPLE REGRESSION: STEPWISE AND OTHER METHODS

In Section 13.6 we saw how the so-called forward inclusion method can be used to determine an optimal combination of independent variables. Nevertheless there are other methods that we could have used to achieve this end.

The clearest alternative to the forward inclusion method is that of backwards elimination. The latter scheme starts with an equation which includes all independent terms. These are taken, in turn, and treated as if they were the last to be included and their respective contributions to the explained sums of squares are determined. The variable with the lowest contribution is eliminated, and the process is repeated with variables being eliminated one at a time until their loss creates a statistically significant loss in explained (regression) sums of squares. At that point the process ceases and the regression model so defined is assumed to be the optimum. Generally this method is no more nor less efficient than that of forward inclusion, though there is something to be said in favour of a strategy that starts with all the variables as this provides a modest overview of the situation which may assist at later stages of interpretation. However, it must not be expected that these two, or any other methods, will necessarily produce the same 'optimum' solution given identical data sets, particularly when multicollinearity in the independent terms is present. Indeed, it is possible that the solutions offered may not even be the optimal ones according to the criterion of the researcher, who should always be prepared to step in and make, perhaps, a non-statistical decision on the 'best' regression model for this purposes. Draper and Smith (1966) have written extensively on the various methods and conclude that there may be no unique solutions; personal judgement may form a necessary part of statistical methods.

The latter point becomes clearer when we consider the next alternative method in which all possible combinations of variables are tested. Clearly this is not a very efficient strategy and is possible only with computational assistance — which, even so, may be severely taxed if there are many variables, bearing in mind that for n variables there are $2^n - 1$ possibilities. The advantage that it has over forwards and backwards methods is due to its immunity to the effects of internal rearrangements of the correlational structure of independent variables as they are added or subtracted from the model. By the former schemes inclusion or elimination is final and irrevocable, yet there are occasions when initially 'good' variables can become redundant at later stages or initially 'poor' ones become better. Whenever multicollinearity is present, which is often, the intercorrelations do change within the regression set as elimination or inclusion proceeds. The 'all possible regressions' method allows us to examine combinations that are not permitted by the earlier schemes but which, nevertheless, may be useful. The price to be paid for this advantage is a demand on time, both on the computer (to prepare the regressions) and the researcher (to interpret them). But in the absence of the procedural aspects of, for example, forward inclusion, selection focuses often on the R^2 values, and there is a tendency to select models with high R^2 estimates but which include many variables some of which may not be contributing very much. But by examining the rate at which explained or regression sums of squares increase with the number of terms, it is usually possible to determine the stage at which 'diminishing returns' set in. This enables us to decide on an optimum number of independent terms at least, though we are still left to decide which combination of the prescribed number of variables is 'best'.

A method that combines the efficiency of forward inclusion (and backwards elimination) with the thoroughness of 'all possible regressions' is stepwise regression. In this scheme the independent variables are re-examined at each stage to identify any that have become superfluous following the introduction of subsequent items, or to permit use of previously rejected variables. In doing so stepwise regression takes particular note of the problems of multicollinearity. At each step both inclusion and exclusion are permitted, and the process continues until neither can be allowed under the selected significance level.

The stepwise method is not, nevertheless, foolproof, and the geographer should be alive to the creation of models (notably when there are a large number of variables) that, while being statistically sound, do not stand up to geographical and theoretical scrutiny. Indeed there is a school of thought that suggests that variable selection should be based on firmly theoretical and deductive considerations, rather than statistical measures such as F ratios. Many of these problems and debates stem from the question of multicollinearity whereby different combinations of variables may give equally good R^2 values. When this is the case the argument for some form of deductive or theoretical framework for the study is strengthened, since without it a sensible interpretation of the results and correct selection becomes hazardous. Sadly this point is often overlooked by geographers who demonstrate a misplaced faith in the ability of stepwise programs to determine the 'best' and most suitable regression equations irrespective of the nature of the raw data.

In general, though not exclusively, multicollinearity becomes more severe with the number of variables. The severity of the problem, as far as it affects the geographer, will depend on why he is regressing at all. There are two principal reasons: either for prediction or explanation. If he is regressing in order to predict he may be less concerned with the delicacies of sensible models than with reducing error terms and raising the highest possible R^2 values. In this way predictions become more accurate and the former (coefficient of explanation) values will wholly determine the selection of the regression model. The geographer seeking an explanation is, almost inevitably, less fortunate. He must be careful of paying too much attention to the R^2 values and, indeed, may prefer a model with a lower degree of explanation but one which is theoretically sounder.

When problems due to multicollinearity become insurmountable, the geographer seeking an explanation, will have to make recourse to one of the strategies discussed in Section 13.4. The replacement of many intercorrelated variables by a far fewer number of orthogonal (uncorrelated) principal components (synthetic composite 'variables' extracted from the raw data) is popular. This has been done successfully by, for example, Wong (1963) and Riddell (1970), but much will depend on the researcher's willingness to sacrifice his original and carefully selected variables to the less precise, but orthogonal, principal components that replace them. Furthermore, though this is more thoroughly and appropriately discussed in Chapter 15, the geographer must always be sure that he can make a successful interpretation of the principal components generated from the raw data in terms of the original variables, and the influences that they portray.

13.8 RESIDUALS IN MULTIPLE REGRESSION

Chapter 11 discussed the conditions of variable and residual behaviour that had to be complied with before the simple regression model could be accepted. All of those conditions apply with equal force to multiple regression models. We shall not repeat these conditions here in any length, but it should be recalled that, as far as residuals are concerned, they should have zero mean, possess a normal distribution and constant variance. On these counts alone it is usual to examine residuals to ensure their complicance, but their study can also yield rewards in terms of improving our understanding of the data and their behaviour. Thus when we check residuals for autocorrelation this ensures that the data conform to our purely statistical needs. Failure to conform might lead us to reject the regression model; but it is more likely to suggest to us that an important variable may have been overlooked whose inclusion would eliminate this difficulty, as well as assist in actually explaining the behaviour of the dependent variable.

For example, if we examine the Durbin–Watson statistics in Table 13.14 for the two regression models for daily evaporation losses discussed earlier, we can see how the addition of the second independent variable greatly reduces the degree of autocorrelation (values close to 2.0 indicate zero autocorrelation — see Section 11.7). Such a condition certainly removes any doubts concerning the statistical character of our data, but we must beware of also concluding that no further variables need be considered — this is an incorrect interpretation of zero autocorrelation. It is quite possible to have such a condition but retain scope for additional items.

Table 13.14 Durbin–Watson statistics for regression models of evaporation

Independent variables	Durbin-Watson d
Sunshine	2.496
Sunshine and maximum daily temperature	1.956

Sections 11.6 and 11.7 have already demonstrated how autocorrelation can be tested on sequences of residuals. When many variables are involved the residuals are best sequenced according to their associated $\hat{Y}$ values, as this provides a good overall impression of their character; but we can also produce a sequence of residuals ordered according to each of the individual independent variables. The latter is far more time-consuming since a sequence is produced for each variable. However, there may be instances when we may be concerned with one particular independent term and its detailed influence on the dependent item.

The use of simple graphical plots of residuals against $\hat{Y}$ can often be most revealing (see Figures 11.12(a) and (b) and may demonstrate residual patterns and inadequacies far more clearly than can numerical processing. When these plots of residuals reveal weaknesses in their adherence to the stipulations of regression modelling, it is not necessary to jettison the data set as some simple measures can be attempted to correct them. These measures take the form of transforming either the regression equation or the raw data. If we take the most common forms of non-adherence we can see how they may be corrected. First the wedge-shaped pattern of residuals shown in Figure 13.3(a) suggests that the data are heteroscedistic (in either all or some of the varables). This is a problem that may be solved by data transformation. The difficulty is that error terms increase proportionally with the magnitude of the predictors which may be multiplicative rather linearly additive in their effects on the dependent variable. Such conditions are not uncommon in geographical data and a solution is to use logarithmic transformations of the raw data. Thus a multiple regression model would become:

$$\log_{10} Y = \log_{10} a + b_1 \log_{10} X_1 + b_2 \log_{10} X_2 \ldots b_n \log_{10} X_n \quad (13.9)$$

By the rules of operations with logarithms introduced in Section 2.3, this equation can also be written as:

$$Y = aX_1^{b_1} X_2^{b_2} \cdot \ldots X_n^{b_n} \quad (13.10)$$

which is, as required, multiplicative in its effect on Y and requires only that operations be performed on the logs of the data. It must, nevertheless, be stressed that the indiscriminate use of such transformations, particularly when mixed with untransformed data, can make interpretation a very difficult job. For purely predictive applications of the model this may be less important.

A second and also common form of residual behaviour is for them to be plotted as curves against $\hat{Y}$ or X (Figure 13.3(b)). The implication is clear: the linear model is inappropriate and the solution here is to change our model to a nonlinear form. The choice and range of curves that may be investigated has already been reviewed in Chapter 12.

Third, there is the case of a linear trend in the residuals (Figure 13.3(c)). This would also be identified by the presence of strong positive autocorrelation and the correction would be to modify the model by the inclusion of another independent variable.

So far we have focused attention on the general character of the residuals, but the individual characters cannot be overlooked. Most important are the so-called outliers. These can best be identified by standardizing the residuals (see Section 11.4) in order that the probabilistic aspect of their deviation is clearer. Thus we may define an outlier as a residual with a z value greater than ± 1.96 (giving a random probability of less than 0.05). Such a degree of departure from the regression line expected value is notable and should be examined. The observation might be found to be an error of measurement or in data preparation and should therefore be excluded and the data set reanalysed. In this form residual analysis is a valuable check. But not all outliers can be so easily explained and dealt with. Some must be due to random variation in the data set, and in this sense there would be nothing 'wrong' with them. Nevertheless we may wish to exclude them and rerun the model to reduce the residual sum of squares. Finally, some outliers may be altogether more constructive. Take the example of a rainfall–altitude regression equation in which rainfall is explained by the altitude of the raingauge.

We might find that a group of outliers cannot be explained by error, but on close inspection possess the common characteristic of valley-bottom sites with unusual degrees of shelter, so that despite their altitudes they receive less than expected rainfall. Rather than reject such points it would be far preferable to add a new variable to the regression set which gives a measure of the degree of exposure of the raingauges. In this way our data set is preserved intact and the model, we hope, improved; although the new variable's contribution may have to undergo the usual F-tests. Because of this feature of residuals it is always a good practice to keep a record of the raw data with notes or a reference of the individual observations.

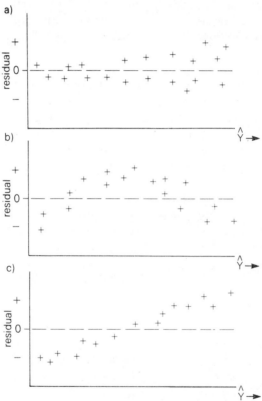

Figure 13.3 Forms of residual behaviour

Clearly much of this checking of residuals can be repetitive. Fortunately some of the larger regression packages contain excellent graphics options that permit the residuals to be plotted in different ways; either against selected variables, or predicted Y values, or even on a probability scale that allows their normality to be assessed very quickly. Both SPSS and BMDP have excellent facilities for such work, which are reviewed in the manuals that accompany these two packages. They require little in terms of specification and full use should always be made of them because, as we have shown, the study of residuals can act as a data check and a means of improving the model.

13.9 PARTIAL CORRELATIONS, PATH ANALYSIS AND CAUSAL MODELS

One of the main aims in many areas of geographical analysis is to establish causal relationships, and for this reason correlation and regression techniques are frequently used by geographers. Usually, the relationships being examined are initially rather simplified and often couched in terms of variable X causing some measureable change in variable Y. Thus, in an earlier example we hypothesized that population controlled the number of stores to be found in particular urban centres (Section 11.1). Unfortunately, in this case and in most other areas of geography such direct causal relationships are very often a gross oversimplification of reality. To help us come to terms with the analysis and testing of these complexities we can make use of partial correlations and path analysis.

Since we have already examined the properties of partial correlation coefficients, let us start here by considering the help that path analysis can bring to bear on the problems of causal modelling. Path analysis was introduced by Wright (1921) as a means of interpreting the causal order among a closed set of variables; and has subsequently been widely used by social scientists through the work of Blalock (1961) and Duncan (1966). The principles of path analysis are discussed in detail by Duncan (1975), and here we shall merely examine some of the more basic elements of the approach.

As we shall show, the examination of causal structures can therefore be undertaken by two methods: one using partial correlations and the other path analysis. First, consider the causal chain presented in Figure 13.4(a); this shows that variable X_1 has a positive effect on X_1, which in turn influences X_3.

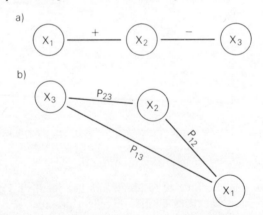

Figure 13.4 Types of casual relationships: (a) causal chain, (b) complex relationships

Indirectly, it also suggests that the partial correlation $r_{32.1}$ should be zero and that X_1 has no independent influence on X_3. Thus, in this simple causal model, X_1 only influences X_3 through variable X_2. Therefore, any correlation between X_1 and X_3

should equal the product of the two correlations between X_2 and X_3 (r_{23}) and X_1 and $X_2(r_{12})$. For example, if $r_{23} = -0.75$ and $r_{12} = +0.65$, then r_{13} should equal $(-0.75)(+0.65) = -0.49$ (Blalock, 1960). This reasoning implies that we can predict one causal relationship (in terms of a correlation) from knowing one set of correlations. We could then go on and test whether such predicted correlations are correct by comparing them with actual ones. If the predicted differ from the actual then we need to go back and search for other causal relationships which we have failed to identify in our model.

Unfortunately, there are few examples of such approaches within the geographical literature and those that do exist have tended to be prone to error. Thus, Taylor (1969) reworked a study by Cox (1968) on the voting behaviour of people living in suburban London to highlight the misconceptions in Cox's hypotheses. However, despite the problems with Cox's work it was nevertheless an early and singular example of the use of causal modelling in geography.

The second approach, that of path analysis, has also received only very limited attention from geographers, compared with the work of sociologists (Duncan, 1966). The aim of the technique is to examine the direct and indirect links in a causal structure. For example, in Figure 13.4(b) we have a path diagram with three variables related through a series of paths. Each of these paths are represented by a path coefficient (p). This coefficient shows the direct influence of one variable on another; thus p_{13} is the path coefficient of X_3 on X_1 (Figure 13.4b) While path coefficients can be represented by ordinary regression coefficients it is customary to use beta values, as discussed by Wright (1960). The real value of path analysis would seem to lie in it being a technique of working out the logical sequences of linear relationships. Those interested in the more detailed practicalities of the method should note that package programs such as SPSS will calculate path coefficients.

13.10 PROBLEMS AND ASSUMPTIONS

Geographers are now more than ever aware of the difficulties inherent not only in regression analysis but also in the use of computers and statistical packages. Thought-provoking studies have appeared from Poole and O'Farrell (1971), Mather and Openshaw (1974) and Gould (1970). Problems have been identified in two areas. First, important assumptions necessary for the validity of regression models may not always be fulfilled. Second, unquestioning dependence on the results of pre-written computer packages can be ill-advised as Longley (1975) and Wempler (1970) have demonstrated.

Take first the question of regression data requirements. It is now clear that the often-quoted need for data normality is subordinate, in regression analysis, to more pressing needs. Indeed, as Mather and Openshaw have indicated, the urge to normalize and transform data can lead to later interpretational difficulties, particularly if different forms of transformation are used within the one regression model. Coefficient reliability appears to be more responsive to other preconditions, and the best linear unbiased estimates are achieved only when these are met. Chapter 11 has dealt with these aspects but we may add others. At the most fundamental level, data accuracy is uncommonly critical because the regression equations permit errors in Y but assume X to be perfectly measured. Rarely, if ever, can this be the case in geography, and the error term e in the regression equation should strictly be expanded to reflect this extra source of error.

The correlational structure of the independent variables also needs close attention, as we have seen. Multicollinearity is a perennial problem, but one that remains soluble by partial correlation methods only as long as the zero-order correlations are not consistently high.

Lastly, it is important that observations used, not only in regression but in many areas of inferential statistics, should be random and independent of one another. This stipulation is perhaps the most difficult for geographers to comply with as their data are often spatially organized and possess a degree of spatial autocorrelation. Under these circumstances, spatially adjacent observations are likely to have similar magnitudes and hence not be random. Much has been written on this important subject (Cliff and Ord, 1970) and the problems are far from being satisfactorily resolved. This situation is particularly vexing to the geographer who, on the one hand, wishes to comply with the conditions of the statistical methods but, on the other, finds that his most problematical area is spatial autocorrelation, the characteristics of which lie at the very core of his discipline.

Happily, it is within the compass of the investigator to check many of these conditions and to take steps to counter some of the non-compliances he encounters. On the question of computer program integrity he is in a less commanding position and is vulnerable to a number of unforseeable pitfalls. Thus, he may avoid the more flagrant statistical transgressions of data collection only to find that the computational procedures leave much to be desired. Computer printouts are not blessed with irrefutable validity and some studies have given grounds for serious disquiet. The work of Wempler is most enlightening in this respect. His paper discusses inconsistencies in results for identical data sets when regressed using a number of different packages, all widely available in North America. The problem is critical when large data sets of widely variable observations are used. The consequent round-off of the substantial sums of squares and covariances within the program can lead to conflicting results and conclusions. The degrees of inconsistency depend on the method of coefficient estimation. A discussion of these issues would go far beyond the scope of this text, and it is sufficient to state that there are many ways in which partial regression coefficients can be estimated. According to Wempler some of these algorithms are more reliable than others. The interested reader is referred to his original paper.

REFERENCES

Blalock, H. M. (1960). 'Correlation analysis and causal inferences', *Am. Anthropologist* **62.** 624–631.

(1961). *Causal Inferences in Non-Experimental Research,* University of N. Carolina Press, Chapel Hill.

Chatfield, C., and Collins, A. J. (1980). *Introduction to Multivariate Analysis,* Chapman and Hall, London.

Cliff, A. D., and Ord, K. (1970). 'Spatial autocorrelation: a review of existing and new measures with applications', *Econ. Geog.* **46,** 269–92.

Cox, K. R. (1968). 'Surburbia and voting behaviour in the London metropolitan area', *Anns Ass. Am. Geogrs* **58,** 111–127.

Draper, N. R., and Smith, H. (1966). *Applied Regression Analysis,* Wiley, Chichester.

Duncan, O. D. (1966). 'Path analysis: sociological examples', *Am. J. Soc.* **72,** 1–16.

(1975). *An Introduction to Structural Education Models,* Academic Press, New York.

Gould, P. (1970). 'Is *statistix inference* the geographical name for a wild goose?', *Econ. Geog.* **46,** 439–48.

Henderson, R. A. (1980). 'The location of immigrant industry within a UK Assisted Area', *Progress in Planning* **14,**2.

Hoerl, A. E., and Kerrnard, R. W. (1970). 'Ridge regression: biased estimation for nonorthogonal problems', *Technometrics* **12,** 55–67.

Johnston, R. J. (1978). *Multivariate Statistical Analysis in Geography,* Longman, London.

Kendall, M. (1975). *Multivariate Analysis,* Griffin, London.

King, L. J. (1961). 'A multivariate analysis of the spacing of urban settlements in the US', *Anns Am. Assoc. Geogrs* **51,** 222–233.

(1969). *Statistical Analysis in Geography,* Prentice-Hall, Englewood Cliffs.

Longley, J. W. (1975). 'An appraisal of least squares programs for the electronic computer from the point of view of the user', *J. Am. Stat. Assoc.* **62,** 819–41.

Mather, P. M., and Openshaw, S. (1974). 'Multivariate methods and geographical data', *The Statistician* **23,** 283–308.

Moriaty, B. M. (1973). 'Causal inference and the problem of nonorthogonal variables', *Geog. Analysis* **5,** 55–61.

Poole, M. A., and O'Farrell, O. N. (1971). 'The assumptions of the linear regression model', *Trans. Inst. Brit. Geogrs* **52,** 145–58.

Riddell, J. B. (1970). 'On structuring a migration model', *Geog. Analysis* **2,** 403–409.

Taylor, P. J. (1969). 'Causal models in geographic research'. *Anns Ass. Am. Geogrs* **59,** 402–404.

Wempler, R. H. (1970). 'A report on the accuracy of some widely used least squares computer programs', *J. Am. Stat. Assoc.* **60,** 549–63.

Wong, S. T. (1963). 'A multivariate statistical model for predicting mean annual flood in New England', *Anns Ass. Am. Geogrs* **53,** 298–311.

Wright, S. (1921). 'Correlation and causation', *J. Agric. Res.* **20,** 557–585.

(1960). 'Path coefficients and path regressions', *Biometrics* **16,** 189–202.

RECOMMENDED READING

Hauser, O. (1974). 'Some problems in the use of stepwise regression techniques in geographical research', *Canadian Geogr* **18,** 148–58. A self-explanatory title for a paper that deals with the problem in a readily-understandable fashion without recourse to lengthy mathematical analysis.

Keeble, D. (1976). *Industrial Location and Planning in the United Kingdom,* London. Chapter 5 of this book provides a good example of regression applications to a geographical problem. It touches on many of the themes adopted in the foregoing chapter and is as informative on the methodology as on the geography.

Mark, D. M., and Peucker, T. K. (1978). 'Regression analysis and geographic models', *Canadian Geogr* **22,** 51–64. This paper reviews the usefulness and limitations of linear and nonlinear regression and its application in gravity flow models. As with many of the items on this list it is appropriate to the geographer with little formal mathematical training.

Mather, P. M. (1976). *Computational Methods of Multivariate Analysis in Physical Geography*, Wiley, Chichester. This has a short section (pp. 73–77) on problems of multicollinearity and the solutions associated with ridge regression.

The geographical literature abounds with good examples of this most popular of techniques. The following references are but the smallest fraction of what can be found.

Boswell, T. D. (1977). 'Inferences concerning intermunicipio migration in Puerto Rico: 1955–1960', *J. Trop. Geog.* **45,** 1–11.

Cruickshank, J. G., and Armstrong, W. J. (1971). 'Soil and agricultural land classification in Co. Londonderry', *Trans. Inst. Brit. Geogrs* **53,** 79–94.

Ferguson, R. I. (1973). 'Channel pattern and sediment type', *Area* **5,** 38–41.

Massey, D. S. (1980). 'Residential segregation and spatial distribution of non-labour forces population: the needy, elderly and disabled', *Econ. Geog.* **56,** 190–199.

Smit, B. (1979). 'Regional employment changes in Canadian agriculture', *Canadian Geogr* **23,** 1–17.

EXERCISES

A random sample of fifteen US states provided data for 6 socioeconomic variables which have been coded here as follows:

X_1 = percentage black population (1980)
X_2 = number of police officers per 1,000 of population (1979)
X_3 = percentage of population completing 1–3 years of college (1980)
X_4 = percentage urban population (1980)
X_5 = median family income in dollars (1979)
Y = crimes per 100,000 people (1980)

These data were processed to give a correlation matrix summarizing the interrelationships between the variables.

1. Use crime rate as the dependent variable and by reference to the correlation matrix construct regression models with the following predictors: (i) variables X_2 and X_4; (ii) two variables of your own choice.

2. Estimate the beta weights of the predictor variables in both cases. Do the contributions to the regression model made by both variables reflect their separate influences as interpreted from their simple correlations?

3. Qualitatively assess the nature of the first of these regression models in the light of both the multicollinearity of the variables and their cause and effect structure.

State	X_1	X_2	X_3	X_4	X_5	Y
Maine	0.3	2.0	29.5	33.0	16208	4243
Connecticut	7.0	2.6	36.9	88.3	23038	5837
New York	13.7	3.7	32.8	90.1	20385	6905
Minnisota	9.9	2.1	27.9	80.3	20710	5447
Kansas	1.3	1.9	34.1	64.6	21217	4737
Delaware	16.2	2.8	30.5	67.0	20658	6689
Maryland	22.7	3.1	34.5	88.8	22850	6558
Florida	13.8	3.0	32.0	87.9	17558	8032
Kentucky	7.1	2.0	21.9	44.5	16399	3532
Arkansas	16.5	1.9	20.7	39.2	14458	3796
Montana	0.3	2.5	37.1	24.0	18839	5019
Idaho	0.3	2.4	37.1	18.3	17278	4531
Nevada	6.4	3.6	35.5	82.0	21666	8592
Washington	2.6	2.1	39.3	80.4	21635	6742

Data source: *Statistical Abstract of the US*, 1982–83, *Washington, DC.*

Correlation matrix

X_2	−0.4416*				
X_3	−0.3763	−0.3091			
X_4	0.5236*	0.5646*	0.1688		
X_5	0.1256	0.4305	0.6308†	0.7047†	
Y	0.3305	0.8072†	0.4636*	0.7441†	0.5820*
	X_1	X_2	X_3	X_4	X_5

*significance at the 0.05 level; †significant at the 0.01 level.

Chapter 14

Methods of Classification

14.1 INTRODUCTION

The term classification covers a wide range of quantitative and qualitative methods. We all classify objects, be they people, places, events or any other of a host of phenomena that enter our consciousness. Classification is a useful and compelling means of describing and summarizing everyday information. We often place people, for example, into groups or classes according to some particular criterion, of which age, sex, income or occupation are common examples — the possibilities are endless. Classification is a natural response to large bodies of information, helping us to comment, describe and analyse what we see. But scientists, too, receive large amounts of data, usually of a specific nature, and like the layman seek to simplify their work and aid their understanding by classifying. Of necessity, their methods can be neither so subjective nor so selective as those of the layman. Scientists prefer more objective and numerical (where possible) approaches to the problem, though the difficulties of subjectivity and selectivity are not necessarily overcome by these adoptions.

Geography has yet to attain the success gained in, for example, biology, where the whole of the known plant or animal kingdoms are organized into a classification scheme. The geographer's subject material is notably diverse in that most geographical individuals can be described by one, or many, of a large number of attributes. Consequently they are not easy to define or measure. Nevertheless, grouping and classification procedures assist in understanding this diversity and have an increasingly important part to play in both description and analysis. Climate is a good example of just such a diverse and multifaceted geographical phenomenon and it is profitable to examine the many attempts at its classification; those by Koppen (Lewis, 1961; Wilcock, 1968), Thornthwaite (1931) and Miller (1931) are typical. In all such cases a recurrent problem emerges: on which of the many aspects and measurable attributes of climate should the classification be based? Rainfall and temperature are, possibly, clear candidates. But what of rainfall intensity, seasonality and duration? Or of temperature range, evaporation, sunshine duration or wind speed? These are all climatic components and different ways of looking at a meteorological whole. It is rarely possible, or even desirable, to include a very large number of attributes, and a major problem is always posed by the need to be selective. The investigator should be guided by the specific needs of his study. Thus, if interest lies with thermal, rather than hydrologic, studies the classification should emphasize the temperature aspects of climate. In this sense there is no need for concern that classification schemes of the same phenomenon may not yield identical results. In all cases the classification should be geared towards the specific needs of the study and should only be undertaken within some framework, and not in any vague or speculative fashion.

14.2 STRATEGIES IN CLASSIFICATION

Within the scope of classification methods are very many alternatives. There are too many to be considered in this chapter and what follows must, of necessity, be highly selective. Nevertheless some common points of issue do emerge and may be dealt with first.

In terms of the raw material the first question to be asked is whether the purpose is to identify distinct or 'natural' groupings of individuals in what might be termed a well-structured data set, or to devise a more arbitrary grouping in what is an homogenous collection of individuals. Figure 14.1 attempts to draw this distinction in graphical terms, though clearly they are ends of a spectrum of possibilities. Harvey (1969) stresses this point, declaring a difference between ordination (the sub-division of a homogeneous set) and classification proper in which distinct groups are defined. Both strategies can be pursued employing the same methods.

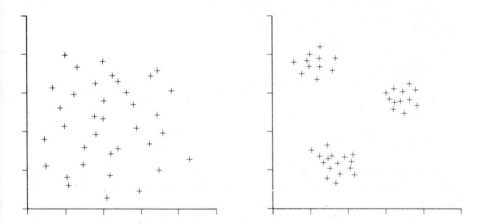

Figure 14.1 Patterns of clustered and dispersed points demonstrating the distinction between real and arbitrary groups

Additionally, and a point stressed in the introduction, these methods may be either quantitative or qualitative. Qualitative classification often proceeds by definition as, for example, in the case of stream segments which can be placed into classes according to some *a priori* definition. In this case Horton's (1945) definition for stream numbering ensures that any channel segment can be accorded a class (stream order). Quantitative classification, however, more often proceeds by enumerating the attributes of a set of individuals, using the enumerations to determine the most similar members of the set. Harvey terms this 'classification by enumeration', and the former 'classification by definition'. Even within the enumeration approach many techniques are encountered. So, for example, groups may be formed by logical division of a complete set, which involves gradually breaking the set down into smaller and smaller groups until only the isolated individuals are left; or the process may be agglomerative whereby the individuals form the starting point and they are successively grouped until they all are embraced within one large group.

Both strategies are hierarchical, with groups of various sizes encountered at different levels. In this connection there is a problem in determining at which level of integration or disintegration the selection of groups is to be made. If the selection is too close to the one large group then there will be too few groups and a conequent loss of information and detail. On the other hand, termination too close to the individual end of the hierarchy causes important generalities and group identities to be lost in the overabundance of small groups.

In all instances the range of detailed methodologies requires that the investigator think carefully about his strategies and purposes. In particular, the selection of appropriate attributes, often from a very large number of possibilities, is critical. This can only be performed if some clear objective is in mind. All too often it is insufficient simply to take a data set, analyse it and see what groups emerge. For unless the data set is well-structured, consisting of natural groups, the choice of variables will predetermine the emergent groups. Such an economy of attributes not only helps to direct the studies effectively, but it also renders the task of data processing more efficient.

14.3 TACTICS IN CLASSIFICATION

Having briefly surveyed the strategies adopted in classification we may take an overview of the tactics by which these strategies are accomplished. Everitt (1974) has made a useful distinction between the following general forms:

(1) Hierarchical methods. Individuals can be formed into groups and these groups, in turn, into larger groups. Alternatively, the process may begin with one large proto-group which is successively sub-divided into smaller groups.

(2) Optimization – partitioning methods. Here the procedures are of a divisive character but, unlike the hierarchical form, permit the relocation of badly grouped individuals at some later stage in the work.

(3) Density of mode-seeking methods. Clusters of objects are formed on the basis of their lying within regions with a high density of individuals, separated from other such clusters by regions of low density, as in Figure 14.1. Such methods are only applicable in the search for 'natural' groupings and not in ordination exercises.

(4) Clumping techniques. These are not yet widely used in geography. Their most important characteristic is that groups are permitted to overlap. Hence their application is restricted to those cases where the user's needs permit such overlaps.

The hierarchical techniques have been most widely applied in geography and form the basis of what follows. Although the detailed considerations are necessarily selective, the exclusion of particular techniques should not be interpreted as suggesting that they are in any way less appropriate or valuable. The final choice belongs with the researcher; but it must be remembered that different methods treat the data in different fashions, not all of which may conform to the investigator's wishes. Some of these differences will be alluded to in the following sections.

Most hierarchical methods, and particularly those to be considered here, operate on the principle of enumeration. The objects to be classified are located in n-dimensional space. The positions of the objects and the dimensionality of the space are dictated by the enumerations and the number of variables respectively. In effect each variable requires a dimension or axis for it to be plotted; the result is that we often find ourselves working in n-dimensional space, where n can be greater than the three which govern our world. While we cannot imagine such a situation our arithmetic is not similarly constrained and can operate over as many dimensions as necessary. The 'distances' between objects located in n-dimensional space can be measured, and those close to one another are assumed to be similar according to the variables used to locate them and may be grouped together to form groups which may, in turn, be linked to others close by.

Because each variable contributes to the individual's location it is important that superfluous elements be excluded. In regression analysis partial coefficients can be used to weed out unwanted variables, but no such possibility exists in classification and the researcher must employ his initiative in this regard. Furthermore, whereas redundant variables in regression models are not detrimental to the model (in the sense of actually reducing explained sums of squares), this is not the case in classification where their inclusion may distort and obscure the groupings that are sought. Many of these issues can be clarified by a detailed example.

14.4 AN EXAMPLE OF AGGLOMERATIVE CLASSIFICATION

In this example West European states are considered on the basis of economic activity. To maintain a convenient level of simplicity attention is confined to two variables only: trading balance expressed in £m Sterling and national income expressed in £ per capita per annum. In most studies many more variables would be used, but the principles remain the same and can be more clearly indicated in the simpler case.

A useful first task, and one possible only for so few variables, is to summarize the data in graphical form, as in Figure 14.2. This simple procedure often provides many benefits and may confirm the presence, or absence, of natural groups or help to determine which precise course of action is to be adopted. It is implicit here that the selected variables have the power to discriminate between the states in a useful manner. Frequently, the validity of this assumption may only be confirmed by the outcome of the analysis. That is to say, do the groups defined stand up to scrutiny? Do they make sense?

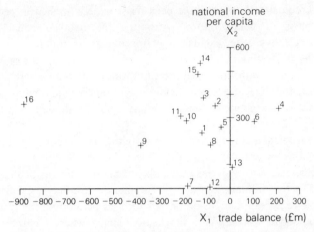

Figure 14.2 Statistical location of European states according
to trade balance and national income per capita; for key see
Figure 14.4

The example uses an agglomerative method, classifying by enumeration. As
Figure 14.2 indicates, the location in this two-dimensional model of the states is
determined by their values (enumerations) along each of the axes. These locations
determine, in turn, the statistical (not the geographical) distances between the
states, and these can be calculated by a straight forward application of
Pythagoras' theorem. In geometrical terms the distance between any two states is
the length of the hypotenuse of a right-angled triangle defined by the location on
the graph of the two states. (Figure 14.3) The lengths of the other two sides of the
triangle then depend on the differences between the two values on each of the two
axes. Distance D is given by:

$$D_{ij} = \sqrt{[(X_{1i} - X_{1j})^2 + (X_{2i} - X_{2j})^2]} \qquad (14.1)$$

Equation 14.1
X_{1i} = individual i's score on variable 1
X_{1j} = individual j's score on variable 1
X_{2i} = individual i's score on variable 2
X_{2j} = individual j's score on variable 2
D_{ij} = statistical distance between i and j

This calculation is repeated for all pairs of individuals and a distance (or
similarity) matrix may be constructed. The matrix for this example is shown in
Table 14.1. It must be remembered that these distances are not expressed in any
recognizable units and are relative only. Notice, too, that this two-dimensional approach is also the basis for the n-dimensional equation, which requires only that
the squared differences along the other axes be also included; hence:

$$D_{ij} = \sqrt{[(X_{1i} - X_{1j})^2 + (X_{2i} - X_{2j})^2 + (X_{3i} - X_{3j})^2 \ldots (X_{ni} - X_{nj})^2]} \quad (14.2)$$

Table 14.1 Distance matrix for the sixteen European states according to GNP per capita and trade balance

	1	2	3	4	5	6	7	8	9	10	11	12	13	14	15
2	127,8														
3	141.9	52.0													
4	333.3	274.3													
5	87.3	95.2	138.2	246.3											
6	217.8	185.5	237.8	119.0	132.7										
7	240.8	368.6	376.8	515.0	305.6	399.6									
8	52.6	164.1	188.3	328.3	91.7	203.4	214.0								
9	279.2	357.8	327.8	610.4	364.6	496.7	271.1	306.1							
10	95.2	132.3	108.2	393.7	157.5	287.8	286.9	146.4	225.5						
11	133.3	147.1	110.3	417.5	189.3	316.3	315.9	184.7	217.0	38.4					
12	230.9	350.9	372.7	445.4	269.5	335.3	100.6	186.9	353.4	304.9	340.2				
13	177.1	271.6	305.5	327.4	178.6	209.3	199.9	124.5	392.7	269.9	308.3	118.1			
14	292.7	177.0	150.9	370.0	272.4	327.2	526.0	337.4	432.6	214.5	227.8	523.2	449.2		
15	259.4	151.2	117.7	368.9	246.5	314.4	490.1	305.8	396.0	267.3	190.1	490.4	420.4	37.7	
16	801.8	826.8	778.3	1011	874.7	1003	811.4	841.3	551.2	717.2	687.3	901.0	941.4	788.9	768.2

Such distances are measures of similarity — the shorter the distance the greater the similarity. The matrix may now be searched to determine the two most similar members which can be merged to form a group. From Table 14.1 it is Sweden (14) and Switzerlend (15) that are merged, their places being taken by a new, single, location mid-way between them (the group centroid). The matrix is now decreased by one order and is recalculated to take account of the new point. The recalculated matrix is searched for its smallest value and the process of merging and recalculation repeated. At each stage the matrix order falls by one and the resulting group is presumed to act as a single point located at its geometric centre. This position, the group centroid, is determined from the mean of the individual points along each of the axes or dimensions. For example, the centre of a cluster of five individuals would be given by the means of the five trade balances and five national incomes. The principle is easily extended into as many dimensions as necessary.

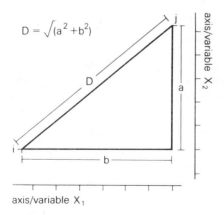

Figure 14.3 Distance D between two locations by use of Pythagoras' theorem

The point is quickly reached when most individuals are integrated into one or other of the emerging groups. But the process of merging continues with the groups themselves until one large proto-group only exists. When groups are merged the convention adopted here has been to estimate the centroid position by reference to all the individuals with the new group. If a group of three is merged with a group of five all eight members are included in the calculations.

The calculations, though not difficult, are repetitive and provide a long sequence of results that may be summarized in the form of a dendrogram. This is a graph in which one axis lists the individuals and the other the levels at which they, and their consequent sub-groups, merge. These levels are expressed by the distances between the groups at the point of merging. The graph is completed by the tree-like structure shown in Figure 14.4, in which the pathways of merging are plotted. Sweden and Switzerlend, for example merged first at a distance of 37.7 units, and they are connected at that point on the dendrogram. Other countries, such as France and Germany or Greece and Portugal, also merge but at higher levels, their distances being 119.0 and 100.6 units respectively. It can also be seen that as the groups become larger the distances between their centroids increases and the mergers occur at ever higher levels.

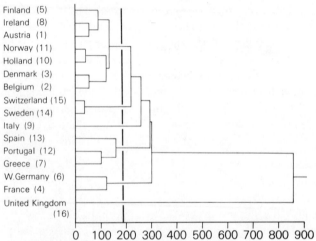

Figure 14.4 Linkage tree of European states. The thicker line shows the sub-divisions at the 6-group level. The code numbers of the states are given in brackets

At this point one of the difficulties of hierarchical procedures, already suggested, emerges very clearly. At what point along the grouping sequence should the process cease? There is little point in pursuing the sequence to its end point. In the absence of natural groups the decision will often be an arbitrary one to be made in the light of the researcher's particular objectives. In the present example it was considered that six groups were sufficient and a line was drawn at the appropriate position on the dendrogram (see Figure 14.4). The appropriate groups may then be abstracted. Those found here are listed in Table 14.2. The groups do, despite the few variables, used, indicate a degree of coherence. The South European states (excluding Italy) emerge as an entity. The traditionally more prosperous states of France and Germany are linked. The two neutral states, Sweden and Switzerland, can be distinguished, while the UK and Italy have such unusual economic characteristics to dispose them to remain as individuals until very close to the end. The largest group is far more difficult to discuss in generalities and may indicate the need for further variables to produce a more decisive grouping. At the same time, the trade balances of both Italy and the UK tend to overwhelm the influence of the one other variable, again emphasizing the need for more variables.

Table 14.2 State groupings classified by GNP per capita and trade balance

Sweden	Portugal
Switzerland	Spain
	Greece
Belgium	
Denmark	France
Holland	West Germany
Norway	
Austria	United Kingdom
Ireland	
Finland	Italy

14.5 PROBLEMS ARISING FROM THE EXAMPLE

In the above example a number of conventions were adopted that require explanation and discussion. The group centroid, for instance, was defined by reference to its members. It could equally have been based on the centroids of the two component groups alone. This would certainly render the calculations less demanding, but would overlook the unequal weighting resulting from the merger of large and small groups. As a result a sub-group of three members would receive weighting equal to a sub-group of, say, ten. Having said this, there are occasions when it might be beneficial to preserve the character of the smaller group and not to allow it to be swallowed up by its larger companion. In this case equal weighting might be quite acceptable. Both methods are widely adopted in computer programs, and it is as well to check the procedures before use as this matter may be critical.

Neither will it have escaped the reader's attention that the distances are estimated on the basis of the numerical values of the attributes or variables. Consequently, attributes with large numerical values will exercise an exaggerated degree of discrimination. Population, for example, may be measured in millions while personal income would, more probably, be in the range of thousands. Any classification based on two such disparate scales may well indicate preferential discrimination by the former.

This difficulty, by no means an uncommon one, is best overcome by standardizing the observations. In this manner all attributes receive the same numerical weighting. In the current example the measurement scales are not vastly different and reclassification using standardized data yielded no appreciable differences. But where such scaling similarities are absent it is always sound practice to adopt standardized data.

A priori assumptions concerning the relative importance of different variables can be made; but they are difficult to substantiate and even more troublesome to translate into numerical terms. Thus, even if it is known that a particular variable is more important than another how can this dominance be expressed as a numerical weighting? Should the scaling factor be 2.0, 3.0, 4.6 or some other quantity?

Attention needs also to be directed towards any correlations between the attributes. In Figure 14.4 the two variables were plotted at right angles to one another. This is the accepted convention but, in the statistical sense, it assumes that the variables are truly orthogonal, i.e. uncorrelated. The more strongly correlated are the two variables the less orthogonal would be the axes, and the angle between the axes should decrease to zero as a perfect positive correlation is approached, and open towards 180° for perfect negative correlations. In the example the correlation (−0.12) was well within the compass of random variation and the problem did not arise. When large numbers of variables are employed such problems are commonplace. One solution is to use a technique known as principal components analysis in order to reduce the number of variables without simply excluding them. Both principal components analysis and its partner factor analysis will be discussed in a later chapter; at this stage it is sufficient to remark that the method replaces a large number of variables with a far smaller number of principal components. The latter should be viewed as composite variables synthesized from the raw data. Furthermore such axes can be made genuinly orthogonal and, hence, zero correlated. Thus the use of principal components analysis overcomes simultaneously the problems of too many variables and their intercorrelations. The replacement of variables by components means that the individuals are enumerated by component scores and not by directly measured quantities. Nevertheless, it must not be expected that these methods will regularly reduce the number of attributes (now components) to a visually manageable three or two, and *n*-dimensional geometry will still be required on most occasions.

A further note of caution is necessary. If the data are in distinct groups then those groups will probably emerge even after such drastic transformations. But if the problem is one of the ordination of a homogeneous data set, then rather different results may be arrived at as a consequence of such data management.

Questions need also to be answered concerning the geometry of group centres. The method used above takes the arithmetic mean location along each of the axes. In mechanical terms such a centroid is analogous to the centre of gravity of a body and as such has an intuitive appeal. But there are alternatives to this. In particular, some procedures measure inter-group distances not from the centroid but from the two nearest members of the two groups, or even the two furthest members. Such conventions are not as bizarre as they appear at first sight. There are occasions when inter-group distances by centroids are highly inappropriate. Thus far the groups have been assumed to be equidimensional, i.e. circular in two dimensions, spherical in three and so on. For many purposes this is adequate, but it is not universally valid and groups arise that are far from equidimensional, but still groups for all that. An example of such a group is suggested in Figure 14.5. Under these circumstances the group centroid might fall beyond the limits of what we might visually perceive as the boundary, and it is important to possess alternative inter-group distance definitions.

Extremes of this character are, thankfully, rare. Furthermore, classification by ordination, the commonly preferred method by geographers, creates groups that are equidimensional, or acceptably so. As a result the group centroid is rarely obsolescent.

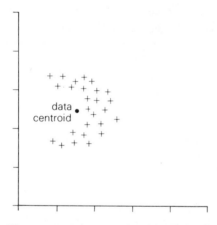

Figure 14.5 Cluster of points the data certroid of which lies beyond the visual limits of the group

14.6 WARD'S GROUPING METHOD

This is an increasingly popular procedure and one which has much in common with the method outlined above. It is hierarchical, agglomerative and employs the group centroid when estimating inter-group distances. There are differences in detail, however, and though these are advantageous they are gained at the expense of yet greater computational effort. The problems of classification are here approached through the medium of sums of squares. Each group possesses a sum of squares provided by the scatter of its members about the centroid. Until now the sum of squares has been regarded largely in a 'non-dimensional' context, i.e. with regard to one variable only. But applications of the Pythagoras method discussed above permit this concept to be extended into n-dimensions (variables). The primary criterion when assembling groups with this method is to ensure that the inevitable increase in the within-groups sum of squares is kept to a minimum. The within-group sum of squares, termed by Ward the 'error sum of squares' or ESS, is accumulated over all the groups at each stage. At the first stage, with no groups, the ESS is zero; but as grouping proceeds the ESS grows ever larger and reflects the loss of information as groups and generalizations replace individuals and detail. The data for the previous example can be reapplied, though the standardized form will be used.

At each level the data are searched to find the merger which gives the minimum increase in ESS and, hence, minimum loss of information. The first merger, then, is between Sweden and Switzerland, giving an ESS of 0.0205. Because of this criterion there is not a steady increase in ESS, but a slow growth over the first few steps and a much accelerated increase over the final stages. Hence, though the second merger (Norway and Holland) raises the ESS by 0.0241 to only 0.0446, the final stage which brings the UK into the group raises it from 19.03 to 30.21. Such a pattern can, however, be put to good use in helping to identify an optimum grouping stage. If a check is kept of the increase in ESS, then the optimum grouping occurs at the point at which ESS begins to accumulate rapidly. A graphical representation such as Figure 14.6 is often valuable. In this example step 10 appears to be critical, with the ESS curve turning sharply upwards at that point. The results are all but identical with those achieved by the other method, as might be expected. Interestingly this, possibly more objective, method of optimum grouping gives an identical arrangement to that indicated in Table 14.2, supporting thereby the purely subjective suggestion for a six-group arrangement. The groups are retabulated in Table 14.3, which also lists the group sums of squares at step 10.

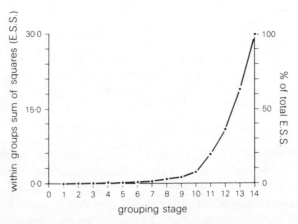

Figure 14.6 Accumulation of within-groups sums of squares as clustering proceeds

The results may also be presented in dedrogram form with ESS replacing pure distance on the horizontal axis. This does, however, lead to a concentration of mergers at the lower end of the scale, with the long 'tail' for the final few steps (see Figure 14.7). One means of overcoming this imbalance would be to adopt a logarithmic scale for ESS, though this has not been done here.

A point that needs to be stressed concerning both the methods thus far discussed, and many others, is that neither can guarantee optimal groupings at each stage. This is particularly important in Ward's method, since at any stage the observed ESS, though arrived at by minimum increments, need not be the minimum at that level. Generally, and fortunately, the differences are minimal and, furthermore, it would be impractical to search the data for possible minor improvements at each step. Most procedures, therefore, do not allow for the subsequent reallocation of badly grouped individuals at later stages. In this simple example the problem did not arise, but this may not always be the case.

Table 14.3 Group sums of squares according to Ward's algorithm

Group composition	Group sum of squares (ESS)
Sweden, Switzerland	0.0205
Belgium, Denmark, Holland, Norway, Austria, Ireland, Finland	1.6067
Greece, Spain, Portugal	0.4990
France, West Germany	0.1682
Italy	0.0
United Kingdom	0.0
	2.2951

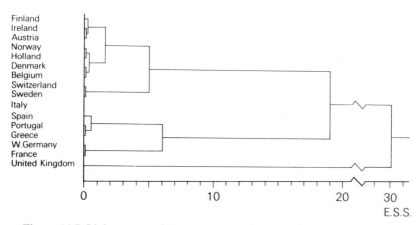

Figure 14.7 Linkage tree of European states by Ward's grouping method

14.7 ASSOCIATION ANALYSIS: AN EXAMPLE OF A DIVISIVE PROCEDURE

Thus far the only hierarchical methods to be considered have been agglomerative. But divisive measures, those that subdivide progressively from a large data set, are also available. Again there is a bewildering variety of techniques of which only one, association analysis, has been selected for detailed examination. As a divisive method it has become popular with geographers and biogeographers especially. It has the additional advantage of using nominal data and dispels the possible notion that numerical classification may proceed only with parametric data.

In principal, association analysis is a direct development from the chi-square test (Chapter 9). The results of two-by-two contingency table tests are used to determine the course of the successive sub-divisions of the data. The raw data consist of, as usual, a set of individuals, each of which is now described by a number of dichotomous variables. But these issues may be best elaborated with an example taken from biogeography, the area where association analysis is most frequently applied and where it has its origins. The data are from a quadrat survey conducted to determine the floristic composition of a small area of heathland. The survey grid was 20 × 20 transects, giving 400 survey points. Each of these points represents an individual for the purposes of this exercise, and the five major plant species of the area form the attributes of variables at each location. The species are recorded simply on a presence of absence basis; hence the dichotomous character of the data. The five species, and the code letters that will be used for them, are given below:

A *Calluna vulgaris*
B *Molinia caerulea*
C *Erica tetralix*
D *E. cinerea*
E *Festuca ovina*

The data are initially prepared in the form of a presence–absence matrix (see Table 14.4 in which part of the matrix is shown). Each row represents a location and each column one of the five species. The cells contain either 1 or 0 according to whether the species is present or absent at that location. In this way each individual or location is enumerated by reference to five species or attributes.

Table 14.4 Part of record sheet for presence and absence of species by quadrat

| Quadrat | Species (by code letter) | | | | |
	A	B	C	D	E
1	0	1	0	0	0
2	1	1	0	0	0
3	0	0	1	1	0
4	0	0	1	0	0
5	1	0	1	0	1
.	.	.	.	.	.
.	.	.	.	.	.

Two-by-two contingency tables may now be prepared for each possible pairing of species. The frequencies to be entered in the tables are derived directly from the presence-absence matrix. The table cells reflect the four possible outcomes for any pair of species at a given location (Table 14.4); both may be present, both may be absent or either one may be present with the other absent. Consultation of the raw data indicates, for each location, which of the four outcomes is registered. The total frequencies over all individuals (survey points) are entered in the contingency tables. The process is repeated for each species pair.

The strength of the association between species is assessed directly by the chi-square statistic (estimated as in Chapter 9). Stronger associations are reflected in higher chi-square values. But there are departures from conventional chi-square methodology. Most importantly, distinctions are drawn between positive and negative associations. The former arise when frequencies concentrate in the presence–presence and absence–absence cells, the latter when they concentrate in the two presence–absence cells (Table 14.5). Positive associations indicate pairs of species with similar ecological demands, negative associations suggest pairs with contrasting requirements. Both, however, are equally informative from a classification point of view. Low or zero associations are demonstrated, as usual, by a more or less equable distribution of frequencies among the four cells.

Table 14.5 Layout of presence–absence
matrix for association analysis

	Species X present	Species X absent
Species Y present		
Species Y absent		

The chi-square statistics are then entered in a summary table (see Table 14.6). The roman typeface indicates significant positive associations, italic script significant negative associations (both at the 0.05 level), with non-significant values denoted by a + and not their numerical values which, by definition, are randomly derived and can offer no interpetative value. This style of presentation was used by Williams and Lambert (1959) in their original studies with this technique.

Table 14.6 Significant X^2 values for inter-species associations (first level)

	A	B	C	D	E
A	—	34.0	13.9	+	+
B	34.0	—	65.2	*27.0*	+
C	13.9	65.2	—	*17.1*	*7.0*
D	+	*27.0*	*17.1*	—	41.5
E	+	+	*7.0*	41.5	—
$\Sigma\chi^2$	47.9	126.6	103.2	85.6	48.5

From the information displayed on the table the first sub-division may be made. Association analysis proceeds by sub-dividing on the basis of the species which, at each level, has the highest total degree of association with its fellows. Quadrats in which that species were registered are separated from those in which it was not, giving thereby, the first two groups. The $\Sigma\chi^2$ values are used to distinguish this species, the highest value indicating the highest degree of total association. In the present example this first sub-division takes place on species B. The two sub-groups of quadrats, those with and those without *M. caerulea,* are re-examined separately and new tables of associations prepared for each based on the four remaining species, but prepared by otherwise identical means. The results are shown in Table 14.7.

Table 14.7 Significant χ^2 values for inter-species associations (second level)

	A	C	D	E		A	C	D	E
A	—	10.7	+	+	A	—	+	+	+
C	10.7	—	+	+	C	+	—	+	+
D	+	+	—	+	D	+	+	—	7.6
E	+	+	+	—	E	+	+	7.6	—
$\Sigma\chi^2$	10.7	10.7	0	0	$\Sigma\chi^2$	0	0	7.6	7.6

sub-group B   sub-group b

The resulting levels of association are inevitably lower. Nevertheless, critical species within each sub-group may again be identified. Further sub-divisions take place on these species to produce four sub-groups. This second-level sub-division need not proceed on the same species for each group. Hence group B (in which *M. Caerulea* is present) sub-divides on *E. tetralix,* and groub b (in which *M. caerulea* is absent) sub-divides on *E. cinerea.* However, in both instances there is what is termed an ambiguity whereby two species share the maximum $\Sigma\chi^2$ value. This difficulty is resolved by reference to the χ^2 values at the previous level; the species recording the higher value being used to make the next division. In group B, therefore, sub-division proceeds on *E. tetralix* and not *C. vulgaris.* For the same reason, *E. cinerea* is preferred to *F. ovina* in group b, and the sub-divisions take place on the presence and absence of those species in the respective groups.

These four groups can, in turn, be examined through the three remaining species in each. However, each of the pairwise associations fails to reach the significance level of 0.05, and no further sub-divisions are possible. The final grouping consists of the four, mutually exclusive, plant associations outlined above. In detail they are:

BC *M. caerulea* present/*E. tetralix* present
Bc *M. caerulae* present/*E. tetralix* absent
bD *M. caerulea* absent/*E. cinerea* present
bd *M. caerulea* absent/*E.cinerea* absent

The results may also be presented in the form of an annotated dendrogram (Figure 14.8). With this grouping scheme each of the 400 quadrat locations may be placed into its appropriate class. The sizes of the classes vary from 189 quadrats in BC down to 38 in bd. Furthermore, each quadrat may be mapped to produce a generalized scheme for the survey area defined in terms of the four classes. Such maps often clarify problems identified in the field and enable verifications to be made of the ecological intepretation of the grouping. They may also emphasize zonations that are far from clear on the ground (Figure 14.9).

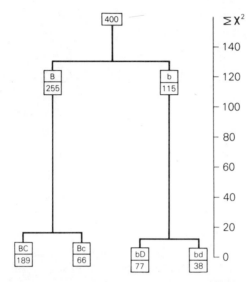

Figure 14.8 Linkage tree of plant species by association analysis (upper case characters indicate species present, and lower case their absence); the number of quadrats within each group are given in the lower section of each box

A valuable test of all derived groupings is to establish their, in this case ecological, justification. In the present case interpretations might run as follows. *M. caerulea* is a plant of moist heaths and damp conditions. Consequently the first, and most important, sub-division may be to distinguish wet and dry ecotypes. In addition, small areas within the survey site are known to be burned over periodically, and although *M. caerulea* recovers rapidly *E. tetralix,* another plant of moist heaths, requires a longer regeneration period. The sub-division of the moist quadrats (B) into those in which *E. tetralix* was and was not registered may well reflect a distinction between areas recently burned and areas that have been left for a number of years.

Quadrats from which *M. caerulea* is absent are interpreted as drier, its place as the dominant species being taken by *E. cinerea,* a heath of dry conditions; and it is in this light that group bC is interpreted. The smallest group, in terms of representative quadrats, is that in which both *M. caerulea* and *E. cinerea* are absent. Such areas may be defined as dry, but the indeterminate floristic composition makes further interpretation difficult. In such cases the map location of quadrat members of the group may shed light on its ecological significance, perhaps identifying zones with particular, and unexpectedly important, conditions.

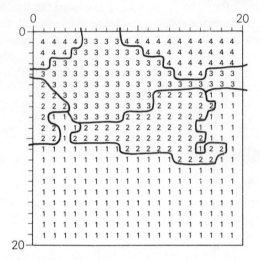

Figure 14.9 Map of quadrats grouped according to the results in Figure 14.8; the groups are BC = 1, Bc = 2, bD = 3 and bd = 4

Association analysis is a direct application of a hypothesis testing technique; but because of its peculiar usage it is not required to test hypotheses in the strict probabilistic sense, although there may well be a working hypothesis in the researcher's mind. Owing to this characteristic there have grown up a number of variants on the theme outlined above. Some workers, for example, prefer to include all chi-square statistics, significant or not, when calculating $\Sigma\chi^2$. Such a recourse certainly has the advantage of dispelling any problems connected with the selection of significance levels, since it should be recalled that different levels may provide different solutions in marginal cases. Having said this, however, the 0.05 level is widely accepted, and although it is admittedly unusual to use significance levels when there is no null hypothesis to test, they do eliminate those random, and consequently meaningless, values that would otherwise contribute towards the $\Sigma\chi^2$ value.

There are also variants on the division criterion. In our example $\Sigma\chi^2$ was used, but the value of $\Sigma\sqrt{~^2}$ is a popular alternative, although one that would not alter the selection of critical species. Others include $E\chi^2/N$ (where N is the number of quadrats) and max $[\Sigma |AD{-}BC|~]$ where A, B, C and D are the symbols we used in Section 9.9 when examining contingency tables.

14.8 CONCLUSIONS

By far the most widespread application of this method has been in biogeography. But it must also be viewed as a valuable non-parametric counterpart to the many other methods in classification. Its most serious problem, and one which it shares with all classification techniques, is the repetitious and voluminous nature of the calculations. This makes computer assistance almost indispensible. Even in this simple case, with only five species, as many as ten possible pairings were created and 400 records had to be searched to complete the contingency tables for each pair. The problem may be alleviated by excluding rarities (those species that are recorded in less than 2 per cent of the quadrats). Even allowing for this very many calculations are necessary. The same arguments apply in the other examples, where, even if the number of variables is reduced, the calculations soon become daunting. As a result many workers may be circumscribed not by their knowledge or objective but by what is available in terms of local computer software and capacity.

The geographer who chooses to write his own programs will encounter problems. In particular, which method should be adopted? There are no easy answers to this question. In the absence of a well-structured data set different methods will yield different groupings. Group stability might well be tested using a number of techniques, but this may only multiply the effort unnecessarily. Ultimately, much will depend on the insight that the researcher can bring to bear on his problem. Classification is not to be undertaken lightly.

REFERENCES

Everitt, B. (1974). *Cluster Analysis,* Heinemann, London.
Harvey, D. (1969). *Explanation in Geography,* Edward Arnold, London.
Horton, R. E. (1945). 'Erosional development of streams and their drainage basins, hydrophysical approach to quantitative morphology', *Bull. Geol. Soc. Am.* **56,** 275–370.
Lewis, P. F. (1961). 'Dichotomous keys to the Koppen system', *Prof. Geogr* **13,** 25–31.
Miller, A. A. (1931). *Climatology,* Methuen, London.
Thornthwaite, C. W. (1931). 'The climates of North America according to a new classification', *Geog. Rev.* **21,** 633–655.
Wilcock, A. A. (1968). 'Koppen after fifty years', *Anns Ass. Am. Geogrs* **58,** 12–28.
Williams, W. T., and Lambert, J. M. (1959). 'Multivariate methods in plant ecology, I: Association analysis in plant commuties', *J. Ecol.* **47,** 83–101.

RECOMMENDED READING

Bunge, W. (1966). 'Gerrymandering, geography and grouping', *Geog. Rev.* **56,** 256–263. A very good introduction for those interested in moving from the field of typological classification into regionalization (spatially controlled classification).
Clark, D. (1973). 'Urban linkage and regional structure in Wales: an analysis of change 1958–68' *Trans. Inst. Brit. Geogrs* **38,** 41–58. This paper provides an example of the use of the 'distance' measure as applied to Ward's algorithm. Data reduction from a large initial set, interestingly, is achieved by canonical vectors. The work is executed in a spatial/regionalizing context.
Everitt, B. (1974). *Cluster Analysis,* Heinemann, London. A non-geographical but thoroughly readable text which explains the principles of a wide range of techniques without ever becoming over-technical.

Fesenmaier, D. R., Goodchild, M. F., and Morrison, S. (1979). 'The spatial structure of the rural–urban fringe: a multivariate approach', *Canadian Geogr.* **23**, 255–265. This is a short paper dealing not only with classification but drawing attention also to the inherent 'fuzziness' of groups determined within essentially homogenous data sets. It includes a good illustration of the manner in which ESS accumulates as agglomeration proceeds.

Harrison, C. M. (1971). 'Recent approaches to the description and analysis of vegetation', *Trans. Inst. Brit. Geogrs* **52**, 113–127. A valuable discussion of developments within biogeographical classification, including association analysis.

Johnston, R. J. (1968). 'Choice in classification; the subjectivity of objective methods', *Anns Ass. Am. Geogrs* **58**, 575–589. An eminently readable and clear exposition of the difficulties of achieving a genuinely objective classification of geographical data. (1976). *Classification in Geography,* Catmog 6, Geobooks, Norwich. A valuable text, concisely written with non-mathematical geographers in mind. By no means as comprehensive as, say, Everitt's text, but very useful.

Parsons, A. J. (1978). 'A technique for the classification of hillslope forms', *Trans. Inst. Brit. Geogrs* (N.S.) **3**, 432–433. Although the methodological explanation is limited in this paper it offers an example of how measured attributes in physical geography may be employed for classification. The paper also discusses the problem of group identification in data sets at different similarity levels.

Sokal, R. R., and Sneath, P. H. A. (1969). *Principles of Numerical Taxonomy,* Freeman, San Francisco. This book is aimed primarily at the life science disciplines. It is highly technical and thorough but well worth consulting by those with greater mathematical expertise.

Williams, W. T., and Lambert, J. M. (1959). 'Multivariate methods in plant ecology, I: Association analysis in plant communities', *J. Ecol.* **47**, 83–101. A detailed and clearly written presentation of the principles and application of association analysis. Further, and closely related, material by the same authors appears in the same journal for 1960 and 1961.

The following references contain applications of classification and clustering techniques. In detail, the approaches differ from those adopted in the text. However, the material offers a valuable selection of the different means by which the same statistical end may be gained.

Austin, M. P., and Orloci, L. (1966). 'Geometric models in ecology, II: An evaluation of some ordination techniques', *J. Ecol.* **54**, 217–227.

Eyles, R. J. (1971). 'A classification of West Malaysian drainage basins', *Anns Ass. Am. Geogrs* **61**, 460–467.

Grimshaw, P. N., Shepherd, M. J., and Willmott, A. J. (1970). 'An application of cluster analysis by computer to the study of urban morphology', *Trans. Inst. Brit. Geogrs* **51**, 143–161.

King, L. J. (1969). *Statistical Analysis in Geography* (Chapter 8), Prentice-Hall, Eaglewood Cliffs.

Mather, P. M., and Doornkamp, J. C. (1970). 'Multivariate analysis in geography with particular reference to drainage basin morphology', *Trans. Inst. Brit. Geogrs* **51**, 163–187.

Yapa, L. S., and Mayfield, R. C. (1978). 'Non-adoption of innovations: evidence from discriminant analysis', *Econ. Geog.* **54**, 145–156.

Chapter 15

Factor Analysis and Related Techniques

15.1 INTRODUCTION

The aim of this chapter is to introduce the practical aspects associated with the application of factor analysis, rather than dealing in detail with any theoretical problems. Consequently, most of the applications and examples are based on the use of readily available computer programs, such as SPSS. We shall start by examining the possible uses of factor analysis, before discussing the variety of techniques that are available to the geographer.

As was suggested in Chapter 13, the most important feature of factor techniques is their ability to reduce a large data set down to a smaller number of factors. This replacement process may be undertaken for a variety of reasons, although at least three main ones can be identified. First, we may wish to produce new combinations of the original data, which may then be used as new variables in some further analysis. For example, we may use factor analysis to combine so-called independent variables in multiple regression to reduce the effect of collinearity. Second, we may wish to reduce the number of variables under investigation. Third, we may use it for exploratory purposes, in an attempt to detect and identify groups of interrelated variables. It is within this last area that most geographical uses of factor techniques are to be found, although as more of these studies are undertaken the exploratory role becomes of less importance. This is partly the case with the work on the factorial ecologies of urban areas. Thus, the earlier exploratory studies revealed three major patterns concerned with socioeconomic status, stage in the life cycle and segregation. These have formed the basis of further work, with factor analysis being used to confirm or reject hypotheses.

At the outset it should be stressed that the term factor analysis does not refer to a single technique, but covers a variety of approaches. Initially, we can recognize three main stages in the application of factor analysis, each of which offers different alternatives to the user (Table 15.1). In the first stage most factor analyses require product-moment correlation coefficients as basic inputs. At this level the alternatives are between using information measuring correlations among different variables for a group of observations, or taking a matrix of correlation coefficients measuring the relationship between a set of individuals. In the first instance we may examine different areas of a city in terms of a variety of census variables measuring social and economic factors — this is called R-mode factoring. By contrast, Q-

273

mode factoring refers to the second approach, in which specified variables are examined over a range of observations. Thus, we may examine the relationships between variables measuring imports and exports for a number of countries. Therefore, in Q-mode analysis the variables form the rows in the correlation matrix; whereas in R-mode factoring the observations form the rows.

Table 15.1 A simplified view of types of factor analysis

Stage	Option types	Terminology
Correlation	(a) Between variables	R–factoring
Matrix	(b) Between individuals	Q–factoring
Extraction of initial factor	(a) Defined factors	Principal components
	(b) Inferred factors	Factor analysis
Rotation to final factors	(a) Uncorrelated	Orthogonal
	(b) Correlated	Oblique

The second major stage is to explore the possibilities of data reduction by constructing a new set of variables based on the interrelationships in the correlation matrix. There are two basic approaches at this level — principal component analysis and factor analysis (Table 15.1).

In both these approaches, new variables are defined as mathematical transformations of the original data. However, in the factor analysis model the assumption is that the observed correlations are largely the result of some underlying regularity in the basic data. Specifically, it assumes that the original variable is influenced by various determinants; a part shared by other variables, known as the common variance; and a unique variance. The latter is the residual from the multiple relationship, and consists of variance accounted for by influences specific to each variable and also that relating to measurement error. In contrast, principal component analysis makes no underlying assumptions about the structure of the original variables, neither does it hypothesize an element of unique variance for individual variables (Figure 15.1). These differences will be examined in more detail in Section 15.3, when the relevance of the two approaches will be discussed.

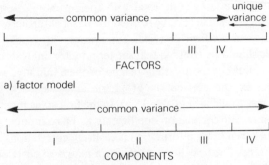

a) factor model

b) principal components model

Figure 15.1 Apportionment of variance in (a) the factor model and (b) the principal components model

The final stage in which variations are possible is the search for interpretable factors. In this search to define the underlying dimensions of a data set a number of solutions are available, which may involve manipulating or rotating the factors, as is shown in Section 15.6.

15.2 THEORETICAL CONSIDERATIONS AND TERMINOLOGY

Before we examine the operational aspects of factor analysis we do need to consider a little of its theoretical background. There are two basic approaches to this, one based on algebraic solutions and the second on geometric interpretations. It is the latter that we shall use here; those readers interested in more detailed explanations should consult Harman (1967).

Let us start by considering how, in an earlier part of this book (Chapter 11), we represented a set of data in terms of their relationships with two variables. In an example we examined settlements in relation to their population totals and numbers of shops. For this simple case the data can be plotted on a two-dimensional scattergraph, with the X and Y axes defining the two variables. As we saw in Chapter 11, in the case of two variables, if all the points fall along a straight line then we have a perfect correlation. Alternatively, if there is no correlation between the two variables the points form a circular scatter on the graph. From the point of view of the geometry of such graphs, these examples are the two limiting and extreme cases, between which all other distributions can be described by ellipses. This point is illustrated in Figure 15.2, which shows the extremes (a) and (c) together with the type of distribution (b) we may find occurring between the two. From this diagram one further important feature can be seen, which is that every distribution or ellipse of points can be defined by two axes. These are a major principal axis running the length of the distribution and a minor principal axis at right angles to the first (Figure 15.2(c)).

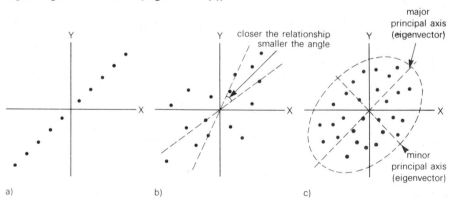

Figure 15.2 Bivariate distributions and types of relationships: (a) perfect correlation, (b) moderate correlation, (c) no correlation

The major principal axis measures the common variance between the variables and may be compared with a regression line for the two variable case. The minor axis indicates the residual variance, i.e. that which is not explained by the relationship. Thus, when we have a perfect relationship, with a correlation coefficient of $+1.0$, the two axes correspond; and the closer the relationship the smaller the angle between the two.

The concern in factor analysis is to define the principal axes for a number of variables, rather than for just two. In this context the axes are defined in n-dimensional space, and obviously cannot be visualized. This is part of the procedure to produce a new set of variables or factors, from the original ones. The aim, in other words, is to find a mean variable that is closely related to our original variables. The process of defining such 'new' variables is carried out by computers and the use of package programs. In the terminology used in factor analysis, the eigenvectors define our principal axes and represent our new variables. A simple example will illustrate the structure of the newly defined factors, although it should be stressed that the mathematical routines used by computer programs are far more accurate (Cooley and Lohnes, 1971).

Let us consider the relationships between the three variables given by the product-moment correlations in Table 15.2. From this information it is possible to calculate a common factor, as a function of the square root of the sum of their mutual correlations. The relationship between each variable with the factor can be calculated by dividing each variable's total correlations by the square root of the total sum of the correlations. These values have been calculated for our simple example and are presented in Table 15.3. In factor analysis these values are known as loadings and they represent the correlations between the original variables and the new factor. As such they can be treated as correlation coefficients, and therefore the square of each value is the proportion of the variance in the individual variable that is associated with the factor. Taken together, these squared factor loadings represent the total variance accounted for by the factor, which is referred to as the eigenvalue.

Table 15.2 Estimation of new mean variable or factor

	X_1	X_2	X_3
X_1	1.0	0.6	0.7
X_2	0.6	1.0	0.8
X_3	0.7	0.8	1.0
Sum of correlations	2.3	2.4	2.5

Factor loadings = sum of each correlation/$\sqrt{}$(total sum of correlations).

Before we proceed to examine the importance of eigenvalues let us first consider the other aspect associated with the extraction of factors. This concerns the concept of communality, which is the proportion of variance accounted for by the common factors. However, the communalities are also the values used in the diagonal of the correlation matrix, i.e. they replace the values of one in Table 15.2. In this matrix, communalities are defined as the correlation of each variable with itself due to the common elements part only, that is the part of the variable common to other variables. (Cattell, 1978). In our simple example we use the value of 1, but Cattell argues against this, and that other estimates should be used. As such this produces circular reasoning, since communality is the sum of the squared factor loadings which are not known until the factors have been determined. However, the factors cannot be calculated from the correlation matrix until the communalities have been inserted. This problem is resolved by package programs through the use of iterative procedures based on estimates of communalities. We need not concern ourselves with the technical details of this problem, but merely register awareness of its existence, and the variety of solutions to it that may be used by different package programs (Cattell, 1978; Cooley and Lohnes, 1971).

Table 15.3 Factor loadings and the eigenvalue

Variable	Loading	Square loading
X_1	0.85	0.72
X_2	0.98	0.89
X_3	0.93	0.86
		Eigenvalue $= 2.37$

Percentage trace of factor $1 = 79$. .

In our example developed in Tables 15.2 and 15.3 we extracted one new factor from our original variables, and measured its total variance or eigenvalue. To consider the overall importance of this eigenvalue we can relate it to the total variance in the correlation matrix of the original variables. This can be measured by using equation 15.1, and is sometimes referred to as the percentage trace of the factor:

$$\text{Percentage of variance} = (\lambda_I/n) \times 100 \qquad (15.1)$$

Equation 15.1
$\lambda_I =$ eigenvalue of factor I
$n =$ number of original variables

From Table 15.2, we can see that in our simple example the first factor accounts for 79 per cent of the total variance. Therefore, this first factor does not account for all the correlation in the original variables.

The next step in the procedure is to derive a second new factor to explain the remaining variance. This is achieved by the subtraction of what can be accounted for by the first factor from the original correlation matrix. Once again we shall not go into the details of this, but those interested are directed to Cattell (1978). However, it should be stressed that the second factor subsumes the parts of the variables not related to the first factor, and is therefore orthogonal or uncorrelated to it. This process can be repeated until we produce a further set of factors that accounts for the correlations in the original variables.

Finally, it is worth repeating that this section has not attempted to explain the mathematical principles of factor analysis, but rather to to review the underlying procedures and basic terminology.

15.3. FACTOR ANALYSIS AND PRINCIPAL COMPONENTS: A COMPARISON

As was explained in Section 15.1 there are two basic factor models: namely, common factor analysis and principal components. Geographers have used both approaches, although a glance through the literature shows that preference has been given to the common factor solution. From our point of view the main points of interest in this comparison between the two models are: first, what essential differences exist between them; and second, how do we decide which one to use.

Let us first consider the main differences between the two in terms of their assumptions and their results. The principal component model assumes a closed system where all the statistical variation in the variables is explained by the variables themselves. In fact, we used a principal component solution in the estimates of the communalities for our example in Table 15.2, when we assumed unity in the diagonal of the correlation matrix. This assumes high correlations between all the variables, with high common variances and low unqiue variances. Statistically, the principal component model is favoured for the easy solution to the communalities problem, and because of this many statistical texts prefer it to common factor analysis (Blackith and Reyment, 1971; Chatfield and Collins 1980).

The common factor model does have problems relating to the estimation of communalities, but unlike principal components it does not assume a closed system. It is this feature which makes it so attractive to geographers, who are dealing with situations where it would be totally unrealistic to assume a closed model. Thus, in most of our studies it is likely that we have not collected all the variables and that some degree of measurement error exists. Common factor analysis enables such problems to be taken into account, and any variance which is unexplained by the factors can be described by a residual error term.

Given these differences we can go on to consider the criteria for selecting one model in preference to the other. First, it should be pointed out that under certain conditions both models would give very similar solutions and work equally well. For example, when all the correlations between the original variables are high and communalities of one are likely estimates, then there may be little difference between the two approaches. Alternatively, if there are some variables in the correlation matrix that have low correlations the substitution of unity in the diagonal will be an overestimate, and the two approaches are likely to give different results.

We shall not concern ourselves too much with the theoretical differences, but rather focus on the practical issues that determine the use of the two models. Thus, we can regard the common factor approach as the most realistic for problems where some measurement error may be involved and some underlying structure is assumed about the relationships between the variables. In contrast, the principal component model deals with a more limiting case, in which no underlying structure is sought and the main aim is purely that of data reduction. The general characteristics of the two approaches are given in Table 15.4, which also hints at the circumstances when one technique may be used in preference to the other. It also indicates the relevant sub-programs for each model contained within standard packages. Clearly, from what we have said common factor analysis provides a more comprehensive model for the needs of most lines of geographical inquiry. For this reason the remainder of this chapter will focus on the application of factor analysis rather than on the more limiting case of principal components.

Table 15.4 Basic characteristics of principal components and common factor analysis

	Principal components	Factor analysis
Character	Assumes a closed system, with no assumptions about underlying structure of variables; identifies only the common variance between variables	Realistic assumptions concerning measurement error; allows search for variable structure; identifies common and unique variance between variables
Best conditions	Usually, if high correlations between variables, large number of variables and only simple data reduction is required	As for principal components, but it will also deal with a smaller variable matrix and allow a wider range of analysis other than simple data reduction
Package programs	SPSS (sub-program PA1) principal factoring without iteration;[1] BMDP program P4M; PCA option	SPSS (sub-program PA2) principal factoring with iteration; BMDP program P4M; PFA option

[1] In SPSS the sub-program allows for a modification of the principal components solution in PA1 by permitting the user to insert estimates of communality into the correlation matrix. This is referred to as the principal factor solution and is not covered in this chapter (Harman, 1967).

15.4 FACTOR ANALYSIS: DATA INPUT

Any factor analysis model, when applied to a specific problem, can be broken down into a series of operational steps (Table 15.1). These start with (1) the original data matrix, (2) the correlation matrix, (3) the initially derived factor matrix, (4) a transformed or rotated factor matrix, and (5) a listing of factor scores. In this section we shall examine the problems concerned with data input, both in terms of the selection of original variables and also the interpretation of the correlation matrix. An analysis of the latter can be important as it gives some idea of the possible linkages within the data and may help in the formulation of hypotheses.

There are two main problems associated with data input into factor analysis programs. The first concerns the possible need to transform the original data, when data normality is required. This may be required when inferential statistics are being used as part of the factor analysis. If this is the case, then ideally each variable should be checked for normality and transformed if necessary. Obviously, if many variables are involved then this type of exercise will be extremely time-consuming, and in such circumstances some geographers have been tempted to employ the same transformation to all the variables. If no inferences are being made then it is probably best not to transform the original data, since such changes often obscure and complicate interpretations.

In the geographical literature the decision as to whether the data should be transformed or not is anything but clear. For example, if we compare studies in one area of research, that concerned with the social structure of urban areas, we can see how a wide variety of solutions have been used. Thus Murdie (1969) was one of the first urban geographers to attempt anything more sophisticated than a blanket logarithmic transformation in his study of Toronto. However, others have been slow to follow his example, and in many cases decisions have been taken against the use of complex transformations. Davies and Lewis (1973), for example, in their study of the urban dimensions of Leicester, used their data in their untransformed format because of potential problems. Indeed, they gave two major reasons for not transforming their data. First, their study of Leicester was just one of a series of comparable studies of large provincial British cities, and they felt that transformations suitable for the distribution of a variable within one city might possibly be quite different from those required by another city. Second, they believed that statistically accurate transformations may tend to complicate the interpretation of the variables.

Perhaps of greater importance is the selection of variables to be used within the factor analysis. Obviously, this selection will be dependent on the type of study being undertaken and the availability of data. It should be stressed, however, that the structure of the final factors will be determined by the quality and type of variables used (Gittus, 1964). For example, in urban social studies, if the original variables were weighted towards those measuring demographic characteristics, clearly this would be reflected in the composition and importance of the resultant factors. In terms of these studies of urban social structure, the selection of input variables is usually based on (1) the experience of other work; (2) the balance of variables between demographic, housing and socioeconomic indices; and (3) the requirements of the specific study. The decision as to which variables to include is obviously difficult and no clear-cut guidelines exist; although when using factor analysis, rather than principal components, there should be some underlying hypothesis.

Finally, we can turn our attention to the question of how many variables should be included. The effect of including a large number of variables that measure similar characteristics and are highly correlated together is merely to increase the total proportion of variance that a particular factor accounts for. Thus, in the social analysis of British towns the actual number of variables used in factor and component studies has varied from 26 in Cardiff and Swansea (Herbert, 1972), to

37 in Hull (Wilkinson *et al.*, 1966) to 60 in Hampshire. In terms of the technique, the only major constraint on the number of variables used within the study is that there should be substantially more cases than variables.

15.5 UNDERSTANDING AND USING THE OUTPUT FROM FACTOR ANALYSIS

There are a number of computer programs available to carry out factor analysis, but in this section attention will be focused on the use of the SPSS system. The emphasis here is placed on understanding the statistical and later the geographical significance of the computer printout. We can use this package to obtain the following information: a variable list, the correlation matrix, communalities, factor loadings, eigenvalues, the percentage of variance explained by each factor and factor scores. We can also obtain factor loadings based on some type of rotational solution (see Section 15.6). Let us consider the use of such information by examining some detailed examples.

The first example concerns an analysis of the morphometric and hydrological characteristics of a sample of thirty British drainage basins (Figure 15.3). The object of the exercise was to examine the underlying physical dimensions of these different river systems and to see whether a smaller number of new variables (three new factors) could be derived. The original variables, listed in Table 15.5, cover a wide range of physical characteristics.

1 Tyne, Bywell
2 Wear, Sunderland Bridge
3 Tees, Broken Scar
4 Tweed, Drybrough
5 Stour, Langham
6 Blackwater, Appleford Bridge
7 Colne, Lexden
8 Ribble, Samlesbury
9 Lune, Halton
10 Duddon, Duddon Hall
11 Derwent, Camerton
12 Wye, Dadora
13 Usk, Chain Bridge
14 Tywi, Ty-Castell Farm
15 Teifi, Glan Teifi
16 Dovey, Dovey Bridge
17 Blithe, Hamstall Ridware
18 Churnet, Rochester
19 Vyrnwy, Vyrnwy Bridge
20 Avon, Melksham
21 Medway, Teston
22 Winterbourne, Bagnor
23 Blackwater, Swallowfield
24 Eye Brook, Eye Brook Reservoir
25 Skerne, South Park
26 Alwen, Druid
27 Dee, Erbistock Reservoir
28 Weaver, Ashbrook Picton
29 Eden, Warwick Bridge
30 Clyde, Hazelbank

Figure 15.3 Map of sample drainage basins used in factor analysis example

Table 15.5 Variables used in the drainage basin study

Variable	Code
Main stream length (km)	MSL
Dry valley factor (ratio)	DVF
Angle of slope, 10–85° (m/km)	S1085
Stream frequency (junction/km^2)	STMFRQ
Mean annual rainfall (mm)	SAAR
Five-year maximum of 2-day rain (mm)	M52D
Effective mean soil moisture deficit (mm)	SMDBAR
Soil index (measure of porosity)	SOIL
Percentage of urban land	URBAN
Mean annual run-off (mm)	RUNOFF
Run-off as a percentage of rainfall	ROPERC
Highest point in basin (m)	MAXELEV
Drainage area (km^2)	DRAREA

Source: *Flood Studies Report*, NERC, London, 1975 (5 volumes).

The first table we can examine is that of the unrotated factor matrix, which contains the loadings between each variable and the new factors. As was stated earlier these loadings represent the correlations between the original variables and the new factors (Table 15.6). With these loadings it is possible to determine which variables 'load' or relate together on the new factors. In our example, for factor 1 we can see that mean annual run-off, mean annual rainfall, run-off as a percentage of rainfall and five-year maximum of two-day rainfall, are all highly positively loaded. In addition we have a few other variables, in particular effective mean soil moisture deficit (negative loading) and stream frequency that also have fairly high loadings. Factor 2 has a lower set of loadings and would seem to only relate to two original variables; namely, mean stream length (negatively) and slope angle (positively). However, once again some other variables may possibly be included within the structure of this factor. In terms of factor 3, the highly related variables are a soil index, the dry valley factor and possibly stream frequency.

Up to this point we have essentially made two decisions. The first concerns how many factors we have examined (in this example three), and the second relates to the composition of the factors. In both cases our decision appears to be somewhat subjective in nature, so can we make these choices in a more objective way? With regard to the selection of the number of factors use can be made of the eigenvalues or the percentage of variance explained by each factor. Thus, an eigenvalue of 1.0 or more indicates that a factor explains more of the total variance than does a single variable. In our example the first three factors had eigenvalues of 6.34, 1.89 and 1.07 respectively, with the remaining factors all having values less than 1.0. A further method of determining the number of factors to be used is through the use of the 'scree-slope' graph as suggested by Cattell (1978). The technique is based on identifying a distinctive break of slope in a plot of the amount of variance explained by each factor. In our drainage basin example this can be identified after three factors, as is shown in Figure 15.4.

Table 15.6 Factor loadings for drainage basin study (unrotated)

Variable	Factor 1	Factor 2	Factor 3
MSL	0.16678	−0.81492	0.02152
DVF	−0.37124	0.33671	0.54190
S1085	0.53171	0.60095	0.01817
STMFRQ	0.83943	0.13912	−0.31507
SAAR	0.94265	0.21618	0.17299
M52D	0.89425	0.28321	0.15472
SMDBAR	−0.85072	0.23387	−0.17695
SOIL	0.45239	0.06927	−0.72139
URBAN	−0.22036	0.03753	−0.01951
RUNOFF	0.96229	0.15782	0.15012
ROPERC	0.94151	−0.13021	0.09665
MAXELE	0.79090	−0.38244	0.03273
DRAREA	0.30600	−0.59917	0.01998

The second problem, that of identifying the structure of each factor, is rather more difficult to solve, although some objectivity can be introduced into the proceedings. For example, a cut-off point can be decided on for the individual loadings, and use can be made of the break-of-slope technique. Some factor ecologists have adopted fairly low cut-off points of 0.3 or 0.4, for distinguishing significant from non-significant factor loadings (Lawton and Pooley, 1975). Alternatively, the composition of each factor can be determined through a use of background knowledge about the variables and the area of study.

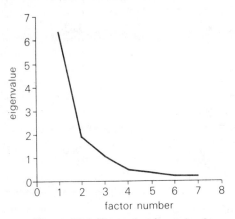

Figure 15.4 'Scree-slope' graph of amount of variance explained by each factor

Let us take a closer look at the problem of interpreting and naming factors in the light of the possible solutions that have been referred to. In many circumstances it is profitable to start by considering not only the variables within each factor, but also their interrelationships. This may be achieved by an examination of the original zero-order correlation matrix, which may lead to the formulation of hypotheses about the importance of groups of variables. To illustrate the usefulness of this approach we can take a second example, concerning an attempt to understand the social dimensions of the nineteenth century city (Pooley and Johnson, 1982), more specifically Exeter in 1871. The variables included in the study are given in Table 15.7, along with their factor loadings.

Table 15.7 Original variables and their factor loadings for Exeter study (varimax rotation)

Variable	Factors		
	1	2	3
Persons in household	0.34	0.66	−0.55
Large households	0.26	0.62	−0.24
Age of head	0.08	−0.09	−0.09
Female heads	−0.17	−0.00	0.07
Children	−0.34	0.10	−0.81
Young adults	0.84	0.10	0.25
Mature adults	0.01	−0.18	0.56
Middle age	−0.06	−0.06	0.00
Old age	−0.26	0.24	−0.13
Females	0.27	0.01	−0.06
Fertile females	0.87	0.05	0.27
Single adults	0.58	0.48	0.12
Nuclear families	−0.03	−0.88	−0.28
Lodgers	0.07	0.79	0.01
Devon born heads	−0.51	−0.31	0.13
Migrant (non-Irish) heads	0.71	0.02	0.17
Irish heads	0.21	0.09	−0.28
High social class	0.46	0.05	0.38
Low social class	−0.05	−0.01	−0.09
Servants	0.58	0.23	0.01
Sharing house	−0.08	−0.55	0.08
Economically active	0.44	0.24	0.63
Workers in manufacturing occupations	−0.05	0.03	−0.07
Workers in dealing occupations	0.27	0.02	0.66
Fertility ratio	−0.82	−0.13	−0.13
Variance explained (per cent)	29.0	15.9	11.0
Cumulative variance explained (per cent)		44.9	55.9

From the correlation matrix two main groups of variables could be identified, as illustrated by the linkage diagrams in Figure 15.5. The first relates to variables measuring socioeconomic status and is fairly clearly identified; the second is rather more complex, but relates to family status. Through the identification of such linkages the task of interpreting factors may become a little easier. Thus, if we examine the factor loadings for this example (Table 15.7) we can see the emergence of these variables relative to specific factors.

The labelling of each factor is therefore related to our original hypotheses and the high loadings on each factor. Thus, for our example of British drainage basins in Table 15.6, factor 1 could be termed the hydrological factor, factor 2 the basin size and factor 3 a measure of porosity. From our second example it can be seen that factor loadings are relatively lower, and we can use three categories of loadings to aid our description of the factors. These three categories are summarized in Table 15.8. In either case it should be stressed that the labelling of variables is always going to be tinged with some degree of subjectivity.

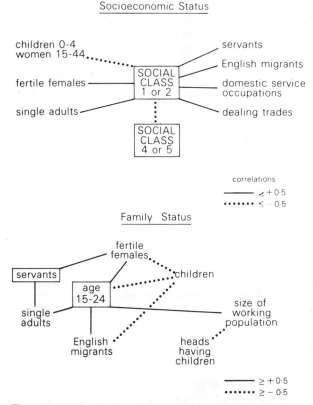

Figure 15.5 Correlation linkage diagrams relating to a study of social structure in 1871 for Exeter, England

15.6 ROTATIONAL SOLUTIONS

To a large extent the aim of factor analysis is to define new variables or factors that adequately and clearly describe the original set of variables. The ideal is to search for a 'simple factor structure' whereby each original variable loads high on one factor and low on the second. Unfortunately, what often happens is that the initial solution derived by a factor analysis program does not provide such a neat and clear factor structure. Thus, in Table 15.6 we have variables that appear to load fairly highly on more than one factor, as is the case with stream frequency. Under this type of solution, then, the factors are not clearly describing the original variables. What we need is an alternative solution in which the factors are rotated to provide a better description of the variable pattern.

Table 15.8 Factor labels and factor structure for Exeter study

Factor 1		
Positive	High	: Young adults, fertile females, migrant (non-Irish) heads
	Medium	: Single adults; servants
	Low	: Persons in household; Economically active High Social class
Negative	Low	: Children
	Medium	: Devon heads
	High	: Fertility ratio
Factor 2		
Positive	High	: Lodgers
	Medium	: Persons in household; large households
	Low	: Single adults
Negative	Low	: Devon heads
	Medium	: Sharing house
	High	: Nuclear families
Factor 3		
Positive	High	: None
	Medium	: Middle-aged; Economically active; workers in dealing occupations
	Low	: High social class
Negative	Low	: None
	Medium	: Persons in household
	High	: Children

Key to categories: Factor loadings $\pm$ 1.00-0.70 high; $\pm$ 0.69-0.50 medium; $\pm$ 0.49-0.30 low.

Rotation of the factors therefore aims at simplifying the factor matrix by separating out significant clusters of variables, without altering their relative positions. In any factor analysis the axes describing the variables can be fixed in an infinite number of positions by geometrically rotating them about their origins. In some ways such indeterminancy is unfortunate since there can be no unique solution, although not all statistical solutions are equally meaningful in terms of our geographical studies.

The major option available when selecting a rotation is whether an orthogonal or oblique solution should be used. Oblique rotations are not frequently used by geographers, since they do not assume independent groups of variables and the results are often rather difficult to interpret. Under this type of rotation the idea of independent or orthogonal factors is replaced, and the assumption is that every variable will have one factor loading of ± 1.0, but that other loadings need not be zero. Two sets of loadings are usually obtained for each variable when using an oblique rotation, and they are called structure loadings and pattern loadings. The difference is indicated in Figure 15.6(a), which also shows the advantages of an oblique solution over an orthogonal one, when the pattern of variables is not totally independent. The structure loadings are used in the basic interpretation of the factors in exactly the same way as are the loadings from an unrotated factor matrix. In contrast pattern loadings are less important for the basic interpretation of factor structure, since they measure the correlation between each variable and a factor, independent of the effect of the other factors. They are, therefore, partial correlation coefficients.

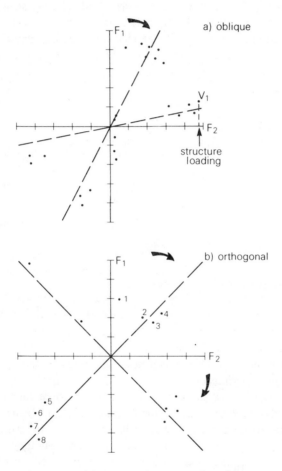

Figure 15.6 (a) Oblique factor rotation; (b) orthogonal (varimax) rotation

To date most geographers have used orthogonal rotational solutions in factor analysis. This is partly because of the wide availability of such rotations in the early package programs, and partly due to the fact that such solutions are often easire to interpret. Orthogonal rotations are based on the assumption that the factors are unrelated, and the ideal is that every variable has a factor loading of

±1.0 on one factor, and zero on all of the other factors. The most widely used orthogonal rotation is that of varimax; although the SPSS package also offers quartimax and equimax solutions (see Table 15.9). Here we shall only consider the varimax technique, which is based on simplifying the columns of a factor matrix, thereby maximizing the sum of the variance of the square loadings in each column; hence the name varimax. The logic of the varimax solution is illustrated graphically in Figure 15.6(b), which shows how the orthogonal rotation of the factors can identify groupings of variables. Thus, the movement of factor 1 highlights the clustering of variables 1–8, which now seem to load more strongly on the rotated factor 1. Compared with oblique solutions, the factor matrix resulting from a varimax rotation contains both the pattern and structure matrix as one set of data. Thus, the values in the rotated factor matrix represent the equivalent of regression weights and correlation coefficients.

Table 15.9 Characteristics of some factor rotations

Orthogonal solutions	
Quartimax	Variable loads high on one factor and almost zero on the others; simplifies the rows of the factor matrix
Varimax	As for quartimax, but simplifies the columns of the factor matrix
Equimax	Seeks a compromise solution, simplifying both rows and columns
Oblique solutions	Factors are allowed to be correlated if correlations exist in the original data

We need not concern ourselves too much with the mathematical background of these techniques, but rather focus on how they can help in the provision of more clearly definable factors. The usefulness of such rotations can be seen by applying the varimax solution to our analysis of drainage basin characteristics. The factor loadings using the varimax rotation are presented in Table 15.10, in which it can be seen that a slightly clearer pattern of variables is to be found.

The choice of whether to rotate the factors, and if so whether to use an oblique or orthogonal solution, may be viewed in two ways. First, we can consider such decisions as inductive processes based on the idea or hypothesis that some distinct pattern of variables does exist and that this may be enhanced by rotating the factors. Under this type of approach the solution of a particular factor rotation may well be based on a trial and error approach. Unfortunately, in the statistical literature there are no worthwhile guidelines available to advise the geographer on which solution is best, although as we have hinted orthogonal solutions continue to receive most attention. How do we know when we have found the 'ideal' rotation? According to Harman (1967) such a condition is achieved when in geometric terms (1) most points lie close to a factor axis, and (2) only a small number of points lie at some distance away from factor axes (see Figures 15.6(a) and (b)).

Table 15.10 Factor loadings for drainage basin study (varimax rotation)

Variable	Factor 1	Factor 2	Factor 3
MSL	0.06498	0.74123	0.15648
DVF	−0.16408	−0.28038	−0.72684
S1085	0.40250	−0.13413	0.08463
STMERQ	0.60867	−0.01272	−0.44937
SAAR	0.94473	0.00353	0.11990
M52D	0.86874	0.01634	0.11568
SMDBAR	−0.83658	−0.26123	−0.15639
SOIL	0.15652	−0.11297	0.77791
URBAN	0.16756	0.09803	0.04499
RUNOFF	0.95407	0.04683	0.15443
ROPERC	0.86762	0.25034	0.21642
MAXELE	0.67329	0.33099	0.16307
DRAREA	0.13526	0.83471	−0.00565

The above inductive process is also termed the 'hypothesis-checking rotation' by some factor analysts, and it contrasts with the second approach of 'hypothesis-creating' or deductive solutions (Cattell, 1978). Under such conditions we would hypothesize that particular groups of variables should exist owing to the nature of our study. This is also known as 'target' rotation, since we are rotating the factors to fit in with a hypothesized or 'target' pattern of variables. There are two basic rotations used in such circumstances: one is the oblique promax solution and the other is termed multiple group factor analysis (Timms, 1971; Johnston, 1978). Such 'target' solutions are obviously only useful when we have sufficient background to draw on from previous studies.

In geography one area where such conditions exist is in the analysis of urban social structures. Both from the theoretical background, initiated by the early work of Shevky and Bell (1955), and from the large number of factorial ecologies undertaken, it is possible to consider the 'target' approach to rotation within this area of research (Berry and Kasarda, 1977). Thus, in a study of urban social areas we could hypothesize that three basic social dimensions or groups of variables should emerge. In North America these are likely to be economic status, family status and ethnic status; while in the UK we may identify housing tenure, socioeconomic status and a mobility factor (Knox, 1982). Given this background we can then decide on the number of factors to be extracted, in this instance three; and then rotate the factors to the required pattern. However, owing to the lack of established theoretical frameworks in many areas of geography, such 'hypothesis-creating' approaches have very rarely been applied.

15.7 USE OF FACTOR SCORES

One important part of the output from factor analysis is the matrix of factor scores, which provides a measure of the relationship between each observation and the new factors. As we shall see this has proved particularly useful for geographers in the analysis of spatial patterns. These scores are values for each observation on the new variables, and as such they reflect to some extent their relationship with the original variables, and the contribution that each new variable makes to their variance. Thus, if an observation has a large value on an original variable, which in turn is highly loaded on a new factor, then it will have a high score for that particular factor.

Strictly speaking, in statistical terms, only scores from a principal component analysis can be directly computed, since it assumes a common variance (see Figure 15.1). The scores for factor analysis can only be estimated given that the model provides for both common and unique variances (see Section 15.1). However, because of the widespread preference for factor analysis, factor scores are frequently used by geographers; consequently many package programs such as SPSS provide this data. The problems associated with the use and interpretation of estimated factor scores, especially when complex factor solutions are involved, were fully discussed by Joshi (1972). One way to tackle the problem of estimated scores from factor analysis is to use 'image factor analysis'. This technique breaks down the data in terms of common and unique variances, the values of which can be estimated by using multiple regression. The residual values from the regression analysis are then taken as the unique part of the variance, while the factor scores are computed from the common variance of the factors (Rees, 1972).

Factor scores can be used in two ways: they can be either plotted and presented in factor space or mapped out in geographical space. Not surprisingly geographers have made most use of the second type of presentation, especially in the study of urban social patterns. The former method merely involves plotting the observations relative to the factor axes, as was done in Figure 15.6 for the original variables. This approach is most useful for examining the grouping of individual observations; and may be used to classify or group particular cases or areas together on the basis of their relationship with each factor. The importance of this grouping procedure can be illustrated with reference to our drainage basin example. Thus, each basin area can be allocated to a particular factor using factor scores. In this way basins that are statistically similar can be identified by reference to their scores on each of the three factors. We might even go so far as to produce a map for each of the factors. This is not done here, but may be carried out for a larger sample of basins to help identify spatial trends. In our present example a simple examination of the factor scores will be sufficient (Table 15.11). From this we find that drainage basins 9 and 10 (the rivers Duddon and Derwent, both of which are in the Lake District; Figure 15.3), score heavily on factor 1, which is the hydrological factor. This may reflect the control exerted by the high rainfall in this area and the generally moist conditions. Drainage basins that score at the other end of factor 1, the rivers Stour and Blackwater (numbers 5 and 6), are both in East England, where rainfall is lower and the hydrological environment far less dynamic.

Factor 2 (basin size) picks out drainage areas 1 and 4, that of the rivers Tyne and Tweed; these are large basins where their size or areal extent controls their behaviour. From Table 5.11 we may also notice how poorly the latter two rivers score on factor 1. That the small basins of the Duddon and Derwent now score at the negative end of factor 2 is also consistent with the character of this factor.

It is once again basin 1, together with 2 and 3, that score towards the positive end of factor three. Basins 2 and 3 are the Wear and Tees respectively; hence all three basins lie in NE England, where we must assume that the soil characteristics and porosity contribute significantly to their behaviour. Clearly, the application of factor analysis can pave the way towards classification; based on a small number of statistically defined factors, rather than a large number of individual variables.

Table 15.11 Factor scores for drainage basin example

Drainage basin	Factor 1	Factor 2	Factor 3
1	−0.4652	0.7332	1.3254
2	−0.8055	0.1221	0.8080
3	−0.0925	0.0275	0.7947
4	−0.0411	1.0271	0.2875
5	−1.3708	−0.3121	0.4511
6	−1.2857	−0.6534	0.4392
7	−1.4121	−0.6776	0.2604
8	0.8191	0.1640	0.7443
9	1.4734	−0.8152	0.5916
10	1.8218	−1.0049	0.1645

Basin numbers refer to locations given in Figure 15.3.

In the remainder of this section attention will be focused on the importance of mapping factor scores, by referring to our earlier example of the social structure of nineteenth-century Exeter. In the same way that contemporary cities have been researched in terms of their social patterns, using factor analysis, so too has the nineteenth-century city. The rationale behind this type of factorial analysis is based on the changing nature of social structure and socio-geographical space as cities evolve (Timms, 1971; Carter and Wheatley, 1982). Furthermore, stimulated by the work of Lawton and Pooley (1975) a number of historical geographers have carried out factor analyses of nineteenth-century cities (Johnson and Pooley, 1982). Such work has provided a fairly sound background and it is within the context of this research that our example from Exter must be viewed.

The data were extracted from a 10 per cent sample of the 1871 census returns (Schofield, 1972), and we have already examined the character of the three main factors to be extracted from the study (see Table 15.8). Using the factor scores we can examine the spatial expression of these social and economic dimensions or factors, in an attempt to describe and understand the structure of nineteenth-century Exeter. To save space, since we are only interested in the use of factor scores, only the scores for factor 1 will be mapped. This first factor is concerned with family composition and age structure, and it also has strong links with variables measuring migrant status and economic status. Consequently, areas in the city which score highly on this factor are characterized by a dominance of young, single adults, non-local born residents and higher than average social status. A distinctive spatial pattern is difficult to discern, but high scores on factor 1 cover part of the old part of the city, as well as new suburban areas (Figure 15.7). Further patterns could obviously be revealed by examining the map of scores for the other two factors and perhaps also by using smaller sub-areas, the importance of which are discussed in Section 15.8.

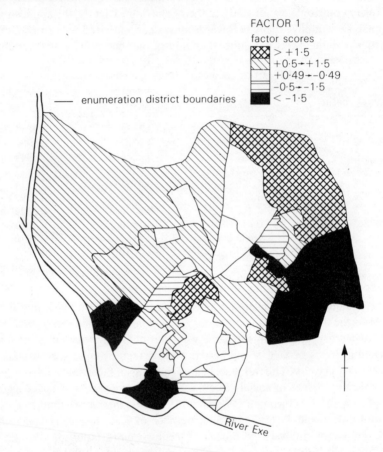

FACTOR 1

factor scores

> +1·5
+0·5 → +1·5
+0·49 → −0·49
−0·5 → −1·5
< −1·5

—— enumeration district boundaries

River Exe

Figure 15.7 Map of factor scores for 1871 Exeter, England

15.8 PROBLEMS IN THE APPLICATION OF FACTOR ANALYSIS

Like most other multivariate statistical techniques applied to geographical problems, factor analysis has attracted its share of criticism. As we have seen, among statisticians factor analysis is often dismissed as an 'elaborate way of doing something which can only ever be crude, namely picking out clusters of interrelated variables' (Hills, 1977). In these debates such people favour the mathematical clarity of principal components. However, countering such criticism is the work of the pro-factor analysts, whose work is mainly rooted within the social sciences (Cattell, 1978). To these people the advantage of factor analysis is its more realistic assumptions, and it is because of such arguments that geographers have come down in favour of a factor approach.

These debates between pro and anti-factor analysts have encouraged some geographers to question the appropriateness of factor techniques (Clark *et al.,* 1974). From these discussions two main issues have emerged. The first concerns how factor analysis should be used by geographers, a problem that to a large extent is created by the flexibility of the technique. A flexibility that has sometimes encouraged people to use the technique but without having any clear reasons for doing so. In basic terms, however, we can say that geographers have used factor and component analyses to (1) attempt to create some order into a large group of variables, as a descriptive device, (2) explore hypothetical constructs and measure related dimensions in a set of variables, and (3) analyse the pattern factor loadings, which may link with the analysis of factor dimensions. The use of factor analyses for these specific tasks is unfortunately further complicated by the realization of the fact that a great variety of factor techniques exist (Johnston, 1978). This awareness of new techniques raises the question of which should be applied to specific problems, since many produce quite different results. Such a confused situation is likely to continue until geographers fully explore the advantages and limitations of different factor models.

Apart from the problems associated with the appropriateness of factor techniques, attention should also be drawn to the problems of interpretation, and specifically to that of over-interpretation. One way in which this problem arises stems from the fact that factor loadings (by which the factors are interpreted) are the square roots of the proportion of standardized variation in a variable which can be accounted for by the factor. Thus the correlations are often not as substantive as their numerical size would show. For example, a factor loading (or correlation coefficient from the correlation matrix) of ± 0.65 shows just 42 per cent agreement in the bivariate distributions of the two items which are being related. This leads to a danger of identifying a factor, for example, by two variables which are themselves only slightly related. As Meyer (1971) observed, the smaller the correlations on which the factoring procedure is based, the greater is the possibility of substantial over-interpretation, even despite apparently high loadings.

Part of the over-interpretation problem, Johnston (1976) believes, stems from a problem of imprecise goals. It may also, as a result of the use of orthogonal rotation methods and of a failure to approximate to the ideal solution (when all loadings are either $\pm$ 0.0 or $\pm$ 1.0), be technically derived.

The selection of variables may lead to a problem of over-interpretation when approximation of factor scores occurs. The scores are produced by multiplying the original standardized loadings; where the factor solution does not approach the ideal one, it is possible that the same factor score can be derived from many combinations of high and low values for individual variables. Joshi (1972), in a study of Katmundu, found that some 'smaller' variables were having more influence on the pattern of scores than were the 'more important' ones by which his socioeconomic status dimension was identified. Joshi subsequently suggested that the troublesome smaller loadings should be removed before computing scores; but as this necessitates some judgement by the analyst it may prove to be difficult. Alternatively, Schmid and Tagashira suggested (1965) that we try to move nearer to the simple structure (that is, the ideal solution) by removal of variables from our analysis, possibly following an examination of the general stability of the results. The above problems are especially important because of the use of the factor score matrix in social area studies.

Other problems in the use of factorial methods in urban areas result from the mismatch between census enumeration districts (the usual unit in British studies) and social groups. This may, owing to unfortunate boundary locations, introduce spurious internal heterogeneity into areas. More importantly, it could bias the correlations on which the factorial ecology is based. To some extent our example of Exeter's social structure in 1871 suffers from this boundary problem, in that we are dealing in some cases with fairly large census areas, which contain a variety of social groups. In terms of factorial ecologies of urban areas it is such problems that have contributed to the declining use of factor methods, to be replaced by less mechanistic approaches (Herbert and Thomas, 1982).

Nevertheless, despite its problems and its changed importance in many areas of geographical inquiry, factor analysis still remains a useful technique. Its flexibility and variety are, given further work, powerful reasons why geographers will continue to apply it to many different types of problems.

REFERENCES

Berry, B. J. L. (1971). 'Introduction: The logic and limitations of comparative factorial ecology', *Econ. Geog.* **47**, 209–219.

Berry, B. J. L., and Kasarda, J.D. (1977). *Contemporary Urban Ecology*, Collier Macmillan, London.

Blackith, R. E., and Reyment, R. A. (1971). *Multivariate Morphometrics*, Academic Press, London.

Carter, M., and Wheatley, S. (1982). *Merthyr Tydfil in 1851: A Study of the Spatial Structure of a Welsh Industrial Town*, University of Wales, Monograph 7.

Cattell, R. B. (1978). *The Scientific Use of Factor Analysis*, Plenum Press, New York.

Chatfield, C., and Collins, A. J. (1980). *Introduction to Multivariate Analysis*, Chapman & Hall, London.

Clark, D., Davies, W. K. D., and Johnston, R. L. (1974). 'The application of factor analysis in human geography', *The Statistician* **23**, 259–281.

Cooley, W. W., and Lohnes, P. R. (1971). *Multivariate Data Analysis*, Wiley, New York.

Davies, W. K. D., and Lewis, G. J. (1973). *The Urban Dimensions of Leicester*, Inst. Brit. Geogrs Spec. Pub. **5**, 71–85.

Gittus, E. (1964). 'An experiment in the definition of urban sub-areas', *Trans. Bartlett Soc.* **2**, 109–135.

Harman, M. H. (1967). *Modern Factor Analysis*, Chicago Univ. Press, Chicago.

Herbert, D. (1972). *Urban Geography: A Social Perspective*, David & Charles, Newton Abbot.

Herbert, D., and Thomas, C. (1982). *Urban Geography: a First Approach*, Wiley, Chichester.

Hills, M. (1977). Book review, *Applied Statistics* **26**, 339–340.

Johnson, J., and Pooley, C. (eds) (1982). *The Structure of Nineteenth Century Cities*, Croom Helm, London.

Johnston, R. J. (1976). 'Residential area characteristics: research methods for identifying urban sub-areas. Social area analysis and factorial ecology', in D. T. Herbert and R. J. Johnston (eds) *Social Areas in Cities* I, Wiley, Chichester.
⸻ (1978). *Multivariate Statistical Analysis in Geography*, Longman, London.
Joshi, T. R. (1972). 'Toward computing factor Scores', in W. P. Adams and F. Helleiner (eds) *International Geography* 2, 906–909.
Knox, P. (1982). *Urban Social Geography*, Longman, London.
Lawton, R., and Pooley, C. G. (1975). *The Urban Dimensions of Nineteenth-Century Liverpool*, Dept. of Geography Working Paper 4, University of Liverpol.
Meyer, E. P. (1973). 'On the relationship between ratio of number of variables to number of factors and factorial determinacy', *Psychometrika* 38, 375–380.
Murdie, R. (1969). *Factorial Ecology of Metropolitan Toronto*, 1951–61, University of Chicago, Dept. of Geography Res. Paper, 116.
Rees, P. H. (1972). 'Problems of classifying sub-areas within cities', in B. J. L. Berry (ed.) *City Classification Handbook*, Wiley, New York, 265–330.
Schmid, C. F., and Tagashira, K. (1965). 'Ecological and demographic indices: a methodological analysis', *Demography* 1, 194–211.
Schofield, R. S. (1972). 'Sampling in historical research', in E. A. Wrigley, (ed.) *Nineteenth Century Society*, Cambridge University Press, Cambridge.
Shevky, E., and Bell, W. (1955). *Social Area Analysis*, Stanford, California.
Timms, D. W. E. (1971). *The Urban Mosaic*, Cambridge University Press, Cambridge.
Wilkinson, H. R., Davidson, N., and Francis, M. K. (1966). *Kingston upon Hull: Social Area Analysis*, Dept. of Geog. Misc. Series 10, University of Hull.

RECOMMENDED READING

Carey, G. W. (1969). 'Principal component factor analysis and its application in geography', *Quantitative Methods in Geography*, Am. Geog. Soc. Symp., New York.
Henshall, J. D. and King, L. J. (1966). 'Some structural characteristics of peasant agriculture in Barbados', *Econ. Geog.* 42, 74–84. An early example of the use of Q-mode factor analysis in human geography.
Kim, J. O., and Mueller, C. W. (1978). *Introduction to Factor Analysis*, Sage University Papers 13 and 14, Sage Publications, London. Provides an introductory explanation, together with examples based in social sciences generally, rather than for geography.
Meyer, D. R. (1971). 'Factor analysis versus correlation analysis: are substantive interpretations congruent', *Econ. Geog.* 47, 336–343. A broad discussion on the relative merits of factor analysis compared with correlation techniques, that emphasizes the choices available to the researcher.
Tinkler, K. J. (1972). 'The physical interpretation of eigenfunctions of dichotomous matrices', *Trans. Inst. Brit. Geogrs* 55, 17–26.

Chapter 16

Spatial Indices & Pattern Analysis

16.1 INTRODUCTION TO SPATIAL TECHNIQUES

Many of the statistical techniques that geographers commonly use tend to be of a non-spatial type, that can just as easily be applied to any other scientific discipline. Indeed, the majority of the techniques mentioned so far in this book fall into this category, where location is but one of many variables under examination. However, there also exists a set of spatial statistics which allow the geographer to both summarize and describe numerically a variety of spatial patterns ranging from simple dot distributions through to contour maps. As Unwin (1981) demonstrates, it is possible to recognize a typology of maps that loosely correspond to the nominal, ordinal, interval and ratio scales of measurement that we discussed in Chapter 4. In this way a picture can be built up of the types of data we are likely to want to analyse using different spatial statistics. To carry out such spatial analysis three major groups of techniques are available, namely: centrographic techniques and related spatial indices, point pattern analysis and measures of spatial surfaces as described in Chapter 17.

16.2 CENTROGRAPHIC TECHNIQUES

Centrographic techniques are an extension of the descriptive statistics discussed in Chapter 5, but applied to data in two-dimensional space. Such techniques, also termed geostatistics, have been used for some considerable time, with the concept of the mean centre being introduced in the USA census of population as early as 1870. Similarly, a school of so-called centrography, based on these methods, was developed in Russia in the early part of the twentieth century (Taylor, 1977). The main measures used within these studies were those concerned with central tendency, in particular the arithmetic mean centre. Geographers have used such spatial statistics since the 1950s (Hart, 1954), and have extended the techniques through the work of Warntz and Neft (1960).

The mean centre can be most easily used to summarize a spatial distribution, either of point patterns or data expressed as aggregate frequences on a map. In both cases the data must first be identified by grid coordinates. If the mean centre of a point distribution is to be calculated, then the first step is to derive the mean of the X and Y coordinates for the point distribution, using equation 16.1:

$$\bar{X} = \frac{\Sigma X_i}{N} \qquad \bar{Y} = \frac{\Sigma Y_i}{N} \qquad (16.1)$$

> Equation 16.1
>
> X_i, Y_i = coordinates of individual points
> N = total number of points

These calculations are demonstrated for a simple, hypothetical example in Table 16.1; while Figure 16.1 shows how the mean centre of this dot distribution is represented by the intersection of $\bar{X}$ and $\bar{Y}$. If the data under analysis are not in the form of a simple point distribution, but rather consist of aggregate data in each grid cell, then the mean centre can be calculated using a formula for grouped data:

$$\frac{\Sigma X_i f_i}{N} \qquad \frac{\Sigma Y_i f_i}{N}$$

This is treated in exactly the same way that grouped data were analysed in Chapter 5, since in this case each set of values along the X and Y coordinates can be envisaged as histograms. It is also possible to calculate the weighted mean centre of a distribution where each dot represents a different value. For example, each dot on the map may represent a factory of different size, and we may want to work out the mean centre of factory floorspace. In this case we could weight our calculations in terms of the size of each factory. The mean centre is then found by multiplying the weights of each occurrence by its X and Y coordinate values, and dividing by the sum of the weights, as in equation 16.2:

$$\bar{X}_w = \frac{\Sigma(X_i W_i)}{\Sigma W_i} \qquad \bar{Y}_w = \frac{\Sigma(Y_i W_i)}{\Sigma W_i} \qquad (16.2)$$

where W are the weighted values.

Table 16.1 Calculation of mean centre

X	Y	Point
1	5	1
2	6	2
2	4	3
2	3	4
3	7	5
3	5	6
3	4	7
4	6	8
4	3	9
5	5	10
29	48	10

$\bar{X} = 29/10 = 2.9;\ \bar{Y} = 48/10 = 4.8.$

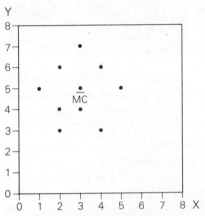

Figure 16.1 Mean centre for a simple
point distribution

As with conventional measures of central tendency, geostatistics can also make use of the median centre and the modal centre. The former, as Ebdon (1977) points out, is unfortunately described in quite different ways by a number of standard statistical texts. For example, Cole and King (1968) and Hammond and McCullagh (1974) define the median centre as the point of intersection of two orthogonal lines from the X and Y axes, which have an equal number of points on each side (Figure 16.2(a)). In contrast, Neft (1966), King (1969) and Smith (1975) refer to the median centre as the point in a distribution at which the sum of the absolute deviations of each point is minimized. That is, distances between the median centre and each point are at a minimum. It therefore represents the point of theoretical 'minimum aggregate travel', and its position can be found by the use of grid overlays and an interative procedure as outlined by Seymour (1965). There are three basic steps involved, in what can be a fairly lengthy process without the aid of a computer. The first stage is to overlay a coordinate grid on the map, the limits of which are set by the four most extreme points in the pattern. Second, for each new grid coordinate point (X_0, Y_0), the square root of the sum of the squared distances to the n points (X_i, Y_i) of the original pattern is calculated (Figure 16.2(b)). The point having the lowest value, using equation 16.3, is then identified:

$$M_c = \sqrt{\Sigma[(X_0 - X_i)^2 + (Y_0 - Y_i)^2]} \qquad (16.3)$$

Equation 16.3
X_0, Y_0 = grid coordinates
X_i, Y_i = original points
M_c = median centre

In the third stage, this point of minimum value is now taken as the centre of a new, finer grid overlay determined in a subjective manner. At this stage, step two is repeated using equation 16.3. A new, more accurate, point of minimum distance is thus established. Such iterations can be carried out as deemed necessary to determine more accurately the 'median centre'.

The concept of the median centre as the point of minimum aggregate travel is of considerable use within the study of economic geography (Smith, 1975). Furthermore this measure, together with a wider set of distance concepts developed by Warntz and Neft (1960), has extended such analysis into the area of spatial modelling.

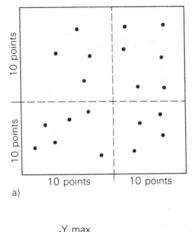

a)

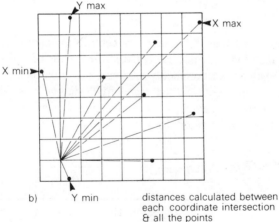

b)   Y min   distances calculated between
                    each coordinate intersection
                    & all the points

Figure 16.2 Two methods of calculating the median centre: (a) the equal frequency approach, (b) the minimum aggregate distance approach

At a more simple, but nevertheless effective, level of analysis the concept of the mean centre has proved extremely useful in studying the changing pattern of distributions over time. These changes can be described by calculating the mean centre of a distribution at different time periods and by plotting such changes, as shown in Figure 16.3. Considerable work has also been carried out by Russian economic geographers during the 1920s and 1930s; when they calculated the mean centre for a variety of economic activities and constructed so-called centrograms (Sviatlovsky and Eells, 1937).

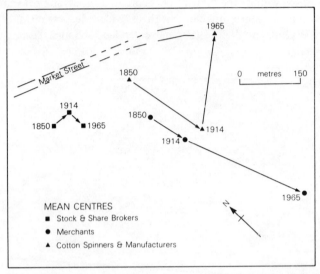

Figure 16.3 Use of the spatial mean to indicate functional change in central Manchester (modified from Varley, 1968)

As was shown in Chapter 5, measures of central tendency were only one way of describing a distribution, since use could also be made of statistics measuring dispersion. In geostatistics a commonly used measure of dispersion is the standard distance, which is analogous to the standard deviation in simple, descriptive statistics. The simplest method of calculating the standard distance of a distribution is shown in equation 16.4:

$$ SD = \sqrt{ \left[\frac{\Sigma(X - \bar{X})^2}{N} + \frac{\Sigma(Y - \bar{Y})^2}{N} \right] } \qquad (16.4) $$

Equation 16.4
N = number of points
$\bar{X}, \bar{Y}$ = mean of X and Y
SD = standard distance

This is based on Pythagoras' theorem, where linear distances between points can be calculated from their X and Y coordinates (Figures 16.4). The method of calculation is shown in Table 16.2 for the same data as in Table 16.1.

Very often it is useful to use the mean centre and the standard distance together in order to describe adequately a spatial distribution. However, a glance through the literature will show that few studies have used both measures.

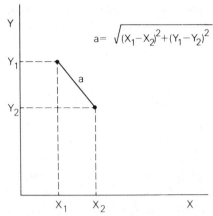

Figure 16.4 Calculation of point to point
distances using Pythagoras' theorem

Table 16.2 Calculation of standard distance

X	Y	$X-\bar{X}$	$(X-\bar{X})^2$	$Y-\bar{Y}$	$(Y-\bar{Y})^2$
1	5	−1.9	3.61	0.2	0.04
2	6	−0.9	0.81	1.2	1.44
2	4	−0.9	0.81	−0.8	0.64
2	3	−0.9	0.81	−1.8	3.24
3	7	0.1	0.01	2.2	4.84
3	5	0.1	0.01	0.2	0.04
3	4	0.1	0.01	−0.8	0.64
4	6	1.1	1.21	1.2	1.44
4	3	1.1	1.21	−1.8	3.24
5	5	2.1	4.41	0.2	0.04
			12.90		15.60

$\bar{X} = 2.9$; $\bar{Y} = 4.8$ (from table 16.1); standard distance on $X = 12.9/10 = 1.29$; standard distance on $Y = 15.60/10 = 1.576$; combined standard distance = 2.87.

Shachar's (1964) analysis of the dispersion of commercial functions in the cities of Tel Aviv, Jerusalem and Rome adequately demonstrated the usefulness of standard distance, while also drawing attention to the problem of using absolute measures of dispersion in a comparative study. Thus, absolute standard distance is affected by the shape of the area under study and also by the size of the study area. For example, in Shachar's results the standard distances of commercial functions in Rome were usually always larger than those found in Tel Aviv or Jerusalem, mainly because it was a larger city. To overcome this problem use needs to be made of relative standard distance measures. There are a number of ways of calculating such relative measures, depending on the phenomena under investigation. In Shachar's study the commercial functions were related to the population distribution and the relative dispersion was calculated by dividing the standard distance of a commercial distribution by the standard distance of the population. In contrast Neft (1966), studying population dispersion in different countries, used a relative measure based on the radius of the area of each country, assumed to be circular.

16.3 SPATIAL INDICES AND THE LORENZ CURVE

Apart from the centrographic techniques discussed in the previous section, distributions can also be compared using spatial indices that relate to the Lorenz curve. The latter is a simple, but effective, means of illustrating graphically the difference between spatial patterns, and in many texts it is introduced as a technique of map comparison (Unwin, 1981). A number of stages are involved in the calculation and construction of the Lorenz curve, as can be illustrated in a simple example which compares the distributions of coloured people and white people in the USA (Table 16.3). First, the ratio between coloureds and whites needs to be calculated for each area (in our example census regions). Second, these areal units are then ranked on the basis of these ratios, from the smallest to the largest. Next, each variable is converted into a percentage of the total for its own area. Thus, in our example, for each census region of the USA the numbers of coloured and white people are expressed as percentages. Finally, as Table 16.3 shows, these percentage values are accumulated, maintaining the ranks and ranked from 1 upwards. These values can then be plotted out as a graph, or Lorenz curve, as shown in Figure 16.5.

Table 16.3 Calculation of Lorenz curve for ethnic segregation in the USA (1981)

Regions	Number (000) of: white	black	X/Y	Rank	X (%)	Y (%)	Cumulation X(%)	Y(%)
N England	11586	475	24.4	8	6.2	1.8	94.7	99.0
Mid Atlantic	30743	4374	7.0	4	16.3	16.5	47.6	69.5
EN Central	36139	4548	7.9	5	19.2	17.2	66.8	86.7
WN Central	16045	789	20.3	7	8.5	3.0	88.5	97.2
S Atlantic	28648	7648	3.7	1	15.2	28.9	15.2	28.9
EC Central	11700	2868	4.1	2	6.2	10.8	21.4	39.7
WS Central	18597	3525	5.3	3	9.9	13.3	31.3	53.0
Mountain	9957	269	37.0	9	5.3	1.0	100.0	100.0
Pacific	24926	1993	12.5	6	13.2	7.5	80.0	94.2
	188341	26489						

In general terms the Lorenz curve has a number of obvious features; one being that if the distributions are proportionally identical in each area then the plot will be a straight line (Figure 16.5). Differences between the distributions will be shown in the form of deviations away from this diagonal line. The extreme case is one of complete separation between two distributions, where the line would follow the X axis and then the vertical boundary of the graph when the value of X reached 100 per cent. In the example given in Figure 16.5 it can be seen that a degree of difference does exist between the two distributions, representing in this case some measure of racial separation at a fairly broad, regional scale within the USA.

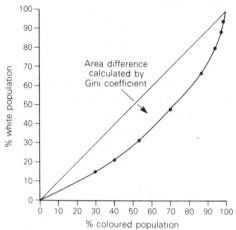

Figure 16.5 The Lorenz curve, based on the data in Table 16.3

One method of measuring the differences revealed by the Lorenz curve is to calculate the index of dissimilarity (D_s). This may be defined as the maximum vertical distances between the diagonal line and the Lorenz curve (Figure 16.5). In fact there are three different ways of obtaining this index. First, it can be calculated from the data, as the maximum difference in the cumulative percentages of the two distributions. Thus, in Table 16.3 D_s would be given by $69.5 - 47.6$, and has the value 21.9. A second method of calculating the index is by measuring it from the Lorenz curve in Figure 16.5, since the area between the two curves is a measure of how poor the fit is. Finally, D_s can be calculated using equation 16.5, which has the advantage that it deals with uncumulated percentages and therefore avoids some of the work in Table 16.3:

$$D_s = \frac{\Sigma |X_i - Y_i|}{2} \tag{16.5}$$

where X_i and Y_i are the individual percentages of each variable. D_s has a range of 0 to 100.

A glance through the literature shows that owing to the flexibility of the index of dissimilarity, it has been applied to a variety of studies. Thus, in our example the index was used as a measure of ethnic segregation, indicating that at a regional scale segregation was not particularly high, with an index of 21.9. Indeed, it is within the context of segregation studies that the index has been widely used. Tauber and Tauber (1965), for example, applied it to over 200 cities in the USA and found that the index varied between 60.4 and 98.1, illustrating high levels of segregation. In Northern Ireland Poole and Boal (1973) applied the technique to a study of segregration tendencies in Belfast. Finally, the more general application of the Lorenz curve to segregation studies was reviewed by Jones and Eyles (1976).

A second type of use for Lorenz curve analysis is within the area of economic geography. In this instance the measure of dissimilarity from the Lorenz curve is known as the 'coefficient of geographical association'. This produces a measure of the extent to which economic activities are concentrated spatially relative to some other form of activity. Smith (1975), for example, presents coefficients of localization for a number of different industries, showing the relative degrees of concentration within the UK.

A further use of the index would be to compare population or employment with land area. Thus, population would be plotted on the Y axis and the size of each areal unit as values of X. A perfectly even distribution of population would therefore have an index of zero, with values approaching 100 (or 1, depending on whether percentages are used) indicating population concentration. Such measures of dispersal or concentration based on the Lorenz curve can obviously be applied to a variety of activities other than population. For example, Wild and Shaw (1974) used a locational index based on the Lorenz curve to rank shop types in order of their locational behaviour and whether they were dispersed or concentrated.

In addition to the measures based on the Lorenz curve, geographers have also used a variety of spatial indices, or coefficients (Isard, 1960). The most extensively used is probably the 'location quotient', which measures the extent to which different areas depart from some norm; for example the national average. The quotient can be calculated using equaion 16.6, or if the data are expressed as percentages, equation 16.7:

$$LQ = \frac{(X_i/X)}{(Y_i/Y)} \qquad (16.6)$$

Equation 16.6

X_i = employment in a given activity i, in an area
X = total employment in an area
Y_i = national employment in activity i
Y = total national employment

$$LQ_j = \frac{X_j}{k} \qquad (16.7)$$

Equation 16.7

X_j = percentage of an activity in j
k = national percentage

In equation 16.6 the location quotient indicates the degree of concentration, with higher values of LQ representing high concentrations, and values of 1 indicating equal distributions. Table 16.4 illustrates how such measures can be derived for two different types of economic activity. Geographical variations in employment concentrations based on the location quotient can be illustrated by mapping the quotients (Figure 16.6).

Table 16.4 Calculation of location quotients for employment in engineering and textile manufacturing in England (1976), thousands.

Standard regions	Engineering	LQ	Textiles	LQ	Total manufacturing
North	192	0.94	52	0.97	438
Yorks and Humberside	242	0.74	149	1.71	711
East Midlands	210	0.78	171	2.38	587
East Anglia	82	0.90	14	0.58	195
South East	943	1.10	108	0.47	1851
South West	215	1.12	36	0.70	420
West Midlands	576	1.27	44	0.36	978
North West	399	0.86	190	1.53	1005
	2860		762		6187

From equation 16.6, LQ for the North (engineering) $= (192/2860)/(438/6187) = 0.067/0.071 = 0.94$.

One of the problems with the location quotient, and indeed with a number of these spatial indices, is that statistically, nothing is known about their sampling distributions (King, 1969). Furthermore, as the scale for the quotient is arranged around unity, values below the national norm are compressed between 0 and 1; but above unity the quotient can rise to any value. One coefficient that does not suffer from these disadvantages is the Gini-coefficient of concentration which can be calculated using equation 16.5, and which varies on a scale of 0 to 100. Both Glasser (1962) and King (1969) illustrate how the sampling distribution of the coefficient can be derived; and that if the sampling is done without replacement from an infinite population then the sampling distribution will approximate to normal.

Finally, some mention must be made of the problems associated with using the Lorenz curve. First, there are data restrictions since variables must be expressed as frequencies for each areal unit, and negative values cannot be included. This therefore makes it rather more difficult to apply the technique to a study of a continuous variable. Second, the index is affected by changes in the spatial boundaries of the study units, and changes in spatial scale. This can be partly illustrated by the example of ethnic segregation, used in Table 16.2, which at a regional scale showed little evidence of segregation. However, the results from the analysis of individual cities by Tauber and Tauber (1965) showed very high levels of segregation at this smaller spatial scale. Indeed there are predictable variations in the index D_s with changes in spatial scale, and lower values of D_s are to be expected in a study with a few areal units. A more sensitive technique in the study of ethnic segregation is Lieberson's isolation index ($P*$), which was fully discussed by Robinson (1980).

Leading on from these difficulties, associated with changing boundaries and scale, is a third problem relating to the fact that the Lorenz curve is insensitive to spatial arrangement, or pattern. Thus, the Lorenz curve and its related indices are an effective way of describing the relationships between distributions, but offer no indication or measure of spatial pattern.

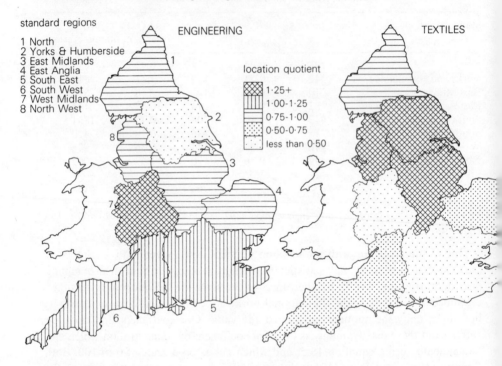

Figure 16.6 Use of location quotients to describe engineering and textile employment in England (1976)

16.4 PATTERN ANALYSIS

In the earlier sections of this chapter, the techniques we discussed were concerned with either measuring spatial distributions or providing summary statistics to describe such distributions. However, in many circumstances geographers may be interested in the locations of individuals relative to each other, often expressed as points on a map. Such work is concerned with the techniques of point pattern analysis. Early approaches to this type of study were subjective and merely involved the mapping of individuals to produce a simple dot map (Monkhouse and Wilkinson, 1963). Very often the interpretation of these patterns lacked objectivity, making accurate description difficult and comparative studies impossible.

The first attempts to study point patterns in an objective fashion were made by plant ecologists in the analysis of plant communities (Clark and Evans, 1954). These studies were soon adapted and used by geographers, initially by Dacey (1960) and King (1962) and then by many others (see Rogers, 1974; and Cliff and Ord, 1981 for a review of this work). All these studies have one important feature in common — they all make use of some type of probability distribution as a means of describing spatial patterns. Thus, a link is established between an observed distribution of points on a map and probability theory as explored in Chapter 7.

One of the most widely used measures is the Poisson probability distribution, which has the following assumptions. First, it postulates the condition of equal probability, which in the context of point patterns refers to the situation where any location on a map has an equal probability of receiving a point For example, if we had a map of a particular woodland, the Poisson theory suggests that any part of the area would have an equal opportunity of having a tree. We may infer from this that the process producing such patterns is therefore a random one. Second, the theory assumes a condition of independence, whereby each of the points located on a map would be independent of one another. Thus, in our woodland example the assumption is that the location of one tree would neither repel nor attract another. Based on these assumptions and the use of the Poisson probability function, as discussed in Section 7.11, it is possible to model a distribution of points on a map. These expected patterns can then be compared with observed or real patterns, and thus used as a standard yardstick, from which we can measure deviations. By applying the Poisson distribution, we are therefore using the concept of randomness as our basic measure.

In the study of point patterns we can recognize two important violations of the assumptions made by the Poisson model, each of which produce non-random processes. The first of these concerns patterns that result from competitive processes. This may be illustrated by the example of food stores competing with each other in a city. Over time, those located close to other food stores may be driven out of business by more powerful competitors, thereby reducing store clusters and producing a regular pattern. The best example of such competitive processes in the geographical literature is that of the evolution of market centres and described by central place theory. A second deviation from the Poisson model is the situation where the locations of existing activies attract others. This is termed a contagious process, and tends to produce a clustered pattern. Harvey (1966) has used the idea of contagious processes in the context of spatial diffusion studies as presented by Hägerstrand (1953).

Given this background it is possible to recognize three basic types of point patterns, namely regular, random, and clustered, each of which can be modelled by a particular probability distribution (Figure 16.7). These distributions provide the geographer with some basic yardsticks with which to compare observed point patterns, and also provide a conceptual framework to account for some of the possible variations from randomness.

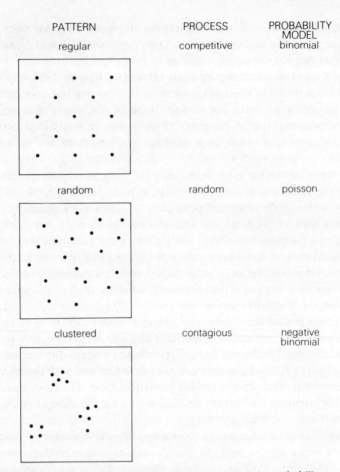

Figure 16.7 Types of point patterns and their probability
distributions

16.5 MEASUREMENT OF POINT PATTERNS

At the start of this section it is important to make clear the distinction that
statisticians draw between 'pattern' and 'dispersion'. Both Dacey (1973) and
Sibley (1976) have urged caution in our use of such terms. By their operational
definitions 'pattern' should be taken to mean the distances between and
arrangements of points in space, and 'dispersion' refers, by contrast, to the areal
extent of a collection of points. Sibley has observed that failure to bear these
distinctions in mind can lead to meaningless and misleading expressions, such as,
'random dispersion'. Attention in this section is devoted to patterns in the sense
that they are defined above.

There are a number of ways in which geographers have studied point patterns.
Later in this section we shall look at 'nearest neighbour analysis', but we shall
start with 'quadrat analysis'. The latter has been popular with both geographers

and botanists but has recently been less widely adopted as an appreciation of some of its problems has grown. Nevertheless it remains an important part of geographical methodology.

In quadrat analysis the study area is overlaid by a grid of lines forming units of equal size, and the number of points in each cell are counted. We have already used this method in Section 7.11 when the spatial distribution of grocers' shops in Sunderland was examined. Traditionally the grid systems are based on squares, although other shapes, such as hexagons, could be used provided that they combine to form a complete cover. Irregular shapes, rectangles or units of different areas should not be used. Squares are clearly the easiest units to construct and measure and will form the basis of what follows, although the theoretical implications apply to all suitable shapes.

The points that can be studied and counted using the quadrat method can consist of any spatially distributed 'point' phenomenon. Shops have already been cited, and there is a substantial literature on their study in this way (Rogers, 1974). But species of plant, cases of disease, industrial locations and even settlement patterns are equally amenable. The points, or events, must be spatially discrete, and continuous variables such as rainfall or altitude are not suitable. We hope that by counting the numbers of point events within each square we can derive a measure of the points' pattern. The guiding principle underlying these attempts is that point patterns can be described according to their location along a continuum which varies from perfectly regular (all points equidistant) at one end to perfectly clustered (all points touching) at the other. The random distribution lies mid-way between theses extremes. Figure 16.7 conveys something of the form of these patterns, and later in this section we shall see how the distances between points can be used to describe the patterns; but another method of description is to compare the quadrat counts with a hypothesized probability distribution. The Poisson distribution (see Section 7.11) has already been examined and has a particular, though limited, application to this problem.

In the earlier example of Sunderland shops we hypothesized that their spatial distribution was random. If that hypothesis was correct then the observed distribution should approximate to the Poisson which is, as we explained in Chapter 7, a random distribution in which events are independent and located without spatial preference. Section 9.7 demonstrated how the differences between observed and hypothesized observations were tested. The distributions are described by the number of quadrats which contain 0, 1, 2, 3 etc. points, and in the latter example the observed data differed so little from that expected from a Poisson distribution that the pattern was concluded to be random.

Another approach to this problem of pattern analysis would have been to use the variance–mean ratio of the observed distribution. The mean, denoted by λ, is given by the observed density of points, which in the earlier example was 0.4 shops per quadrat. The variance of the observed distribution is given by:

$$\sigma^2 = \frac{\Sigma X^2}{\Sigma X} - \frac{\Sigma X}{N} \tag{16.8}$$

where X is the number of points in each quadrat and N the number of quadrats.

One of the hallmarks of the Poisson distribution is the equality of mean and variance. Thus, if the variance–mean ratio is unity or close to it, we may conclude the distribution to be Poisson and, consequently, spatially random.

Departures from unity in the ratio reflect tendencies towards either clustering or regularity. Regularly located points yield very low variances because most quadrats record a similar number of points, with the result that the variance–mean ratio is less than 1.0. On the other hand clustered point patterns give variances that are very high, because a few quadrats have many points and the majority have very few or none, and the resulting variance–mean ratios are greater than 1.0. The degree of departure from 1.0 can be converted to a z score after calculating the standard error of the difference (SE_X) from:

$$SE_X = \sqrt{[2/(N - 1)]} \qquad (16.9)$$

in which N is the total number of quandrats, and:

$$z = \frac{\text{observed ratio} - \text{expected ratio}}{\text{standard error}} \qquad (16.10)$$

Given this information we can rework the shops example. The mean of the spatial pattern we have already established to be 0.4 shops per quadrat, and its variance we estimate from equation 16.8. The constituent observations (X) consist of the number of shops counted in each of the 135 quadrats. In this case there were 96 quadrats with zero shops, 27 with one, 9 with two, 2 with three and only 1 with four. We may prepare our data for equation 16.8 in the form shown in Table 16.5, taking care to keep the distinction between N (the number of quadrats) and ΣX (the total number of shops) clearly in mind. We find that $\sigma^2 = 1 \cdot 418$ and the variance–mean ratio is therefore $1.418/0.4 = 3.545$.

Table 16.5 Method of calculating the variance of spatial patterns using quadrat counts

Number of shops per quadrat (n)	Number of quadrats with n shops (q)	Number of shops by quadrats (X)	X^2
0	96	0	0
1	27	27	27
2	9	18	36
3	2	6	18
4	1	4	16
	135	53	96

In their original form the X data would consist of 135 observations corresponding to the number of shops in each of the 135 quadrats. Here they are treated in a simpler, condensed, form in which $X = nq$ and $X^2 = n^2q$.

We can move on to examine the degree to which the variance–mean ratio of the shop pattern differs from the hypothesized value of 1.0. The first step is to calculate the standard error of the difference from equation 16.9:

$$SE_X = \sqrt{[2/(135 - 1)]}$$
$$= \sqrt{0.01493} = 0.1222$$

From this the z value (t value if ΣX is less than 30) is calculated by dividing the standard error into the difference between the observed and expected variance–mean ratios:

$$z = \frac{3.545 - 1.0}{0.1222} = 20.827$$

Thus, at the 0.01 significance level, for which the critical z values are ± 2.58, the differences are sufficiently great for us to conclude that the distribution is, by this method, non-random. Moreover, if we pay attention to the sign of the z value we can see, because of the nature of equation 16.10, that positive results show a tendency towards clustering while negative values arise from a tendency towards regularity. The positive result here suggests a degree of clustering of the points.

Hence we come, by different routes, to contradictory conclusions concerning this distribution. Such difficulties are not uncommon in point pattern analysis and warn us to be on our guard. There are, unfortunately, further problems that have to be considered. Most importantly, both the observed and hypothesized (Poisson) distributions are density-dependent as by varying the grid-size placed over the study area or its limits the measured density can be drastically altered. Figure 7.17 has already shown the dramatic effect that variations in density (which provides the Poisson mean λ) can have on the character of the distribution. It is unfortunate that pattern is wholly independent of density, yet the distributions used here to analyse them are strongly density-dependent. Equally patterns are independent of scale, and those shown in Figure 16.7 are as likely to be encountered on a scale of kilometres as a scale of centimetres. Nevertheless, variations in quadrat size can still lead to unrepresentative contrasts between the observed distributions that they generate.

Clearly we need to select the boundaries of the study areas with care. Less easy to accomplish is the choice of a suitable quadrat size. Greig-Smith (1964) has pointed out that by the judicious selection of quadrat size we can simulate a Poisson distribution where none exists. The effect of quadrat size has been widely examined by plant ecologists (Greig-Smith, 1964; Pielou, 1969). Their work has shown that, in visually clustered patterns such as that in Figure 16.7, very small quadrats usually produce results suggesting randomness; since it is likely that quadrats would only contain small numbers of points and not measure the clusters. Similarly, very large quadrats in such circumstances produce results that seem to indicate a regular pattern, as most quadrats would contain similar numbers of points. If, on the other hand, we are examining a pattern that appears to be visually regular, then this characteristic will be shown as quadrat size increases (Greig-Smith, 1952).

There are two ways in which this problem of quadrat size can be tackled. The first is to derive some method of determining an ideal quadrat size for particular point distributions. Ecologists have examined such a notion and define quadrat size as $2A/N$, where A is the area of study and N is the number of points, i.e. twice the mean area around each point. Taylor (1977), however, suggests that such quadrats are probably too large for geographical studies, particularly where spatial competition is important. In these conditions he recommends the use of quadrats determined by the area of the map divided by the number of points (A/N).

A further, though time-consumming, solution is to test for Poisson randomness over a range of quadrat sizes. If it is present throughout we might be confident in claiming its existence. The problem of appropriate quadrat size has yet to be resolved and the final choice may well be determined by the nature of the project rather than the character of the data. But we should, despite these warnings, note the advice of Harvey (1966) who reminds us that 'the quadrat sampling has considerable potential for testing models of location'.

In addition we can use the effect of quadrat size variations in a positive fashion. Thus, by using quadrats of different sizes to examine the same pattern we can explore variations in scale or 'grain' of point distributions. This has been discussed by Kershaw (1964), who plotted quadrat size against variance of the distributions and argued that peaks of high variance on the resulting graphs could be interpreted as point clusters.

There is, however nothing intrinsically sacred about the Poisson distribution and its applicability of spatial observations. Scatters of points in space can be described by other distributions whose exclusion from this text should not be interpreted as a relegation of their importance. Most of these distributions make allowances for a 'contagion' effect in which occurences influence the probability of other events. The Poisson distribution specifically forbids any such effect and requires the events to be wholly independent. Contagion effects can often be very important in describing point patterns. Some services tend, by their nature, to be largely randomly located over wide areas of towns and cities. On the other hand there are other retail functions, such as banks and department stores, that tend to cluster together in city centres; it would be inappropriate to attempt to describe their patterns in the urban field by the Poisson distribution.

There are some probability distributions that can take specific account of contagion effects, and they can be applied to cases in which the Poisson distribution is theoretically imprecise, i.e. if it is thought that the point pattern is not random. These distributions possess rather exotic titles such as 'Neyman's type A' and the 'Polya–Aeppli distribution'. They are less widely known than those reviewed in Chapter 7, but are particularly useful when studying patterns generated by clustering and contagion. A good example might be the pattern of locations created by car component factories which tend to cluster around the car factories themselves. However, the equations describing these probability distributions are not easy to evaluate. The interested reader is referred to more detailed discussions (e.g. Rogers, 1974).

Despite the mathematical elegance of these distributions they remain, like the simpler Poisson distribution, highly dependent on quadrat size. As a result the failure of a point pattern to conform to one of them does not imply that the pattern is not one of contagion. Neither does a correspondence provide irrefutable proof that contagion is present. Indeed, any point pattern may be approximated by two or three different probability distributions. Here, then, is another reason why quadrat analysis is now less popular; because different theoretical interpretations, based on approximations to different distributions, can be made for each case. Nevertheless the thoughtful use of quadrat analysis for point pattern description or for testing *a priori* assumptions against real-world observations remains a valuable geographical tool.

An alternative method to quadrat analysis is to use measures based on the spacing between points, by taking the distance of each point to its nearest neighbour. The pioneer work was carried out by plant ecologists Clark and Evans (1954), who developed a measurement index and linked it to the Poisson probability distribution (the role of this distribution is discussed in Chapter 7). The test requires a knowledge of the population density, and the analysis involves a comparison between an observed spacing of a point distribution and the spacing expected in a random pattern. The average expected distances are calculated using equation 16.11, the derivation of which was presented by Clark and Evans:

$$\bar{r}_e = 0.5 \sqrt{(A/N)} \qquad (16.11)$$

Equation 16.11
$\bar{r}_e$ = average expected distance
A = area of study region
N = number of points

Once again, such expected values are described by the Poisson function. The nearest neighbour statistic (R) is derived by dividing the observed distances by the expected, with the results falling within a range of values from 0 to 2.1491:

$$R = \bar{r}_a/\bar{r}_e \qquad (16.12)$$

Equation 16.12
$\bar{r}_a$ = average observed distance
$\bar{r}_e$ = average expected distance

This index shows how more, or less, spaced the observed distribution is compared with a random one.

Under conditions of maximum aggregation all the individuals in a point distribution occupy the same locus and the distance to the nearest neighbour is therefore zero. At the other extreme, conditions of maximum spacing, the individuals will be distributed in an even hexagonal pattern (Figure 16.7). Consequently, every point will be equidistant from six other individuals, so that the mean distance to the nearest neighbour is maximized and $R = 2.149$. If $R = 1$, then the observed and expected distances are equal, thus indicating a random pattern. Therefore, when values of R are less than 1 this suggests distributions tending toward a clustered pattern, while values above 1 describe tendencies towards dispersion.

The concept of randomness has traditionally been applied as the base measure in point pattern analysis. In theory a random pattern is one in which the location of each point is totally uninfluenced by the remaining points. However, in practice it is much more useful to view patterns as deviating from clustered or regular, with the results falling on a continuum between the two. In reality, locational forces are unlikely to operate randomly, but are more often capable of transforming either of the extreme conditions towards a random pattern. The applicability of this approach can be improved with the application of a test assessing the significance of the R value, which takes into account possible variations in the random processes. Thus, the probability that an R value could have arisen by chance can be established, using the standard error and z scores (see Chapter 8). The standard error of the expected average nearest neighbour distance can be calculated using equation 16.13:

$$SE_{r_e} = \frac{0.26136}{\sqrt{[N(N/A)]}} \qquad (16.13)$$

Equation 16.13
N = number of points
A = area of study
SE_{r_e} = standard error of expected average distance

The standard error can be then used in the normal fashion, with 95 per cent of the expected values falling within $\pm 1.96 SE_{r_e}$ of the average computed distance. Getis (1964), in some early work on the use of nearest neighbour techniques, extended the use of standard errors to test the significance of patterns by computing the z value (see Section 9.5), using equation 16.14:

$$z_R = \frac{|\bar{r}_e - \bar{r}_a|}{SE_{r_e}} \qquad (16.14)$$

Equation 16.14
$\bar{r}_e$ = average expected distance
$\bar{r}_a$ = average observed distance
SE_{r_e} = standard error of expected distance

These inferential statistics can in turn be used to delimit a range of random matching conditions, as suggested by Pinder and Witherick (1973).

The implication of the nearest neighbour technique is that the area under study is an isotropic surface. Attention therefore needs to be given to defining the actual study area in terms of a 'biotope space' (Hudson, 1969); which, for example, in an urban environment involves measuring the built-up area. In addition, consideration needs to be given to the definition of any study area, since both the size and shape of the area may influence the results. As a general rule the study area should be defined relative to the problem under investigation. For example, as Getis (1964) showed in his study of urban retail patterns, the most appropriate study area was a circle, enclosing most of the built-up area.

The basic nearest neighbour technique can be extended to take into account the scale elements in a point pattern. This can be achieved by measuring order neighbour distances up to the Nth value, and by calculating the corresponding values of R. The formula for such measures was derived by Thompson (1956):

$$\text{Expected distance to } n\text{th neighbour} = \frac{1}{\sqrt{M}} \frac{(2n)!n}{(2^n n!)^2} \tag{16.15}$$

Equation 16.15

$M = $ density of points per unit area
$n\ = $ order of nearest neighbour

If measurements are taken to the nth order neighbour, then clearly some idea of the scale at which point patterns are occuring can be gained. In theory, by plotting rank order of the nearest neighbour against the corresponding R values a measure of the 'grain' or scale of pattern intensity can be achieved. Thus, in terms of mean distance between points, if the pattern were of a 'fine-grained' nature then a plot of order neighbour against mean distance would give a curve of only gradual increase, with a smooth profile. Conversely, in a 'coarse-grained' pattern such a curve would be less smooth with sharp increases, as distance measurements become of an inter-group type. One problem in applying this extension concerns the number of neighbours to which measurements should be taken, for as Cowie (1968) states 'there is no objective method to indicate how many neighbours should be employed'. However, in the few attempts at this type of analysis the number selected has been at least of the order of three (Cowie, 1968; Sibley, 1971). The calculation of the technique and the extension proposed by Thomspon are obviously time-consuming by hand, especially for large numbers of points; and in such circumstances the measurement and calculations can be speeded up using a computer program.

The technique can be illustrated by an examination of patterns of retail change, which has been a frequent use of such spatial statistics (Kivell and Shaw, 1980). The example presented here relates specifically to changes in the pattern of footwear retailers in Kingston upon Hull from 1880 to 1950. For the purposes of the analysis two distinct retail organizational types were recognized, multiples and independents, that had different locational requirements. The form refers to those firms having five or more branch shops, and they were often controlled by national companies.

Table 16.6 Trends in the pattern of footwear retailers

	Independents			Multiples		
Time	Nearest neighbour value	Pattern type	Shop	Nearest neighbour value	Pattern type	Shops
1880	1.053	regular	256	nil	nil	nil
1890	1.264	regular	275	0.915	random	19
1900	0.975	random	235	0.630	clustered	16
1910	0.919	random	189	0.684	clustered	22
1920	0.991	random	124	0.873	clustered	25
1930	0.923	random	100	0.640	clustered	32
1940	0.863	clustered	52	0.426	clustered	29
1950	0.781	clustered	22	0.591	clustered	19

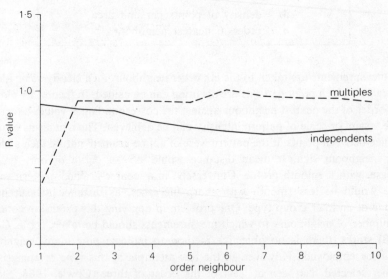

Figure 16.8 Nearest neighbour values against order neighbour for stores in Hull

An analysis of first-order neighbours using equation 16.12 gives an indication of pattern intensity, and clear differences can be recognized between the two shop types (Table 16.6). Despite differences in basic patterns, the overall trend for both types was towards a clustered state, a condition that took the independents longer to achieve. Such results essentially represent the local variations in the retail pattern, and the spatial scale of the analysis can be extended using higher-order neighbours. In this case, calculations were made to the tenth-order neighbour to measure the grain of the pattern. An example of a graph, plotting rank order neighbour against corresponding values of R, is illustrated in Figure 16.8.

The independent retailers exhibited a fine-grained pattern, and the resultant curves decrease in a monotonic fashion towards a general clustered pattern. At high-order neighbours the R value is low, indicating a high degree of large-scale clustering. In contrast, the multiples had a more complex pattern, with marked spatial variations occurring at different scales of measurement. Indications are that they had a coarser-grained pattern with a much stronger element of localized clustering. The step-like nature of the rank order curves suggests that the multiples occured in losely associated clumps of stores, probably at points of relatively high accessibility.

16.6 PROBLEMS IN THE INTERPRETATION OF POINT PATTERNS

Point pattern analysis, in addition to having a number of methodological problems, also suffers from difficulties in the interpretation of results. From the previous two sections in this chapter it is clear that the analysis of point patterns is based on a comparison of the observed distribution with some known theoretical one. If there is a close fit between the two then we may describe a pattern as either random, clustered or dispersed. However, we should be very cautious about inferring processes from point patterns. First, as Harvey (1966) and Rogers (1969) have shown, there may be a number of different probability models that fit our observed points, and we therefore must in some cases use other background evidence to discriminate between them. Second, even if we only have one clear model that fits our data, this is not sufficient evidence to infer conclusively that a specific process produced the pattern.

It is because of such inferential problems that since the mid-1970s point pattern analysis has rather fallen from favour in many areas of geographical research. Thus, while the technique can help us objectively to describe and classify point patterns, it fails to provide the necessary information about spatial processes. This is especially the case in human geography with its increasing emphasis on the study of behavioural processes.

REFERENCES

Clark, P. J., and Evans, F. C. (1954). 'Distance to nearest neighbour as a measure of spatial relations in populations', *Ecology* **35**, 445–453.

Cliff, A. D., and Ord, J. K. (1981). *Spatial Processes: Models and Applications*, Pion, London.

Cole, J. P., and King C. A. M. (1968). *Quantitative Geography*, Wiley, London.

Court, A. (1964). 'The elusive point of minimum travel', *Anns Ass. Am. Geogrs* **54**, 400–403.

Cowie, S. R. (1968). 'The cumulative frequency nearest neighbour method for the identification of spatial patterns', Seminar Paper Ser. 10, Dept. of Geog., University of Bristol.

318

Dacey, M. F. (1960). 'A note on the derivation of nearest-neighbour distances', *J. Reg. Sci.* **2**, 81–87.

—— (1966). 'A county-seat model for the areal pattern of an urban system', *Geog. Rev.* **56**, 527–542.

—— (1973). 'Some questions about spatial distributions', in R. J. Chorley (ed.) *Directions in Geography*, Methuen, London.

Dice, L. R. (1952). 'Measures of spacing between individuals in a population', *Contrib. Lab. Vert. Biology* **55**, Univ. of Michigan.

Ebdon, D. (1977). *Statistics in Geography: A practical approach*, Blackwell, Oxford.

Getis, A. (1964). 'Temporal land use pattern analysis with the use of nearest neighbour and quadrat methods', *Anns Ass. Am. Geogrs* **54**, 391–399.

Glasser, G. J. (1962). 'Variance formulas for the mean difference and coefficient of concentration', *J. Am. Stats Assoc.* **57**, 648–654.

Greig-Smith, P. (1952). 'The use of random and contiguous quadrats in the study of the structure of plant communities', *Anns Botany* (*N.S.*) **16**, 293–316.

—— (1964). *Quantitative Plant Ecology*, Butterworths, London.

Hägerstrand, T. (1953). 'On Monte Carlo simulation of diffusion', reprinted in W. L. Garrison and D. F. Marble (eds) *Quantitative Geography* (1967), Northwestern Univ., Evanston.

Hammond, R., and McCullagh, P. S. (1978). *Quantitative Techniques in Geography: an Introduction*, 2nd edn, Clarendon, Oxford.

Hart, J. F. (1954). 'Central tendency in areal distributions', *Econ. Geog* **30**, 48–59.

Harvey, D. W. (1966). 'Geographic processes and the analysis of point patterns', *Trans. Inst. Brit. Geogrs* **40**, 81–95.

Hudson, J. C. (1969). 'A location theory for settlement', *Anns. Ass. Am. Geogrs* **59**, 365–381.

Hudson, J. C., and Fowler, P. M. (1966). 'The concept of pattern in geography', Discussion Paper Series 1, Dept. of Geog., Univ. of Iowa.

Isard, W. (1960). *Methods of Regional Analysis*, Wiley, New York.

Jones, E., and Eyles, J. (1976). *An Introduction to Social Geography*, Oxford Univ. Press, Oxford.

Kershaw, K. A. (1964). *Quantitative and Dynamic Ecology*, Arnold, London.

King, L. J. (1962). 'A quantitative expression of the pattern of urban settlements in the USA', *Tijds. voor Econs. en Sociale Geog.* **53**, 1–7.

—— (1969). *Statistical Analysis in Geography*, Prentice-Hall, Englewood Cliffs.

Kivell, P. T., and Shaw, G. (1980). 'The study of retail location', in J. A. Dawson (ed.) *Retail Geography*, Croom Helm, London.

Monkhouse, F. J., and Wilkinson, H. R. (1971). *Maps and Diagrams*, 3rd edn, Methuen, London.

Morisita, M. (1959). 'Measuring the dispersion of individuals and analysis of the distributional pattern', Memorial Fac. of Science Series E, No. 2, Kyushu University, 215–235.

Neft, D. (1966). *Statistical Analysis for Areal Distributions*, Reg. Sci. Inst. Monograph 2, Philadelphia.

Pielou, E. C. (1959). 'The use of point-to-plant distances in the study of the pattern of plant populations', *J. Ecol.* **45**, 607–613.

—— (1969). *An Introduction to Mathematical Ecology* Wiley, New York.

Pinder, D. A., and Witherick, M. E. (1972). 'The principles, practice and pitfalls of nearest neighbour analysis', *Geography* **57**, 277–88.

Poole, M. A., and Boal, F. (1973). 'Segregation in Belfast', in B. D. Clarke and M. B. Cleave (eds) *Social Patterns in Cities*, IBG Special Publication.

Porter, P. W. (1963). 'What is the point of minimum aggregate travel', *Anns Ass. Am. Geogrs* **54**, 403–406.

Robinson, V. (1980). 'Lieberson's isolation index', *Area* **12** (4), 307–312.

Rogers, A. (1974). *Statistical Analysis of Spatial Dispersion: the Quadrat Method* Pion, London.

Seymour, D. R. (1965). 'IBM 7090 program for locating bivariate means and bivariate medians', Tech. Rep. 16, Dept of Geog., Northwestern University.

Shacher, A. (1964). 'Some applications of geo-statistical methods in urban research', *Papers and Proceedings, Reg. Sci. Ass.* **18**, 197–202.

Sibley, D. (1971). 'A temporal analysis of the distribution of shops in British cities', unpublished PhD thesis, Univ. of Cambridge.

(1976). 'On pattern and dispersion', Area **8**, 163–165.

Smith D. M. (1975). *Patterns in Human Geography,* Penguin, Harmondsworth.

Sviatlovsky, E. E., and Eells, W. C. (1937). 'The centrographical method and regional analysis', *Geog. Rev.* **27**, 240–254.

Tauber, K. E., and Tauber, A. F. T. (1965). *Negroes in Cities,* Aldine, Chicago.

Taylor, P. J. (1977). *Quantitative Methods in Geography,* Houghton-Miffin, Boston.

Thompson, H. R. (1965). 'Distribution of distance to Nth neighbour in a population of randomly distributed individuals', *Ecology* **37**, 391–394.

Unwin, D. (1981). *Introductory Spatial Analysis,* Methuen, London.

Varley, R. (1968). 'Land use analysis in the city centre with special reference to Manchester', Uni. of Wales unpub. M.A.

Warntz, W., and Neft, D. (1960). 'Contributions to a statistical methodology for areal distributions', *J. Reg. Sci.* **2**, 47–66.

Wild, M. T., and Shaw, G. (1974). 'Locational behaviour of urban retailing during the nineteenth century', *Trans. Inst. Brit. Geogrs* **61**, 101–118.

RECOMMENDED READING

Rogers, A. (1974). *Statistical Analysis of Spatial Dispersion,* Pion, London. Although this text is complicated in parts there is a great deal to interest any geographer concerned with point patterns and the ways in which we can describe and interpret them. Several probability distributions are covered in great detail, with examples.

Harvey, D. W. (1966). 'Geographical processes and the analysis of point patterns', *Trans Inst. Brit. Geogrs* **40**, 81–95. Without ever becoming too complicated this paper discusses the many problems of point pattern analysis together with the applications of various probability distributions. Useful references for further mathematical studies.

Unwin, D. (1981). *Introductory Spatial Analysis,* Methuen, London. A generally well explained text that covers most aspects of spatial statistics, and contains useful worksheets.

Chapter 17

Trend Surface Analysis

17.1 INTRODUCTION

Trend surface analysis has had a relatively long, if controvertial, geographical history. It is used to fit a mathematically definable three-dimensional 'surface' through a set of spatially scattered observations of some given variable. Point locational data of, for example, soil pH within a prescribed area, or individual house prices within a town, could both be treated by trend surface analysis. In purpose and execution it is analogous to regression procedures and can be envisaged as a combination of nonlinear and multiple methods. But, whereas regression permits the dependent variable to be calculated by reference to the numerical value of one, or more, predictor variables, in trend surface analysis estimates are made on the basis of spatial location; the latter defined as Cartesian (rectangular grid) coordinates, termed here M and N. In the example which follows a grid was created specifically for the study (Figure 17.2), but frequently the British Ordnance Survey grid system is perfectly adequate.

We may now compare the simple regression and linear trend surface expressions in Table 17.1 in which linear variations in Y are accounted for by the behaviour of predictor variable X, in the case of regression, and by location along M and N in trend surface analysis.

Table 17.1 Polynomial two- and three-dimensional expressions

Order	Two-dimensional (regression) expression	Three-dimensional (trend surface) expression
Linear	$Y = a + bX$	$Y = a + bM + cN$
Quadratic	$Y = a + bX + cX^2$	$Y = a + bM + cN + dM^2 + eMN + fN^2$
Cubic	$Y = a + bX + cX^2 + dX^3$	$Y = a + bM + cN + dM^2 + eMN + fN^2$ $+ gM^3 + hM^2 N + iMN^2 + jN^3$

All lower case characters are the coefficients. Upper case characters M and N are the spatial coordinates of the variable Y under study.

The criterion for defining the best-fit surface is that of least-squares, i.e. the surface about which the sum of squared deviations of the observed data is reduced to a minimum. Despite the fact that the question of statistical significance in trend surface analysis is an area of dispute, the least-squares criterion has the virtue of precedence while leaving open the way of significance testing if it is required.

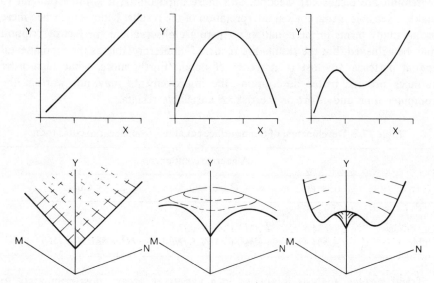

Figure 17.1 Graphical representations of two— and three-dimensional polynomial functions (linear, quadratic and cubic)

Clearly, few phenomena are generalized quite so simply as might be implied by the linear surface, and more diverse forms of variation can be approximated by higher orders of surface. Because there are, however, only two spatial dimensions, these developments can only be achieved through additional, and nonlinear, terms in M and N. Traditionally, though by no means exclusively, the non-linear element has been introduced by polynomial expressions. This is now almost a matter of convention but should not be assumed to indicate anything concerning the character of the study variable. The corresponding regression (two-dimensional curves) and trend surface (three-dimensional surfaces) expressions are indicated in Table 17.1 and, diagramatically, in Figure 17.1.

The flexibility of the two-dimensional curves can, by these algebraic means, be extended to the three-dimensional surface. The latter are lengthier because they need to take into account the two spatial dimensions as well as the combinations of M and N that may arise. The cubic surface, for example, includes in addition to M^3 and N^3 the terms M^2N and MN^2. Consequently, the cubic equation is, superficially, most intimidating but should not cause undue concern to the student and is best viewed as a logical development from the linear and quadratic models (Table 17.2). The degrees of curvature and slope, and the location of troughs and peaks on the trend surface, vary according to the values attached to the equation coefficients. For comparison the reader should examine the theoretical and symmetrical surfaces of Figure 17.1 with the two-dimensional counterparts in Figure 12.9 and the real surfaces used later in this chapter.

Higher orders of surface exist, but are generally not used for two reasons. First, estimation of the least-squares coefficients rapidly becomes difficult and the results questionable (see Chapter 13 in which the corresponding positions for multiple regression are discussed). Second, and more importantly, it is often difficult to make a 'sensible' geographical interpretation of the results. Ultimately, when there are as many terms in the equation as there are observations, perfect description may be achieved; but the results are rendered uninterpretable as the fundamental spatial patterns are lost in a welter of detail. Furthermore, while high-order surfaces provide better descriptions the improvements may not warrant the computer time and effort or the risk of unreliable results.

Table 17.2 Development of trend surface equations from linear to cubic form

	Algebraic components		
	Linear	Quadratic	Cubic
Linear	$Y = a + bM + cN$		
Quadratic	$Y = a + bM + cN$	$+ dM^2 + eMN + fN^2$	
Cubic	$Y = a + bM + cN$	$+ dM^2 + eMN + fN^2$	$+ gM^3 + hM^2N + iMN^2 + jN^3$

Trend surface analysis is applied in a variety of areas; most commonly to determine the existence of spatial trends and to identify, by the use of residuals about the surface, the areas of locally organized behaviour where alternative controls override the regional influences. The identification of these regional and local components may lead to an increased understanding of the character of the study variable. In a similar vein, Chorley and Haggett (1965) have suggested that trend surfaces may be used to test process–response models in which the spatial variation in Y represents changes in the balance of the controlling forces. Such hypothesis testing represents the most sophisticated level of application, but it is also effective as a means of concisely describing spatially organized phenomena. Other applications have included the study of residuals in geological data in order to identify possible oilfield locations (Merrian and Harbaugh, 1964) or to isolate planation surfaces from the general irregularity of surface features (Rodda, 1970).

17.2 ESTIMATION OF LEAST-SQUARES TREND SURFACES

The trend surface equations are comprised of three parts: the predicted value of the variable Y(termed $\hat{Y}$), the spatial location for Y expressed in terms of coordinates M and N, and the least-squares coefficients a, b, c etc. If M, N and the coefficient values are known, then $\hat{Y}$ can be estimated with no great difficulty, even for the cubic surface. It is the evaluation of the least-squares coefficients that throws up the greatest problems. In particular, the large numbers of them to be determined (3, 6 and 10 for the linear, quadratic and cubic surfaces respectively) require that computers are all but indispensable.

Unlike many of the techniques thus far reviewed trend surface analysis is not covered in any of the commercially available computer packages, although some, such as **SPSS**, can be adapted to deal with this problem. This lack of a conveniently available program can be a problem for undergraduate geographers. At present there are a large number of trend surface programs written specifically for individuals or departments, although some are more generally available. Little publicity surrounds the majority of the programs and students are advised to consult staff in their own departments to see if suitable material exists before taking the hazardous course of writing their own or adapting another's program for their own system.

There are many ways in which the least-squares coefficients can be calculated, and many of them are identical to those used in multiple regression and, as a result, will suffer from the same shortcomings (see Chapter 13). There are, however, a number of simple steps that can be taken to reduce some of these difficulties, the most effective of which is the careful 'management' of the data. Rounding errors are commonplace when the numbers are large, and many trend-surface programs treat the raw data in such a way that observations may be raised to the fourth, fifth or even sixth powers before summing over all observations (the details need not concern us). Consequently, if our original values of M, N and Y are measured in, say, hundreds, it will not be long before quite difficult numbers are being processed in the execution of the program. For this reason it is advisable to scale the units of the coordinates with some care. For example an Ordnance Survey six-figure reference of 362 273 might best be treated as 3.62 and 2.73, with other references scaled accordingly. The same principles can be applied to the study-specific grids such as that used later in this chapter, in which M ranges from 0 to 12 and N from 0 to 18. These ranges are far preferable to those of 0 to 120 and 0 to 180. At the same time it must not be forgotten that values of less than 1.0 should be similarly avoided as their products and powers will become impossibly small. Hence a range of 0.01 to 0.1 would be better multiplied by 100 to give 1.0 to 10. It hardly requires emphasizing that such 'juggling' with the coordinate values, provided it is consistently applied, will not affect the shape or statistical significance of the surfaces.

We can now examine the application of these principles in a detailed worked example.

17.3 EXAMPLE OF TREND SURFACE ANALYSIS

The spatially distributed variable under study is house price, sampled within the built-up area of Newcastle-under-Lyme, Staffordshire. The house prices (Y) are measured in £000s. Each observation is also located by reference to M and N coordinates. The raw data may be presented as in Table 17.3. In this instance the reference grid (Figure 17.2) is arbitrary in orientation, origin and scaling. Frequently, the Ordnance Survey grid offers a ready-made system which dispenses with the need for study-specific grids.

The data were processed using a program written specifically for the task. The least-squares estimates were found by employing a branch of mathematics known as matrix algebra, which though not examined in this text is often used to supply answers to such problems. In this connection many computer systems offer pre-written sub-routines which can be incorporated into simple programs in order to overcome the more thorny algebraic issues associated with such work.

Table 17.3 Part of data set used in trend surface analysis of house prices

Y value (£000s)	M coordinate	N coordinate
2.4	3.4	17.6
2.2	4.5	17.5
10.9	5.4	5.8
10.2	9.1	3.3
.	.	.
.	.	.

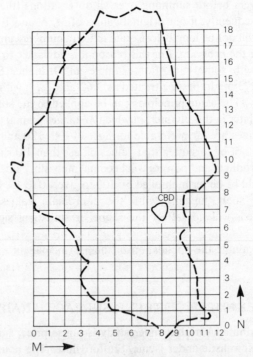

Figure 17.2 Arbitrary grid for locating houses within the borough (pecked-line) of Newcastle-under-Lyme, England

The derived least-squares trend surface equations are given in Table 17.4, and goodness-of-fit summaries in Table 17.5. As the technique is one of least-squares the total sum of squares of Y, i.e. variations about $\bar{Y}$, can, as with the regression models, be regarded as consisting of two parts — that due to the variation of the surface itself and that due to the residual's variation about the surface. The R^2 (coefficient of explanation) values are expressed by the ratio of the surface to the total sum of squares. Specifically, R^2 represents the proportion of variation in Y 'explained' by the regional trends. The linear surface, for example, accounts for 58.6 per cent of the variation in Y, the quadratic for 76.1 per cent and the cubic for 79.5 per cent. This explanation is by spatial location and not, as in regression, by variation in another, predictor, variable.

Table 17.4 Least-squares estimates for the coefficients of the three trend surfaces.

Surface	Least-squares equation
Linear	$Y = 10.951 + 0.238M - 0.692N$
Quadratic	$Y = 23.515 - 0.2792M - 3.1491N - 0.047M^2 + 0.1053MN + 0.0930N^2$
Cubic	$Y = 44.17 - 3.2770M - 8.3441N + 0.0746M^2 + 0.6039MN + 0.4895N^2$ $- 0.0047M^3 - 0.0047M^2N - 0.0024MN^2 - 0.0085N^3$

Table 17.5 Trend surfaces of house prices by component sums of squares

Surface	Surface sum of squares	Residual sum of squares	Coefficient of explanation
Linear	1067.16	752.93	0.586
Quadratic	1384.21	435.88	0.761
Cubic	1447.63	372.46	0.795

Total sum of squares = 1447.63 + 372.46 = 1820.09

At this stage it is helpful to take some sample points, together with the least-squares surfaces, and to examine how the equations can be used to provide estimates of the spatially distributed variable. The obvious starting point is with the linear surface which has only three coefficients. Taking the point defined by $M = 3.5$ and $N = 11.5$, we have a house price value of 2.0 thousands of pounds. If the former two values are now substituted into the linear surface's equation, we have:

$$\hat{Y} = 10.951 + 0.238 \times 3.5 - 0.692 \times 11.5 = 3.826$$

which gives a residual term of 1.826 thousands of pounds. Values of $\hat{Y}$ can be produced for many locations, and when sufficient have been calculated the contours of the surface can be plotted. This has been done in Figure 17.3, where the character of all linear surfaces can be seen — straight, parallel and equidistant contours. In this example the slope is from south-east to north-west. Such surfaces are notably inflexible and in all but the most uncomplicated of variable behaviour will yield relatively low R^2 values. In order to permit a degree of flexibility in the description we may turn to the quadratic surface. The same sample point can be re-used and the estimates compared. Substitution of these values into the quadratic equation gives a much longer expression in which we may note that the a, b and c coefficients are different from their linear counterparts; thus:

$$\hat{Y} = 23.515 - 0.2792 \times 3.5 - 3.1491 \times 11.5 - 0.047 \times 3.5^2$$
$$+ 0.1053 \times 3.5 \times 11.5 + 0.0930 \times 11.5^2 = 2.2850$$

$$Y = 10{\cdot}951 + 0{\cdot}238M - 0{\cdot}692N$$

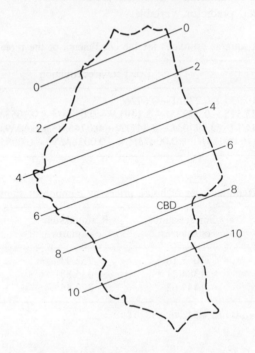

Figure 17.3 Contours of the linear trend
surface for Newcastle house prices (in 1000s)

With this value the residual is now reduced to 0.2850 thousands of pounds and the overall R^2 rises substantially from 0.586 (for the linear surface) to 0.751. Once again, when several $\hat{Y}$ values are obtained the contours of the surface may be plotted (Figure 17.4). But even this degree of curvature is limited as the equation permits the surface contours to flex in only one direction. Finally, the cubic surface can be examined which, with its extra terms, allows the contours to bend in more than one direction. Despite these extra terms and the rather fearsome appearance of the equation, its evaluation is a simple, though tedious, task. Using the same M and N values we have:

$$\hat{Y} = 44.17 - 3.277 \times 3.5 - 8.3441 \times 11.5 + 0.0746 \times 3.5^2$$
$$+ 0.6039 \times 3.5 \times 11.5 + 0.4895 \times 11.5^2 - 0.0047 \times 3.5^3$$
$$- 0.0047 \times 3.5^2 \times 11.5 - 0.024 \times 3.5 \times 11.5^2 - 0.0085 \times 11.5^3$$
$$= 1.801$$

The residual term at this point is now reduced to −0.199 thousands of pounds and the overall R^2 value up to 0.795. Figure 17.5 shows the character of this cubic trend surface and how it is able to vary its shape to account for an even greater proportion of the spatial variability of the house price variable. The preparation of trend surface contour maps can be a very time-consumming business, both in its plotting and in preparing the necessary $\hat{Y}$ values; once again computational aid in data preparation and mapping is invaluable.

$$Y = 23{\cdot}515 - 0{\cdot}2792M - 3{\cdot}1491N - 0{\cdot}0474M^2$$
$$+ 0{\cdot}1053MN + 0{\cdot}0930N^2$$

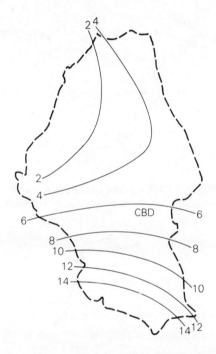

Figure 17.4 Contours of the quadratic trend
surface for Newcastle house prices (in £000s)

In all cases the R^2 values suggest the presence of important spatial trends, with house prices declining towards the north and east of the study area. Notice also the over-generalizing tendency of the linear surface, which leads to expected house prices of below £0.0; a misleading characteristic for which allowances can be made with the higher order surfaces. To complete the picture, the residuals about these surfaces may also be plotted. The residual patterns are, almost inevitably, less organized, though some 'grouping' of values does occur. The plot of the residuals from the cubic surface is shown in Figure 17.6. The larger positive and negative values identify the areas where behaviour deviates more markedly from the regional pattern.

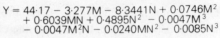

$$Y = 44\cdot17 - 3\cdot277M - 8\cdot3441N + 0\cdot0746M^2$$
$$+ 0\cdot6039MN + 0\cdot4895N^2 - 0\cdot0047M^3$$
$$- 0\cdot0047M^2N - 0\cdot0240MN^2 - 0\cdot0085N^3$$

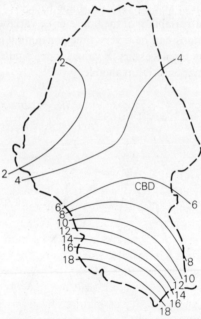

Figure 17.5 Contours of the cubic trend
surface for Newcastle house prices (in £000s)

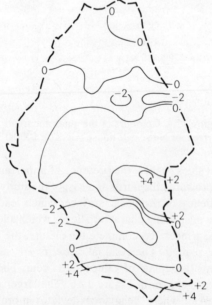

Figure 17.6 Residuals from the cubic trend
surface (£000s)

17.4 SIGNIFICANCE TESTING OF TREND SURFACES

The course of this chapter brings us now to the vexed question of significance testing of trend surfaces. In all cases the basic requirements for regression model testing can be extended to include trend surfaces. In particular, residuals should have zero mean, a constant variance irrespective of location, and should be uncorrelated. The latter condition poses the most serious difficulties for the trend surface analyst, and the problem of spatial autocorrelation, of perennial concern to the statistical geographer, comes clearly to the fore. In trend surface analysis both the raw data and the residuals may well be spatially autocorrelated. Indeed, we may go further and suggest that if some form of spatial autocorrelation of the raw data did not exist then there would be little point in pursuing such an analysis. At the same time its presence, particularly in the residuals, is a matter of concern, since it represents an infringement of a major precondition in hypothesis testing. From Figure 17.6 the degree of spatial autocorrelation in the residuals may be inferred, and the observed grouping of positive and negative residuals might not be expected if they were randomly distributed about the surface.

Although degrees of spatial autocorrelation can be measured (Cliff and Ord. 1970) the debate concerning its precise implications for trend surface analysis remains unresolved and the resilience of the technique to these impositions undertermined. Bearing in mind, however, that the R^2 values provide no information on the genuine reliability of trend surfaces, which can be viewed as sample estimates of unknown population conditions, there is a need to consider how their statistical significance can be assessed.

It has been shown that the total sum of squares can be apportioned between that due to the surface and that due to residuals about the surface. Division of these quantities by their respective degrees of freedom permits the variance-ratio F test to be applied. Variance, it will be recalled, is the quotient of sum of squares and degrees of freedom. The latter are here determined by the number of observations n and the number of terms k in the expression (Table 17.6).

Table 17.6 Degrees of fredom in trend surface significance testing

Surface	Surface variance degrees of freedom (k)	Residual variance degrees of freedom ($n - k - 1$)
Linear	2	107
Quadratic	5	104
Cubic	9	100

k = number of terms in the surface equation, excluding the intercept term a.
n = number of observations (in the current example 110).

If no spatial trends are present the estimated coefficients will, as in regression, tend to zero, producing a horizontal plane; and the constant term a will tend to $\bar{Y}$.

Thus, the H_o under test is that all the coefficients equal zero (H_o: $b = c = d = e... = 0.0$), though this is usually expressed more qualitatively as a null hypothesis of no spatial trend in the nth order.

Surface, residual and total sums of squares are estimated as if they were the dependent term of a regression equation (see Section 11.2), and the by now familiar equations can be used to establish these quantities:

$$\text{Total sum of squares} = \Sigma(Y - \bar{Y})^2 \qquad (17.4)$$
$$\text{Surface sum of squares} = \Sigma(\hat{Y} - \bar{Y})^2 \qquad (17.5)$$
$$\text{Residual sum of squares} = \Sigma(\hat{Y} - Y)^2 \qquad (17.6)$$

Equations 17.4 — 17.6

Y = observations of variable Y

$\bar{Y}$ = mean of Y

$\hat{Y}$ = predicted values of Y at given M,N location

To obtain the necessary quantities the process of estimation of $\hat{Y}$ outlined above should be repeated for each known point in the sample to give a $\hat{Y}$ value corresponding to each observed value. This process is repeated once again for each surface under study.

Using the data from Tables 17.5 and 17.6 an ANOVA table can be drawn up (Table 17.7). Details for the linear trend surface are given in full, but those for the quadratic and cubic forms in summary only. In all three cases H_o may be firmly rejected and significant spatial trends in house prices variation are confirmed to exist. This is in accord with the widely held intuitive notion that urban organization does not create random assemblages of housing but that in some ways regularity and order may be detected.

Table 17.7 Linear trend surface ANOVA table

Source of variation	Sum of squares	Degrees of freedom	Variance	F ratio
Linear surface	1067.16	2	533.58	
Residual	752.93	$110 - 2 - 1$	7.037	75.8
	1820.09	$110 - 1$		

$F_{crit, 0.01}$(with 2 and 107 degrees of freedom) = 4.75.
Quadratic surface F ratio = 65.42.
$F_{crit, 0.01}$ (with 5 and 104 degrees of freedom) = 3.24.
Cubic surface F ratio = 42.78.
$F_{crit, 0.01}$ (with 9 and 100 degrees of freedom) = 2.61.

As in the current example, it is a common experience that significance in the linear surface 'imposes' significance on the immediately higher order surfaces. Consequently, researchers frequently need to determine the optimum surface, as clearly there is no justification for using the most complex model if a simpler one will suffice. Neither should the high-order surfaces always be abandoned out of hand, as they may offer substantial descriptive advantages over simpler, less flexible, forms. Selection is achieved by regarding the sequence of surfaces as analogous to a multiple regression expression into which terms are successively added. Following principles established in Chapter 13, attention is concentrated not on the surfaces, known to be significant, but on the improvement gained by the addition of extra terms. The additional terms will, inevitably, increase the surface (explained) variance, and it is the investigator's task to assess the significance of this change. The H_o now under test is that of no improvement in surface description, and the F ratio is derived from the variance of the change in surface 'explanation' divided by the residual variance remaining with the higher order of the two surfaces. The associated degrees of freedom remain $n - k - 1$ for the error variance, but correspond to the number of extra terms for the variance of the increase in explanation (change in surface sum of squares); 3 for the linear to quadratic step and 4 for the quadratic to cubic step.

The results shown in Table 17.8 indicate that at both steps a significant improvement in description is gained and H_o is rejected at the 0.01 level. As a result the cubic surface is adopted as providing the most efficient approximation to the spatial variations in Y. However, the reader will already be aware that a measure of spatial autocorrelation may still persist in the residuals about the surface (Figure 17.6). Its principal effect is to underestimate the error sum of squares and in doing so raises the probability of wrongly rejecting H_o. The risk of this type I error is best reduced by increasing the significance level (α) to 0.01 or higher when testing the individual surfaces. As Unwin (1975) has recorded, 'surfaces that give poor — just significant — fits and the residuals associated with them should be treated with extreme caution'.

Table 17.8 ANOVA table for testing sequential improvements of description by trend surfaces

Source of variation	Sum of squares	Degrees of freedom	Variance	F ratio
Changes in surface sum of squares (linear–quadratic)	317.05	3	105.68	
Residuals (quadratic)	435.88	$110 - 5 - 1$	4.19	25.22[1]
Changes in surface sum of squares (quadratic-cubic)	63.42	4	15.86	
Residuals (cubic)	372.46	$110 - 9 - 1$	3.72	4.26[2]

[1] $F_{crit, 0.01}$ (with 3 and 104 degrees of freedom) = 4.01.
[2] $F_{crit, 0.01}$ (with 4 and 100 degrees of freedom) = 3.54.

17.5 DATA REQUIREMENTS OF TREND SURFACE ANALYSIS

Overlooking the problems of spatial autocorrelation and the implicit specifications acquired from regression analysis, there are other aspects of data preconditions specifically relating to trend surface analysis. This example uses values, selected at random, which have point references. These points, however, are area-based in that they represent values associated not with points in the strict geometrical sense, but with discrete areas — the plots of land on which the houses were built. Specifically, trend surface analysis should employ only observations of a variable which changes continuously, and not discontinuously, in space. Altitude of the land surface, rainfall and soil pH are good examples of variables which are continuous and, within obvious constraints, measureable at any location. In the present case one may acknowledge the data limitations but, simultaneously, recognize that the relatively small aerial units come close to a quasi-continuous distribution. This approach echoes the sentiments of Chorley and Haggett's (1965) paper in which they suggest that rigid attitudes in this respect may be unnecessary. Nevertheless, the use of observations taken from aerial units large in relation to the study area are best avoided as the results may well depend on the size and shape of the units rather than the intrinsic spatial behaviour of Y. It is typical of this technique, that contradictory statements will be encountered in the literature, and King (1967), for example, has suggested the use of trend surface methods to

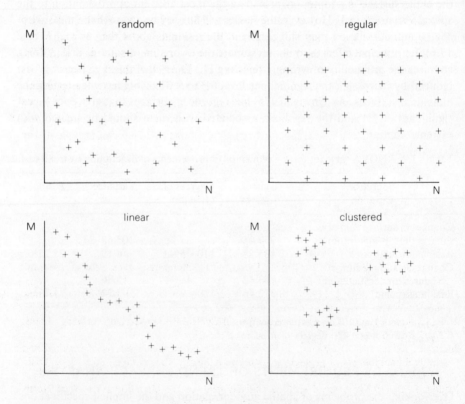

Figure 17.7 Different types of sampling methods for data points in trend surface analysis; the random and regular systems are acceptable, the linear and clustered are not

reconstruct, from highly discontinuous and fragmented data, the character of planation surfaces.

Other aspects can also be isolated for attention. Having regard to the data points themselves, random or regularly distributed sampling locations are usually acceptable, but clustered or linearly organized points (Figure 17.7) will distort the form of the fitted surface and render it sensitive to slight inacuracies or additions to the sparsely sampled zones.

At the same time significance testing becomes all but impossible because the effective degrees of freedom cannot be determined. Neither can regular sampling frameworks guarantee avoidance of these problems, and if spatial periodicity coincides with the sample grid spacing, then important aspects of the variables' behaviour may be entirely overlooked.

The problem of surface instability at the edges can be overcome by using a boundary of sample points extending beyond the area defined for study. It hardly requires emphasizing that, while interpolation within the sampled area can be useful, extrapolations beyond it may produce (at best) unreliable or (at worst) nonsensical conclusions. The shape of the study area can also be influential and is best kept equidimensional to avoid the tendency for the surface contours to orient themselves parallel to the major axis.

Finally, it will be noted that the computer's efficiency and the results' reliability may be enhanced by the simple expedient of constraining not only M and N but also Y values. Hence, in the present example house prices were scaled in £000s. Such manipulations in no way affect the shape of the trend surface or its significance though, clearly, the magnitudes of the coefficients are scale-dependent.

Trend surface analysis, so popular with geographers, appears to have a large number of pitfalls awaiting the unwary. Yet if these are acknowledged and avoided the judicious application of trend surface methods may yield enormous benefits for the geographer.

REFERENCES

Chorley, R. J., and Haggett, P. (1965). 'Trend surface mapping in geographical research', *Trans. Inst. Brit. Geogrs* **37**, 47–67.

Cliff, A. D., and Ord, K. (1970). 'Spatial autocorrelation: a review of existing and new measures with applications', *Econ. Geog.* **46**, 269–292.

King, C. A. M. (1969). 'Trend surface analysis of Central Pennine erosion surfaces', *Trans. Inst. Brit. Geogrs* **47**, 47–59.

Merriam, D. F., and Harbaugh, J. W. (1964). 'Trend surface analysis of regional and residual components of geologic structure in Kansas', *Kansas Geol. Surv. Spec. Dist.*, Publ. ll.

Rodda, J. C. (1970).'A trend surface analysis trial for the planation surfaces of North Cardiganshire', *Trans. Inst. Brit. Geogrs* **50**, 107–114.

Unwin, D. J. (1975). *An Introduction to Trend Surface Analysis*, Catmog 5, Geobooks, Norwich.

RECOMMENDED READING

Aburmere, S. (1979). 'The diffusion of economic development in Bendel State of Nigeria', *Geog. Ann* **61B,** 103–111. For the benefit of those who might imagine trend surface analysis to be wholly within the domain of physical geography, here is a successful and informative application from the human side of the discipline.

Chorley, R. J., and Haggett, P. (1965). 'Trend surface mapping in geographical research', *Trans. Inst. Brit. Geogrs* **37,** 47–67. A seminal work which, though largely theoretical in its treatment, was to pave the way for many later analytical studies. Longer than many papers, but worth reading as an item of historical and intrinsic value.

Cliff, A. D., and Ord, K. (1970). 'Spatial autocorrelation: a review of existing and new measures with applications', *Econ. Geog* **46,** 269–292.

Doornkamp, J. C. (1972). 'Trend-surface analysis of planation surfaces with an East African case study', in R. J. Chorley (ed.) *Spatial Analysis in Geomorphology,* Methuen, London. A thorough and detailed example of the application of trend surface analysis which balances very well for the reader the theoretical and practical aspects of the technique.

Gwebu, T. D. (1979). 'Population change determinants in the early transitional society: the Western Sierra Leone example', *Geog. Ann* **61B,** 91–101. This paper, in addition to offering an interesting case study, also provides an unusual degree of detail in respect of the statistical analysis. It is worth consulting on both counts.

King, C. A. M. (1969). 'Trend surface analysis of Central Pennine erosion surfaces', *Trans. Inst. Brit. Geogrs* **47,** 47–59. See Smith et al. (1969) for comments.

Mather, P. A. (1977). 'Clustered data-point distributions in trend surface analysis', *Geog. Analysis* **9,** 84–93. See Robinson (1972) for comments.

Merriam, D. F., and Harbaugh, J. W. (1964). 'Trend surface analysis of regional and residual components of geologic structure in Kansas', Kansas Geol. Surv. Spec. Dist., Publ. 11.

Robinson, G. (1972). 'Trials on trends through clusters of cirques', *Area* **4,** 104–113. This paper and that by Mather (1977) deal specifically with the problem of data-point clustering; one which is common throughout geography but notably pertinent in this context.

Robinson, G. and Fairburn, K. J. (1969). 'An application of trend surface mapping to the distribution of residuals from a regression', *Anns Ass. Am. Geogrs* **59,** 158–170. How to use trend surface methods in association with other techniques is demonstrated.

Rodda, J. C. (1970). 'A trend surface analysis trial for the planation surfaces of North Cardiganshire', *Trans. Inst. Brit. Geogrs* **50,** 107–114. This paper represents an attempt to turn the 'patterns' and groupings of residuals about the trend surface to the investigator's advantage, using them to identify a stepped landscape dominated by erosion surfaces.

Smith, D. E., Sissons, J. B., and Cullingford, R. A. (1969). 'Isobases for the Main Perth raised shoreline in South-Eastern Scotland as determined by trend surface analysis', *Trans. Inst. Brit. Geogrs* **46,** 45–56. Together with the paper by King (1969) we have here two examples of the manner in which trend surface analysis has been used to reconstruct formerly continuous features from now fragmentary evidence.

Unwin, D. J. (1975). *An Introduction to Trend Surface Analysis,* Catmog 5, Geobooks, Norwich. A highly readable and clear expression of the principles and problems of trend surface analysis. The bibliography is also comprehensive and useful.

Chapter 18

Computer Programs and Manuals

18.1 PROGRAMS

The following list gives basic information on the availability of SPSS, BMDP and Minitab programs for the analytical techniques described in this book. The reader is referred to the companion volume (see the Preface) for further study.

Techniques		Packages	
	SPSS	*BMDP*	*MINITAB*
Descriptive statistics			
Frequency tables	Frequencies	P2D, P4D	TABLE
Histograms	Frequencies	P2D, P5D	HIST
Scatterplots	Scattergram	P6D	PLOT
Central tendency	Condescriptive	P1D, P2D	AVERAGE, MEDIAN
Dispersion	Condescriptive	P1D, P2D	STANDARD DEVIATION
Probability distributions			
Standardized data	—	P1S	—
Binomial probabilities	—	—	BINOMIAL
Poisson probabilities	—	—	POISSON
Hypothesis testing			
One-sample χ^2	Chi-square	—	—
Kolmogorov/Smirnov test	K–S	—	—
Two-sample and k-sample χ^2	Crosstabs	P1F	CHISQ
Mann–Whitney	M–W	P3S	MANN WHIT
t-test difference between means	T-test	P3D	TTEST
Kruskall–Wallis	K–W	P3S	—
One-way analysis of variance	Oneway	P1V	ONEWAYZ
Correlation			
Product-moment	Pearson Corr	P1R	CORRELATION
Spearman's rank	Nonpar Corr	P1F	—
Point biserial	—	—	—
Phi-coefficient	Crosstabs	P1F	—

Techniques	SPSS	Packages BMDP	MINITAB
Regression analysis			
Linear regression	Regression	P1R	REGRESS
Nonlinear regression	Regression	P3R	—
Multiple regression	Regression	P1R	REGRESS
Stepwise regression	Regression	P2R	STEP
Statistical classification			
Cluster analysis	—	P1M, P2M, P3M—	
Association analysis	—	P3F	—
Factor analysis			
Principal components	Factor	P4M	—
Factor analysis	Factor	P4M	—

18.2 MANUALS

Dixon, W. J. (1981). *BMDP Statistical Software*, University of California Press, Berkeley.

Hull, C. H., and Nie, N. H. (1981) *SPSS Update 7-9*, McGraw-Hill, New York.

Nie, N. H., and Hull, C. H. (1975) *Statistical Packages for the Social Sciences (SPSS)*, 2nd edn, McGraw-Hill, New York.

Ryan, T. A., Joiner, B.L., and Ryal, B. F. (1976). *Minitab Student Handbook*, Duxbury Press, North Scituate, Mass.

(1982) *Minitab Reference Manual*, Duxbury Press, Boston, Mass.

SPSS Inc. (1983) *SPSS-X, User's Guide*, McGraw-Hill, New York.

This presents information on the new version of SPSS, which contains some major differences from previous updates.

Appendices

APPENDIX I

Areas beneath the standard normal curve

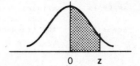

z	p	z	p	z	p	z	p
0.00	0.00000	0.50	0.19146	1.00	0.34134	1.50	0.43319
0.01	0.00399	0.51	0.19497	1.01	0.34375	1.51	0.43448
0.02	0.00798	0.52	0.19847	1.02	0.34614	1.52	0.43574
0.03	0.01197	0.53	0.20194	1.03	0.34849	1.53	0.43699
0.04	0.01595	0.54	0.20540	1.04	0.35083	1.54	0.43822
0.05	0.01994	0.55	0.20884	1.05	0.35314	1.55	0.43943
0.06	0.02392	0.56	0.21226	1.06	0.35543	1.56	0.44062
0.07	0.02790	0.57	0.21566	1.07	0.35769	1.57	0.44179
0.08	0.03188	0.58	0.21904	1.08	0.35993	1.58	0.44295
0.09	0.03586	0.59	0.22240	1.09	0.36214	1.59	0.44408
0.10	0.03983	0.60	0.22575	1.10	0.36433	1.60	0.44520
0.11	0.04380	0.61	0.22907	1.11	0.36650	1.61	0.44630
0.12	0.04776	0.62	0.23237	1.12	0.36864	1.62	0.44738
0.13	0.05172	0.63	0.23565	1.13	0.37076	1.63	0.44835
0.14	0.05567	0.64	0.23891	1.14	0.37286	1.64	0.44950
0.15	0.05962	0.65	0.24215	1.15	0.37493	1.65	0.45053
0.16	0.06356	0.66	0.24537	1.16	0.37698	1.66	0.45154
0.17	0.06750	0.67	0.24857	1.17	0.37900	1.67	0.45254
0.18	0.07142	0.68	0.25175	1.18	0.38100	1.68	0.45352
0.19	0.07535	0.69	0.25490	1.19	0.38298	1.69	0.45449
0.20	0.07926	0.70	0.25805	1.20	0.38493	1.70	0.45543
0.21	0.08317	0.71	0.26115	1.21	0.38686	1.71	0.45637
0.22	0.08706	0.72	0.26424	1.22	0.38877	1.72	0.45728
0.23	0.09095	0.73	0.26730	1.23	0.39065	1.73	0.45818
0.24	0.09483	0.74	0.27035	1.24	0.39251	1.74	0.45907
0.25	0.09871	0.75	0.27337	1.25	0.39435	1.75	0.45994
0.26	0.10257	0.76	0.27637	1.26	0.39617	1.76	0.46080
0.27	0.10642	0.77	0.27935	1.27	0.39796	1.77	0.46164
0.28	0.11026	0.78	0.28230	1.28	0.39973	1.78	0.46246
0.29	0.11409	0.79	0.28524	1.29	0.40147	1.79	0.46327
0.30	0.11791	0.80	0.28814	1.30	0.40320	1.80	0.46407
0.31	0.12172	0.81	0.29103	1.31	0.40490	1.81	0.46485
0.32	0.12552	0.82	0.29389	1.32	0.40658	1.82	0.46562
0.33	0.12930	0.83	0.29673	1.33	0.40824	1.83	0.46638
0.34	0.13307	0.84	0.29955	1.34	0.40988	1.84	0.46712

APPENDIX I (continued)

Areas beneath the standard normal curve

z	p	z	p	z	p	z	p
0.35	0.13683	0.85	0.30234	1.35	0.41149	1.85	0.46784
0.36	0.14058	0.86	0.30511	1.36	0.41309	1.86	0.46856
0.37	0.14431	0.87	0.30785	1.37	0.41466	1.87	0.46926
0.38	0.14803	0.88	0.31057	1.38	0.41621	1.88	0.46995
0.39	0.15173	0.89	0.31327	1.39	0.41774	1.89	0.47062
0.40	0.15542	0.90	0.31594	1.40	0.41924	1.90	0.47128
0.41	0.15910	0.91	0.31859	1.41	0.42073	1.91	0.47193
0.42	0.16276	0.92	0.32121	1.42	0.42220	1.92	0.47257
0.43	0.16640	0.93	0.32381	1.43	0.42364	1.93	0.47320
0.44	0.17003	0.94	0.32639	1.44	0.42507	1.94	0.47381
0.45	0.17364	0.95	0.32894	1.45	0.42647	1.95	0.47441
0.46	0.17724	0.96	0.33147	1.46	0.42785	1.96	0.47500
0.47	0.18082	0.97	0.33398	1.47	0.42922	1.97	0.47558
0.48	0.18439	0.98	0.33646	1.48	0.43056	1.98	0.47615
0.49	0.18793	0.99	0.33891	1.49	0.33189	1.99	0.47670
2.00	0.47725	2.50	0.49379	3.00	0.49865	3.50	0.49977
2.05	0.47982	2.55	0.49461	3.05	0.49886	3.55	0.49981
2.10	0.48214	2.60	0.49534	3.10	0.49903	3.60	0.49984
2.15	0.48422	2.65	0.49598	3.15	0.49918	3.65	0.49987
2.20	0.48610	2.70	0.49653	3.20	0.49931	3.70	0.49989
2.25	0.48778	2.75	0.49702	3.25	0.49942	3.75	0.49991
2.30	0.48928	2.80	0.49744	3.30	0.49952	3.80	0.49993
2.35	0.49061	2.85	0.49781	3.35	0.49960	3.85	0.49994
2.40	0.49180	2.90	0.49813	3.40	0.49966	3.90	0.49995
2.45	0.49286	2.95	0.49841	3.45	0.49972	3.95	0.49996
						4.00	0.49997

Columns headed p give the probabilities of an event within the range of zero
to z. The standard normal curve is perfectly symmetrical and the table is
equally applicable to negative z values.
For probabilities of events beyond the specified z subtract p from 0.50000

APPENDIX II

Critical values on Student's t distribution

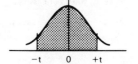

$-t \quad 0 \quad +t$

confidence limits

	0.90	0.95	0.98	0.99	0.999

two-tailed significance levels (one-tailed levels bracketed)

v	0.10 (0.05)	0.05 (0.025)	0.02 (0.01)	0.01 (0.005)	0.001 (0.0005)
1	6.31	12.71	31.81	63.66	636.6
2	2.92	4.30	6.97	9.93	31.60
3	2.35	3.18	4.54	5.84	12.92
4	2.13	2.78	3.75	4.60	8.61
5	2.02	2.57	3.37	4.03	6.86
6	1.94	2.45	3.14	3.71	5.96
7	1.90	2.37	3.00	3.50	5.41
8	1.86	2.31	2.90	3.36	5.04
9	1.83	2.26	2.82	3.25	4.78
10	1.81	2.23	2.76	3.17	4.59
11	1.80	2.20	2.72	3.11	4.44
12	1.78	2.18	2.68	3.06	4.32
13	1.77	2.16	2.65	3.01	4.23
14	1.76	2.15	2.62	2.98	4.14
15	1.75	2.13	2.60	2.95	4.07
16	1.75	2.12	2.58	2.92	4.02
17	1.74	2.11	2.57	2.90	3.97
18	1.73	2.10	2.55	2.88	3.92
19	1.73	2.09	2.54	2.86	3.88
20	1.73	2.09	2.53	2.85	3.85
21	1.72	2.08	2.52	2.83	3.82
22	1.72	2.07	2.51	2.82	3.79
23	1.71	2.07	2.50	2.81	3.77
24	1.71	2.06	2.49	2.80	3.75
25	1.71	2.06	2.49	2.79	3.73
26	1.71	2.06	2.48	2.78	3.71
27	1.70	2.05	2.47	2.77	3.69
28	1.70	2.05	2.47	2.76	3.67
29	1.70	2.05	2.46	2.76	3.66
30	1.70	2.04	2.46	2.75	3.65
40	1.68	2.02	2.42	2.70	3.55
60	1.67	2.00	2.39	2.66	3.46
over 60	approximates to the normal distribution				
z	1.64	1.96	2.33	2.58	3.29

The critical t value is found by reference to the appropriate degrees of freedom (v) and the selected significance level or confidence limit. In the former case the values for two-tailed tests should be read as + and - t and the null hypothesis is rejected if the test statistic falls within either rejection region. Equivalent one-tailed critical values are found under the appropriate bracketed headings and should be assigned to either + or - t.

APPENDIX III

Critical values on the chi-square distribution

significance level

v	0.10	0.05	0.01	0.005	0.001
1	2.71	3.84	6.64	7.88	10.83
2	4.60	5.99	9.21	10.60	13.82
3	6.25	7.82	11.34	12.84	16.27
4	7.78	9.49	13.28	14.86	18.46
5	9.24	11.07	15.09	16.75	20.52
6	10.64	12.59	16.81	18.55	22.46
7	12.02	14.07	18.48	20.28	24.32
8	13.36	15.51	20.29	21.96	26.12
9	14.68	16.92	21.67	23.59	27.88
10	15.99	18.31	23.21	25.19	29.59
11	17.28	19.68	24.72	26.76	31.26
12	18.55	21.03	26.22	28.30	32.91
13	19.81	22.36	27.69	30.82	34.53
14	21.06	23.68	29.14	31.32	36.12
15	22.31	25.00	30.58	32.80	37.70
16	23.54	26.30	32.00	34.27	39.29
17	24.77	27.59	33.41	35.72	40.75
18	25.99	28.87	34.80	37.16	42.31
19	27.20	30.14	36.19	38.58	43.82
20	28.41	31.41	37.57	40.00	45.32
21	29.62	32.67	38.93	41.40	46.80
22	30.81	33.92	40.29	42.80	48.27
23	32.01	35.17	41.64	44.18	49.73
24	33.20	36.42	42.98	45.56	51.18
25	34.38	37.65	44.31	46.93	52.62
26	35.56	35.88	45.64	48.29	54.05
27	36.74	40.11	46.96	49.65	55.48
28	37.92	41.34	48.28	50.99	56.89
29	39.09	42.56	49.59	52.34	58.30
30	40.26	43.77	50.89	53.67	59.70
40	51.81	55.76	63.69	66.77	73.40
50	63.17	67.51	76.15	79.49	86.66
60	74.40	79.08	88.38	91.95	99.61
70	85.53	90.53	100.43	104.22	112.32
80	96.58	101.88	112.33	116.32	124.84
90	105.57	113.15	124.12	128.30	137.21
100	118.50	124.34	135.81	140.17	149.45

The critical values are determined by reference to the sample
degrees of freedom (v) and the selected significance level.
If the test statistic equals or exceeds the critical value
then the null hypothesis is rejected.

APPENDIX IV

Critical values of the Kolmogorov-Smirnov statistic (D)

significance level

n	0.20	0.15	0.10	0.05	0.01
1	0.900	0.925	0.950	0.975	0.995
2	0.684	0.726	0.776	0.842	0.929
3	0.565	0.597	0.642	0.708	0.828
4	0.494	0.525	0.564	0.624	0.733
5	0.446	0.474	0.510	0.565	0.669
6	0.410	0.436	0.470	0.521	0.618
7	0.381	0.405	0.438	0.486	0.577
8	0.358	0.381	0.411	0.457	0.543
9	0.339	0.360	0.388	0.432	0.514
10	0.322	0.342	0.368	0.410	0.490
11	0.307	0.326	0.352	0.391	0.468
12	0.295	0.313	0.338	0.375	0.450
13	0.284	0.302	0.325	0.361	0.433
14	0.274	0.292	0.314	0.349	0.418
15	0.266	0.283	0.304	0.338	0.404
16	0.258	0.274	0.295	0.328	0.392
17	0.250	0.266	0.286	0.318	0.381
18	0.244	0.259	0.278	0.309	0.371
19	0.237	0.252	0.272	0.301	0.363
20	0.231	0.246	0.264	0.294	0.356
25	0.210	0.220	0.240	0.270	0.320
30	0.190	0.200	0.220	0.240	0.290
35	0.180	0.190	0.210	0.230	0.270
over 35	$\dfrac{1.07}{\sqrt{n}}$	$\dfrac{1.14}{\sqrt{n}}$	$\dfrac{1.22}{\sqrt{n}}$	$\dfrac{1.36}{\sqrt{n}}$	$\dfrac{1.63}{\sqrt{n}}$

The null hypothesis is rejected if the observed D statistic
exceeds the critical value for that sample size (n) at the
selected significance level.
For n of greater than 35 the equations should be used to
determine the critical value.

APPENDIX Va

Probabilities for the Mann-Whitney test statistic (U)

	n2 = 3		
n1	1	2	3
U			
0	0.250	0.100	0.050
1	0.500	0.200	0.100
2	0.750	0.400	0.200
3		0.600	0.350
4			0.500
5			0.650

	n2 = 4			
n1	1	2	3	4
U				
0	0.200	0.067	0.028	0.014
1	0.400	0.133	0.057	0.029
2	0.600	0.267	0.114	0.057
3		0.400	0.200	0.100
4		0.600	0.314	0.171
5			0.429	0.243
6			0.571	0.343
7				0.443

	n2 = 5				
n1	1	2	3	4	5
U					
0	0.167	0.047	0.018	0.008	0.004
1	0.333	0.095	0.036	0.016	0.008
2	0.500	0.190	0.071	0.032	0.016
3	0.667	0.286	0.125	0.056	0.028
4		0.429	0.196	0.095	0.048
5		0.571	0.286	0.143	0.075
6			0.393	0.206	0.111
7			0.500	0.278	0.155
8			0.607	0.365	0.210
9				0.452	0.274
10				0.548	0.345
11					0.421
12					0.500
13					0.579

	n2 = 6					
n1	1	2	3	4	5	6
U						
0	0.143	0.036	0.012	0.005	0.002	0.001
1	0.286	0.071	0.024	0.010	0.004	0.002
2	0.428	0.143	0.048	0.019	0.009	0.004
3	0.571	0.214	0.083	0.033	0.015	0.008
4		0.321	0.131	0.057	0.026	0.013
5		0.429	0.190	0.086	0.041	0.021
6		0.571	0.275	0.129	0.063	0.032
7			0.357	0.176	0.089	0.047
8			0.452	0.238	0.123	0.066
9			0.548	0.305	0.165	0.090
10				0.381	0.214	0.120
11				0.457	0.268	0.155
12				0.545	0.331	0.197
13					0.396	0.242
14					0.465	0.294
15					0.535	0.350
16						0.409
17						0.469

Consult the notes at the foot of Appendix Vc for details of how to use these tables of critical values and their associated random probabilities.

APPENDIX Vb

Probabilities for the Mann-Whitney test statistic (U)

n2 = 7

n1	1	2	3	4	5	6	7
U							
0	0.125	0.028	0.008	0.003	0.001	0.001	0.000
1	0.250	0.056	0.017	0.006	0.003	0.001	0.001
2	0.375	0.111	0.033	0.012	0.005	0.002	0.001
3	0.500	0.167	0.058	0.021	0.009	0.004	0.002
4	0.625	0.250	0.092	0.036	0.015	0.007	0.003
5		0.333	0.133	0.055	0.024	0.011	0.006
6		0.444	0.192	0.082	0.037	0.017	0.009
7		0.556	0.258	0.115	0.053	0.026	0.013
8			0.333	0.158	0.074	0.037	0.019
9			0.417	0.206	0.101	0.051	0.027
10			0.500	0.264	0.134	0.069	0.036
11			0.583	0.324	0.172	0.090	0.049
12				0.394	0.216	0.117	0.064
13				0.464	0.265	0.147	0.082
14				0.538	0.319	0.183	0.104
15					0.378	0.223	0.130
16					0.438	0.267	0.159
17					0.500	0.314	0.191
18					0.562	0.365	0.228
19						0.418	0.267
20						0.473	0.310
21						0.527	0.355
22							0.402
23							0.451
24							0.500

Consult the notes at the foot of Appendix Vc for details of how
to use this table of critical values and their associated
random probabilities.

APPENDIX Vc

Probabilities for the Mann-Whitney test statistic (U)

n2 = 8

n1 U	1	2	3	4	5	6	7	8
0	0.111	0.022	0.006	0.002	0.001	0.000	0.000	0.000
1	0.222	0.044	0.012	0.004	0.002	0.001	0.000	0.000
2	0.333	0.089	0.024	0.008	0.003	0.001	0.001	0.000
3	0.444	0.135	0.042	0.014	0.005	0.002	0.001	0.001
4	0.556	0.200	0.067	0.024	0.009	0.004	0.002	0.001
5		0.267	0.097	0.036	0.015	0.006	0.003	0.001
6		0.356	0.139	0.055	0.023	0.010	0.005	0.002
7		0.444	0.188	0.077	0.033	0.015	0.007	0.003
8		0.556	0.248	0.107	0.047	0.021	0.010	0.005
9			0.315	0.141	0.064	0.030	0.014	0.007
10			0.387	0.184	0.085	0.041	0.020	0.010
11			0.461	0.230	0.111	0.054	0.027	0.014
12			0.539	0.285	0.142	0.071	0.036	0.019
13				0.341	0.177	0.091	0.047	0.025
14				0.404	0.217	0.114	0.060	0.032
15				0.467	0.262	0.141	0.076	0.041
16				0.533	0.311	0.172	0.095	0.052
17					0.362	0.207	0.116	0.065
18					0.416	0.245	0.140	0.080
19					0.472	0.286	0.168	0.097
20					0.528	0.331	0.198	0.117
21						0.377	0.232	0.139
22						0.426	0.268	0.164
23						0.475	0.306	0.191
24						0.525	0.347	0.221
25							0.389	0.253
26							0.433	0.287
27							0.478	0.323
28							0.522	0.360
29								0.399
30								0.439
31								0.480
32								0.520

Appendices Va - Vc are arranged to give probabilities for
each U statistic determined from the larger and smaller group
sizes (n1 and n2 respectively). The null hypothesis is
rejected if the tabled probability of U is less than the
selected significance level.

APPENDIX Vd

Critical values of the Mann-Whitney test statistic (U)

significance level = 0.10 (two-tailed) or 0.05 (one-tailed)

n1	9	10	11	12	13	14	15	16	17	18	19	20
n2												
1											0	0
2	1	1	1	2	2	2	3	3	4	4	4	4
3	3	4	5	5	6	7	7	8	9	9	10	11
4	6	7	8	9	10	11	12	14	15	16	17	18
5	9	11	12	13	15	16	18	19	20	22	23	25
6	12	14	16	17	19	21	23	25	26	28	30	32
7	15	17	19	21	24	26	28	30	33	35	37	39
8	18	20	23	26	28	31	33	36	39	41	44	47
9	21	24	27	30	33	36	39	42	45	48	51	54
10	24	27	31	34	37	41	44	48	51	55	58	62
11	27	31	34	38	42	46	50	54	57	61	65	69
12	30	34	38	42	47	51	55	60	64	68	72	77
13	33	37	42	47	51	56	61	65	70	75	80	84
14	36	41	46	51	56	61	66	71	77	82	87	92
15	39	44	50	55	61	66	72	77	83	88	94	100
16	42	48	54	60	65	71	77	83	89	95	101	107
17	45	51	57	64	70	77	83	89	96	102	109	115
18	48	55	61	68	75	82	88	95	102	109	116	123
19	51	58	65	72	80	87	94	101	109	116	123	130
20	54	62	69	77	84	92	100	107	115	123	130	138

The null hypothesis is rejected if the test statistic (U) is
less than or equal to the critical value for the larger and
smaller group sizes (n1 and n2 respectively) at the selected
significance level.

APPENDIX Ve

Critical values of the Mann-Whitney test statistic (U)

significance level = 0.02 (two-tailed) or 0.01 (one-tailed)

n1 n2	9	10	11	12	13	14	15	16	17	18	19	20
2					0	0	0	0	0	0	1	1
3	1	1	1	2	2	2	3	3	4	4	4	5
4	3	3	4	5	5	6	7	7	8	9	9	10
5	5	6	7	8	9	10	11	12	13	14	15	16
6	7	8	9	11	12	13	15	16	18	19	20	22
7	9	11	12	14	16	17	19	21	23	24	26	28
8	11	13	15	17	20	22	24	26	28	30	32	34
9	14	16	18	21	23	26	28	31	33	36	38	40
10	16	19	22	24	27	30	33	36	38	41	44	47
11	18	22	25	28	31	34	37	41	44	47	50	53
12	21	24	28	31	35	38	42	46	49	53	56	60
13	23	27	31	35	39	43	47	51	55	59	63	67
14	26	30	34	38	43	47	51	56	60	65	69	73
15	28	33	37	42	47	51	56	61	66	70	75	80
16	31	36	41	46	51	56	61	66	71	76	82	87
17	33	38	44	49	55	60	66	71	77	82	88	93
18	36	41	47	53	59	65	70	76	82	88	94	100
19	38	44	50	56	63	69	75	82	88	94	101	107
20	40	47	53	60	67	73	80	87	93	100	107	114

The null hypothesis is rejected if the test statistic (U) is
less than or equal to the critical value for the larger and
smaller group sizes (n1 and n2 respectively) at the selected
significance level.

APPENDIX Vf

Critical values of the Mann-Whitney test statistic (U)

significance level = 0.05 (two-tailed) or 0.025 (one-tailed)

n1	9	10	11	12	13	14	15	16	17	18	19	20
n2												
2	0	0	0	1	1	1	1	1	2	2	2	2
3	2	3	3	4	4	5	5	6	6	7	7	8
4	4	5	6	7	8	9	10	11	11	12	13	13
5	7	8	9	11	12	13	14	15	17	18	19	20
6	10	11	13	14	16	17	19	21	22	24	25	27
7	12	14	16	18	20	22	24	26	28	30	32	34
8	15	17	19	22	24	26	29	31	34	36	38	41
9	17	20	23	26	28	31	34	37	39	42	45	48
10	20	23	26	29	33	36	39	42	45	48	52	55
11	23	26	30	33	37	40	44	47	51	55	58	62
12	26	29	33	37	41	45	49	53	57	61	65	69
13	28	33	37	41	45	50	54	59	63	67	72	76
14	31	36	40	45	50	55	59	64	67	74	78	83
15	34	39	44	49	54	59	64	70	75	80	85	90
16	37	42	47	53	59	64	70	75	81	86	92	98
17	39	45	51	57	63	67	75	81	87	93	99	105
18	42	48	55	61	67	74	80	86	93	99	106	112
19	45	52	58	65	72	78	85	92	99	106	113	119
20	48	55	62	69	76	83	90	98	105	112	119	127

The null hypothesis is rejected if the test statistic (U) is
less than or equal to the critical value for the larger and
smaller group sizes (n1 and n2 respectively) at the selected
significance level.

APPENDIX VIa

Critical values on the F distribution

significance level = 0.10

v_1	1	2	3	4	5	6	8	12	24	inf
v_2										
1	39.86	49.50	53.59	55.83	57.24	58.20	59.44	60.71	62.00	63.33
2	8.52	9.00	9.16	9.24	9.29	9.33	9.37	9.41	9.45	9.49
3	5.54	5.46	5.39	5.34	5.31	5.28	5.25	5.22	5.18	5.13
4	4.54	4.32	4.19	4.11	4.05	4.01	3.95	3.90	3.83	3.76
5	4.06	3.78	3.62	3.52	3.45	3.40	3.34	3.27	3.19	3.10
6	3.78	3.46	3.29	3.18	3.11	3.05	2.98	2.90	2.80	2.72
7	3.59	3.26	3.07	2.96	2.88	2.83	2.75	2.67	2.58	2.47
8	3.46	3.11	2.92	2.81	2.73	2.67	2.59	2.50	2.40	2.29
9	3.36	3.01	2.81	2.69	2.61	2.55	2.47	2.38	2.28	2.16
10	3.29	2.92	2.73	2.61	2.52	2.46	2.38	2.28	2.18	2.06
11	3.23	2.86	2.66	2.54	2.45	2.39	2.30	2.21	2.10	1.97
12	3.18	2.81	2.61	2.48	2.39	2.33	2.24	2.15	2.04	1.90
13	3.14	2.76	2.56	2.43	2.35	2.28	2.20	2.10	1.98	1.85
14	3.10	2.73	2.52	2.39	2.31	2.24	2.15	2.05	1.94	1.80
15	3.07	2.70	2.49	2.36	2.27	2.21	2.12	2.02	1.90	1.76
16	3.05	2.67	2.46	2.33	2.24	2.18	2.09	1.99	1.87	1.72
17	3.03	2.64	2.44	2.31	2.22	2.15	2.06	1.96	1.84	1.69
18	3.01	2.62	2.42	2.29	2.20	2.13	2.04	1.93	1.81	1.66
19	2.99	2.61	2.40	2.27	2.18	2.11	2.02	1.91	1.79	1.63
20	2.97	2.59	2.38	2.25	2.16	2.09	2.00	1.89	1.77	1.61
21	2.96	2.57	2.36	2.23	2.14	2.08	1.98	1.87	1.75	1.59
22	2.95	2.56	2.35	2.22	2.13	2.06	1.97	1.86	1.73	1.57
23	2.94	2.55	2.34	2.21	2.11	2.05	1.95	1.84	1.72	1.55
24	2.93	2.54	2.33	2.19	2.10	2.04	1.94	1.83	1.70	1.53
25	2.92	2.53	2.32	2.18	2.09	2.02	1.93	1.82	1.69	1.52
26	2.91	2.52	2.31	2.17	2.08	2.01	1.92	1.81	1.68	1.50
27	2.90	2.51	2.30	2.17	2.07	2.00	1.91	1.80	1.67	1.49
28	2.89	2.50	2.29	2.16	2.06	2.00	1.90	1.79	1.67	1.48
29	2.89	2.50	2.28	2.15	2.06	1.99	1.89	1.78	1.65	1.47
30	2.88	2.49	2.28	2.14	2.05	1.98	1.88	1.77	1.64	1.46
40	2.84	2.44	2.23	2.09	2.00	1.93	1.83	1.71	1.57	1.38
60	2.79	2.39	2.18	2.04	1.95	1.87	1.77	1.66	1.51	1.29
120	2.75	2.35	2.13	1.99	1.90	1.82	1.72	1.60	1.45	1.19
inf	2.71	2.30	2.08	1.94	1.85	1.77	1.67	1.55	1.38	1.00

The critical F value is assessed by reference to the degrees of freedom
for the greater and lesser variance estimates (v_1 and v_2 respectively).
The null hypothesis is rejected when the calculated F value equals or
exceeds the critical value. Missing values should be interpolated.

APPENDIX VIb

Critical values on the F distribution

significance level = 0.05

v_1	1	2	3	4	5	6	8	12	24	inf
v_2										
1	161.4	199.7	215.7	224.6	230.2	234.0	238.9	243.9	249.0	254.3
2	18.51	19.00	19.16	19.25	19.30	19.33	19.37	19.41	19.45	19.50
3	10.13	9.55	9.28	9.12	9.01	8.94	8.84	8.74	8.64	8.53
4	7.71	6.94	6.59	6.39	6.26	6.16	6.04	5.91	5.77	5.63
5	6.61	5.79	5.41	5.19	5.05	4.95	4.81	4.68	4.53	4.36
6	5.99	5.14	4.76	4.53	4.39	4.28	4.15	4.00	3.84	3.67
7	5.59	4.74	4.35	4.12	3.97	3.87	3.73	3.57	3.41	3.23
8	5.32	4.46	4.07	3.84	3.69	3.58	3.44	3.28	3.12	2.93
9	5.12	4.26	3.86	3.63	3.48	3.37	3.23	3.07	2.90	2.71
10	4.96	4.10	3.71	3.48	3.33	3.22	3.07	2.91	2.74	2.54
11	4.84	3.98	3.59	3.36	3.20	3.09	2.95	2.79	2.61	2.40
12	4.75	3.88	3.49	3.26	3.11	3.00	2.85	2.69	2.50	2.30
13	4.67	3.80	3.41	3.18	3.02	2.92	2.77	2.60	2.42	2.21
14	4.60	3.74	3.34	3.11	2.96	2.85	2.70	2.53	2.35	2.13
15	4.54	3.68	3.29	3.06	2.90	2.79	2.64	2.48	2.29	2.07
16	4.49	3.63	3.24	3.01	2.85	2.74	2.59	2.42	2.24	2.01
17	4.45	3.59	3.20	2.96	2.81	2.70	2.55	2.38	2.19	1.96
18	4.41	3.55	3.16	2.93	2.77	2.66	2.51	2.34	2.15	1.92
19	4.38	3.52	3.13	2.90	2.74	2.63	2.48	2.31	2.11	1.88
20	4.35	3.49	3.10	2.87	2.71	2.60	2.45	2.28	2.08	1.84
21	4.32	3.47	3.07	2.84	2.68	2.57	2.42	2.25	2.05	1.81
22	4.30	3.44	3.05	2.82	2.66	2.55	2.40	2.23	2.03	1.78
23	4.28	3.42	3.03	2.80	2.64	2.53	2.38	2.20	2.00	1.76
24	4.26	3.40	3.01	2.78	2.62	2.51	2.36	2.18	1.98	1.73
25	4.24	3.38	2.99	2.76	2.60	2.49	2.34	2.16	1.96	1.71
26	4.22	3.37	2.98	2.74	2.59	2.47	2.32	2.15	1.95	1.69
27	4.21	3.35	2.96	2.73	2.57	2.46	2.30	2.13	1.93	1.67
28	4.20	3.34	2.95	2.71	2.56	2.44	2.29	2.12	1.91	1.65
29	4.18	3.33	2.93	2.70	2.54	2.43	2.28	2.10	1.90	1.64
30	4.17	3.32	2.92	2.69	2.53	2.42	2.27	2.09	1.89	1.62
40	4.08	3.23	2.84	2.61	2.45	2.34	2.18	2.00	1.79	1.51
60	4.00	3.15	2.76	2.52	2.37	2.25	2.10	1.92	1.70	1.39
120	3.93	3.07	2.68	2.45	2.29	2.17	2.02	1.83	1.61	1.25
Inf	3.84	2.99	2.60	2.37	2.21	2.09	1.94	1.75	1.52	1.00

The critical F value is assessed by reference to the degrees of freedom
for the greater and lesser variance estimates (v_1 and v_2 respectively).
The null hypothesis is rejected when the calculated F value equals or
exceeds the critical value. Missing values should be interpolated.

APPENDIX Vic

Critical values on the F distribution

significance level = 0.01

v_1	1	2	3	4	5	6	8	12	24	Inf
v_2										
1	4052	4999	5403	5625	5764	5859	5981	6106	6234	6366
2	98.49	99.01	99.17	99.25	99.30	99.33	99.36	99.42	99.46	99.50
3	34.12	30.81	29.46	28.71	28.24	27.91	27.49	27.05	26.60	26.12
4	21.20	18.00	16.69	15.98	15.52	15.21	14.80	14.37	13.93	13.46
5	16.26	13.27	12.06	11.39	10.97	10.67	10.27	9.89	9.47	9.02
6	13.74	10.92	9.78	9.15	8.75	8.47	8.10	7.72	7.31	6.88
7	12.25	9.55	8.45	7.85	7.46	7.19	6.84	6.47	6.07	5.65
8	11.26	8.65	7.59	7.01	6.63	6.37	6.03	5.67	5.28	4.86
9	10.56	8.02	6.99	6.42	6.06	5.80	5.47	5.11	4.73	4.31
10	10.04	7.56	6.55	5.99	5.64	5.39	5.06	4.71	4.33	3.91
11	9.65	7.20	6.22	5.67	5.32	5.07	4.74	4.40	4.02	3.60
12	9.33	6.93	5.95	5.41	5.06	4.82	4.50	4.16	3.78	3.36
13	9.07	6.70	5.74	5.20	4.86	4.62	4.30	3.96	3.59	3.16
14	8.86	6.51	5.56	5.03	4.69	4.46	4.14	3.80	3.43	3.00
15	8.68	6.36	5.42	4.89	4.56	4.32	4.00	3.67	3.29	2.87
16	8.53	6.23	5.29	4.77	4.44	4.20	3.89	3.55	3.18	2.75
17	8.40	6.11	5.18	4.67	4.34	4.10	3.79	3.45	3.08	2.65
18	8.28	6.01	5.09	4.58	4.25	4.01	3.71	3.37	3.00	2.57
19	8.18	5.93	5.01	4.50	4.17	3.94	3.63	3.30	2.92	2.49
20	8.10	5.85	4.94	4.43	4.10	3.87	3.56	3.23	2.86	2.42
21	8.02	5.78	4.87	4.37	4.04	3.81	3.51	3.17	2.80	2.36
22	7.94	5.72	4.82	4.31	3.99	3.76	3.45	3.12	2.75	2.31
23	7.88	5.66	4.76	4.26	3.94	3.71	3.41	3.07	2.70	2.26
24	7.82	5.61	4.72	4.22	3.90	3.67	3.36	3.03	2.66	2.21
25	7.77	5.57	4.68	4.18	3.86	3.63	3.32	2.99	2.62	2.17
26	7.72	5.53	4.64	4.14	3.82	3.59	3.29	2.96	2.58	2.13
27	7.68	5.49	4.60	4.11	3.78	3.56	3.26	2.93	2.55	2.10
28	7.64	5.45	4.57	4.07	3.75	3.53	3.23	2.90	2.52	2.06
29	7.60	5.42	4.54	4.04	3.73	3.50	3.20	2.87	2.49	2.03
30	7.56	5.39	4.51	4.02	3.70	3.47	3.17	2.84	2.47	2.01
40	7.31	5.18	4.31	3.83	3.51	3.29	2.99	2.66	2.29	1.80
60	7.08	4.98	4.13	3.65	3.34	3.12	2.82	2.50	2.12	1.60
120	6.85	4.79	3.95	3.48	3.17	2.96	2.66	2.34	1.95	1.38
Inf	6.64	4.60	3.78	3.32	3.02	2.80	2.51	2.18	1.79	1.00

The critical F value is assessed by reference to the degrees of freedom
for the greater and lesser variance estimates (v_1 and v_2 respectively).
The null hypothesis is rejected when the calculated F value equals or
exceeds the critical value. Missing values should be interpolated.

APPENDIX VII

Critical values of the Kruskall-Wallis test statistic (H)

sample sizes			significance level		
n1	n2	n3	0.10	0.05	0.01
2	2	2	4.571		
3	2	1	4.286		
3	2	2	4.470	4.714	
3	3	1	4.571	5.143	
3	3	2	4.556	5.210	
3	3	3	4.622	5.600	6.489
4	2	1	4.199	4.822	
4	2	2	4.458	5.125	6.000
4	3	2	4.444	5.400	6.300
4	3	3	4.700	5.727	6.746
4	4	1	4.066	4.966	6.667
4	4	2	4.555	5.300	6.875
4	4	3	4.477	5.586	7.144
4	4	4	4.581	5.692	7.490
5	2	1	4.170	5.000	
5	2	2	4.342	5.071	6.373
5	3	1	3.982	4.915	6.400
5	3	2	4.495	5.205	6.822
5	3	3	4.503	5.580	7.030
5	4	1	3.974	4.923	6.885
5	4	2	4.522	5.268	7.118
5	4	3	4.542	5.631	7.440
5	4	4	4.619	5.618	7.752
5	5	1	4.056	5.018	7.073
5	5	2	4.508	5.339	7.269
5	5	3	4.545	5.642	7.543
5	5	4	4.521	5.643	7.791
5	5	5	4.560	5.720	7.980

The null hypothesis is rejected if the calculated H value
equals or exceeds the critical value. For some combinations
of small sample sizes (n1, n2 and n3 denote sample sizes) there
are no sensible values within this range of significance
levels.

APPENDIX IX

Critical values of the Spearman rank correlation coefficient

two-tailed significance levels (one-tailed in brackets)

n	0.1(0.05)	0.05(0.025)	0.02(0.01)	0.01(0.005)
5	0.900	1.000	1.000	
6	0.829	0.886	0.943	1.000
7	0.714	0.786	0.893	0.929
8	0.643	0.738	0.833	0.881
9	0.600	0.700	0.783	0.833
10	0.564	0.648	0.745	0.794
11	0.536	0.618	0.709	0.755
12	0.503	0.587	0.678	0.727
13	0.484	0.560	0.648	0.703
14	0.464	0.538	0.626	0.679
15	0.446	0.521	0.604	0.654
16	0.429	0.503	0.582	0.635
17	0.414	0.488	0.566	0.618
18	0.401	0.472	0.550	0.600
19	0.391	0.460	0.535	0.584
20	0.380	0.447	0.522	0.570
21	0.370	0.436	0.509	0.556
22	0.361	0.425	0.497	0.544
23	0.353	0.416	0.486	0.532
24	0.344	0.407	0.476	0.521
25	0.337	0.398	0.466	0.511
26	0.331	0.390	0.457	0.501
27	0.324	0.383	0.449	0.492
28	0.318	0.375	0.441	0.483
29	0.312	0.368	0.433	0.475
30	0.306	0.362	0.425	0.467
31	0.301	0.356	0.419	0.459
32	0.296	0.350	0.412	0.452
33	0.291	0.345	0.405	0.446
34	0.287	0.340	0.400	0.439
35	0.283	0.335	0.394	0.433
40	0.264	0.313	0.368	0.405
45	0.248	0.294	0.347	0.382
50	0.235	0.279	0.329	0.363
55	0.224	0.266	0.314	0.346
60	0.214	0.255	0.301	0.331
65	0.206	0.245	0.291	0.322
70	0.198	0.236	0.280	0.310
80	0.185	0.221	0.262	0.290
90	0.174	0.208	0.247	0.273
100	0.165	0.197	0.234	0.259

The observed correlation coefficient is significant if it
exceeds the critical value at the selected significance
level for sample size n.

APPENDIX VIII

Critical values of the Pearson product-moment correlation coefficient

two-tailed significance levels (one-tailed in brackets)

n	0.1(0.05)	0.05(0.025)	0.02(0.01)	0.01(0.005)
3	0.988	0.997	1.000	1.000
4	0.900	0.950	0.980	0.990
5	0.805	0.878	0.934	0.959
6	0.729	0.811	0.882	0.917
7	0.669	0.754	0.833	0.875
8	0.621	0.707	0.789	0.834
9	0.582	0.666	0.750	0.798
10	0.549	0.632	0.715	0.765
11	0.521	0.602	0.685	0.735
12	0.497	0.576	0.658	0.708
13	0.476	0.553	0.634	0.684
14	0.458	0.532	0.612	0.661
15	0.441	0.514	0.592	0.641
16	0.426	0.497	0.574	0.623
17	0.412	0.482	0.558	0.606
18	0.400	0.468	0.543	0.590
19	0.389	0.456	0.529	0.575
20	0.378	0.444	0.516	0.561
21	0.369	0.433	0.503	0.549
22	0.360	0.423	0.492	0.537
23	0.352	0.413	0.482	0.526
24	0.344	0.404	0.472	0.515
25	0.337	0.396	0.462	0.505
26	0.330	0.388	0.453	0.496
27	0.323	0.381	0.445	0.487
28	0.317	0.374	0.437	0.479
29	0.311	0.367	0.430	0.471
30	0.306	0.361	0.423	0.463
31	0.301	0.355	0.416	0.456
32	0.296	0.349	0.409	0.449
33	0.291	0.344	0.403	0.443
34	0.287	0.339	0.397	0.436
35	0.283	0.334	0.391	0.430
40	0.264	0.312	0.367	0.403
45	0.249	0.294	0.346	0.380
50	0.235	0.279	0.328	0.361
55	0.224	0.266	0.313	0.345
60	0.214	0.254	0.300	0.330
65	0.206	0.244	0.288	0.317
70	0.198	0.235	0.278	0.306
75	0.191	0.227	0.268	0.296
80	0.185	0.220	0.260	0.286
85	0.180	0.213	0.252	0.278
90	0.174	0.207	0.245	0.270
95	0.170	0.202	0.238	0.263
100	0.165	0.197	0.232	0.256
150	0.135	0.160	0.190	0.210
200	0.117	0.139	0.164	0.182

The observed correlation coefficient is significant if it
exceeds the critical value at the selected significance
level for sample size n.

APPENDIX Xa

Critical bounds of the Durbin-Watson statistic

significance level = 0.05

	k = 1		k = 2		k = 3		k = 4		k = 5	
n	dL	dU	dL	dU	dL	dU	dL	dU	dL	dU
15	1.077	1.361	0.946	1.543	0.814	1.750	0.685	1.977	0.562	2.220
16	1.106	1.371	0.982	1.539	0.857	1.728	0.734	1.935	0.615	2.157
17	1.133	1.381	1.015	1.536	0.897	1.710	0.779	1.900	0.664	2.104
18	1.158	1.391	1.046	1.535	0.933	1.696	0.820	1.872	0.710	2.060
19	1.180	1.401	1.074	1.536	0.967	1.685	0.859	1.848	0.752	2.023
20	1.201	1.411	1.100	1.537	0.998	1.676	0.894	1.828	0.792	1.991
21	1.221	1.420	1.125	1.538	1.026	1.669	0.927	1.812	0.829	1.964
22	1.239	1.429	1.147	1.541	1.053	1.664	0.958	1.797	0.863	1.940
23	1.257	1.437	1.168	1.543	1.078	1.660	0.986	1.785	0.895	1.920
24	1.273	1.446	1.188	1.546	1.101	1.656	1.013	1.775	0.925	1.902
25	1.288	1.454	1.206	1.550	1.123	1.654	1.038	1.767	0.953	1.886
25	1.302	1.461	1.224	1.553	1.143	1.652	1.062	1.759	0.979	1.873
27	1.316	1.469	1.240	1.556	1.162	1.651	1.084	1.753	1.004	1.861
28	1.328	1.476	1.255	1.560	1.181	1.650	1.104	1.747	1.028	1.850
29	1.341	1.483	1.270	1.563	1.198	1.650	1.124	1.743	1.050	1.841
30	1.352	1.489	1.284	1.567	1.214	1.650	1.143	1.739	1.071	1.833
31	1.363	1.496	1.297	1.570	1.229	1.650	1.160	1.735	1.090	1.825
32	1.373	1.502	1.309	1.574	1.244	1.650	1.177	1.732	1.109	1.819
33	1.383	1.508	1.321	1.577	1.258	1.651	1.193	1.730	1.127	1.813
34	1.393	1.514	1.333	1.580	1.271	1.652	1.208	1.728	1.144	1.808
35	1.402	1.519	1.343	1.584	1.283	1.653	1.222	1.726	1.160	1.803
36	1.411	1.525	1.354	1.587	1.295	1.654	1.236	1.724	1.175	1.799
37	1.419	1.530	1.364	1.590	1.307	1.655	1.249	1.723	1.190	1.795
38	1.427	1.535	1.373	1.594	1.318	1.656	1.261	1.722	1.204	1.792
39	1.435	1.540	1.382	1.597	1.328	1.658	1.273	1.722	1.218	1.789
40	1.442	1.544	1.391	1.600	1.338	1.659	1.285	1.721	1.230	1.786
45	1.475	1.566	1.430	1.615	1.383	1.666	1.336	1.720	1.287	1.776
50	1.503	1.585	1.462	1.628	1.421	1.674	1.378	1.721	1.335	1.771
55	1.528	1.601	1.490	1.641	1.452	1.681	1.414	1.724	1.374	1.768
60	1.549	1.616	1.514	1.652	1.480	1.689	1.444	1.727	1.408	1.767
65	1.567	1.629	1.536	1.662	1.503	1.696	1.471	1.731	1.438	1.767
70	1.583	1.641	1.554	1.672	1.525	1.703	1.494	1.735	1.464	1.768
75	1.598	1.652	1.571	1.680	1.543	1.709	1.515	1.739	1.487	1.770
80	1.611	1.662	1.586	1.688	1.560	1.715	1.534	1.743	1.507	1.772
85	1.624	1.671	1.600	1.696	1.575	1.721	1.550	1.747	1.525	1.774
90	1.635	1.679	1.612	1.703	1.589	1.726	1.566	1.751	1.542	1.776
95	1.645	1.687	1.623	1.709	1.602	1.732	1.579	1.755	1.557	1.778
100	1.654	1.694	1.634	1.715	1.613	1.736	1.592	1.758	1.571	1.780

The upper (dU) and lower (dL) bounds are given by reference to the number of observations (n) and the number of independent variables (k).

APPENDIX Xb

Critical bounds of the Durbin-Watson statistic

significance level = 0.01

n	k = 1		k = 2		k = 3		k = 4		k = 5	
	dL	dU	dL	dU	dL	dU	dL	dU	dL	dU
15	0.811	1.070	0.700	1.252	0.591	1.464	0.488	1.704	0.391	1.967
16	0.844	1.086	0.737	1.252	0.633	1.446	0.532	1.663	0.437	1.900
17	0.874	1.102	0.772	1.255	0.672	1.432	0.574	1.630	0.480	1.847
18	0.902	1.118	0.805	1.259	0.708	1.422	0.613	1.604	0.522	1.803
19	0.928	1.132	0.835	1.265	0.742	1.415	0.650	1.584	0.561	1.767
20	0.952	1.147	0.863	1.271	0.773	1.411	0.685	1.567	0.598	1.737
21	0.975	1.161	0.890	1.277	0.803	1.408	0.718	1.554	0.633	1.712
22	0.997	1.174	0.914	1.284	0.831	1.407	0.748	1.543	0.667	1.691
23	1.018	1.187	0.938	1.291	0.858	1.407	0.777	1.534	0.698	1.673
24	1.037	1.199	0.960	1.298	0.882	1.407	0.805	1.528	0.728	1.658
25	1.055	1.211	0.981	1.305	0.906	1.409	0.831	1.523	0.756	1.645
26	1.072	1.222	1.001	1.312	0.928	1.411	0.855	1.518	0.783	1.635
27	1.089	1.233	1.019	1.319	0.949	1.413	0.878	1.515	0.808	1.626
28	1.104	1.244	1.037	1.325	0.969	1.415	0.900	1.513	0.832	1.618
29	1.119	1.254	1.054	1.332	0.988	1.418	0.921	1.512	0.855	1.611
30	1.133	1.263	1.070	1.339	1.006	1.421	0.941	1.511	0.877	1.606
31	1.147	1.273	1.085	1.345	1.023	1.425	0.960	1.510	0.897	1.601
32	1.160	1.282	1.100	1.352	1.040	1.428	0.979	1.510	0.917	1.597
33	1.172	1.291	1.114	1.358	1.055	1.432	0.996	1.510	0.936	1.594
34	1.184	1.299	1.128	1.364	1.070	1.435	1.012	1.511	0.954	1.591
35	1.195	1.307	1.140	1.370	1.085	1.439	1.028	1.512	0.971	1.589
36	1.206	1.315	1.153	1.376	1.098	1.442	1.043	1.513	0.988	1.588
37	1.217	1.323	1.165	1.382	1.112	1.446	1.058	1.514	1.004	1.586
38	1.227	1.330	1.176	1.388	1.124	1.449	1.072	1.515	1.019	1.585
39	1.237	1.337	1.187	1.393	1.137	1.453	1.085	1.517	1.034	1.584
40	1.246	1.344	1.198	1.398	1.148	1.457	1.098	1.518	1.048	1.584
45	1.288	1.376	1.245	1.423	1.201	1.474	1.156	1.528	1.111	1.584
50	1.324	1.403	1.285	1.446	1.245	1.491	1.205	1.528	1.164	1.587
55	1.356	1.427	1.320	1.466	1.284	1.506	1.247	1.548	1.209	1.592
60	1.383	1.449	1.350	1.484	1.317	1.520	1.283	1.558	1.249	1.598
65	1.407	1.468	1.377	1.500	1.346	1.534	1.315	1.568	1.283	1.604
70	1.429	1.485	1.400	1.515	1.372	1.546	1.343	1.578	1.313	1.611
75	1.448	1.501	1.422	1.529	1.395	1.557	1.368	1.587	1.340	1.617
80	1.466	1.515	1.441	1.541	1.416	1.568	1.390	1.595	1.364	1.624
85	1.482	1.528	1.458	1.553	1.435	1.578	1.411	1.603	1.386	1.630
90	1.496	1.540	1.474	1.563	1.452	1.587	1.429	1.611	1.406	1.636
95	1.510	1.552	1.489	1.573	1.468	1.596	1.446	1.618	1.425	1.642
100	1.522	1.562	1.503	1.583	1.482	1.604	1.462	1.625	1.441	1.647

The upper (dU) and lower (dL) bounds are given by reference to the number of observations (n) and the number of independent variables (k).

APPENDIX XI

Table of random numbers

19223	95034	05756	27813	96409	12531	42544	82853
73676	47150	98400	01927	27764	42648	84225	36290
45467	71709	77588	00095	32863	29485	82226	90056
52711	38889	93074	60227	40011	85848	48767	52573
95592	94007	69971	91481	60779	53971	17297	59335
68417	35013	15829	72765	85089	57067	50211	47487
82739	57890	20807	47511	81767	55330	94383	14893
60940	72042	17868	24943	61790	90565	87964	18883
36009	19365	15412	39638	85453	46816	84385	41979
38448	48789	18338	24697	39364	42006	76688	08708
81486	69487	60513	09297	00412	71238	27499	39950
59636	88804	04643	71197	18352	73089	84898	45785
62568	70206	40235	03699	71080	22553	11486	11776
45159	32992	75750	66280	03819	56202	02938	70915
61041	77684	94322	24709	73689	14526	31893	32593
38162	98532	61283	70632	23417	26185	41448	75532
73190	32533	04470	29669	84407	90785	65496	86382
95857	07118	87664	92099	58806	66979	98624	84826
35476	55975	39421	65850	04266	35435	43742	11937
71487	09984	29077	14863	61683	47052	62224	51025
13873	81598	95052	90908	73592	75186	87136	95761
54580	81507	27102	56027	55893	33063	41842	81868
71035	09001	43367	49497	72719	96758	27611	91965
96927	19931	36089	74192	77567	88741	48409	41903
43909	99477	25330	64359	40085	16925	85117	36071
15689	14227	06565	14374	13352	49367	81982	87209
36759	58984	68288	22913	18638	54303	00795	07827
69051	64817	87174	09751	84534	06489	87201	97245
05007	16332	81194	14873	04197	85576	45195	96565
68732	55259	84292	08796	43456	93739	31865	97150
45740	41807	65561	33302	07051	93623	18132	09547
66925	55685	39100	78458	11206	19876	87151	31260
08421	44753	77377	28744	75592	08563	79140	94254
53645	66812	61421	47836	12609	15374	98481	14592
66831	68908	40772	21558	47781	33586	79117	06928
55588	99404	70708	41098	43563	56934	48394	51719
12975	13258	13048	45114	72321	81940	00360	02428
96767	35964	23822	96012	94591	65194	50842	53372
72829	50232	97892	63408	77919	44575	24870	04178
88656	42628	17787	49376	61762	16953	88604	12724
62974	88145	83083	69543	46109	59505	69680	00900
19687	12633	57857	95806	09931	02150	43163	58636
37609	59057	66979	83401	60705	02384	90597	93600
54973	86278	88737	74351	47500	84552	19909	67181
00694	05977	19664	65441	20903	62371	22725	53340
71546	05233	53946	68743	72460	27601	45403	88692
07511	88915	41276	16853	84509	79367	32337	03316

Author Index

Subject Index

364